DREUX DE NETTANCOURT ▦ **Incompatibility and Incongruity
in Wild and Cultivated Plants**

2nd Edition

Springer
*Berlin
Heidelberg
New York
Barcelona
Hong Kong
London
Milan
Paris
Singapore
Tokyo*

DREUX DE NETTANCOURT

Incompatibility and Incongruity in Wild and Cultivated Plants

Second, totally revised and enlarged edition

With 41 Figures in 67 separate Illustrations and 19 Tables

Springer

Professor DREUX DE NETTANCOURT
Université Catholique de Louvain
Faculté des sciences agronomiques
Unité de biochimie physiologique
Croix du Sud 2/20
B-1348 Louvain-la-Neuve, Belgium

Cover illustration: Cross section of an incompatible pollen tube in the intercellular space of the stylar conducting tissue of *Lycopersicum peruvianum* (courtesy University of Siena). The endoplasmic reticulum shows a concentric parallel configuration (de Nettancourt et al. 1973a)

The first edition was published under the title
"Incompatibility in Angiosperms" as Vol. 3 of the Series
"Monographs on Theoretical and Applied Genetics", Springer 1977.

ISBN 3-540-65217-5 Springer-Verlag Berlin Heidelberg New York

Library of Congress Cataloging-in-Publication Data
De Nettancourt, D., 1933 – Incompatibility and incongruity in wild and cultivated plants/Dreux de Nettancourt. – 2nd, totally rev. and enlarged ed. p. cm. Rev. ed. of: Incompatibility in angiosperms. 1977. Includes bibliographical references (p.). ISBN 3540652175 (alk. paper) 1. Angiosperms. 2. Plant genetics. 3. Pollination. 4. Plant breeding. I. De Nettancourt, D., 1933 – Incompatibility in angiosperms. QK495.A1 D46 2001 581.3'5 – dc21

Springer-Verlag Berlin Heidelberg New York
a member of BertelsmannSpringer Science+Business Media GmbH
© Springer-Verlag Berlin Heidelberg 2001
   Printed in Germany

The use of general descriptive names, registered names, trademarks, etc. in this publication does not imply, even in the absence of a specific statement, that such names are exempt from the relevant protective laws and regulations and therefore free for general use.

Production: Klemens Schwind
Cover design: design & production GmbH, D-69121 Heidelberg
Typesetting: K+V Fotosatz GmbH, Beerfelden

SPIN 10695077   31/3136   5 4 3 2 1 0 – Printed on acid-free paper

To Gabrielle, my wife,
for her encouragement, patience
and everlasting guidance
regarding the word processor

Preface

The aim of this book is to provide a picture, as complete as possible, of the current state of knowledge of pollen-pistil barriers in flowering plants and of the ability of man to silence, mutate or transfer the genes that control them. The work was conducted in two phases. The first one essentially consisted of a review of early work, such as it had been summarized in a monograph I wrote in 1977 regarding the main features, origin and classification of self-incompatibility systems, their distribution among the angiosperms, the modalities of their inheritance at the scale of individuals and populations, the mutability of SI genes, and their possible involvement in the control of barriers between species. The major part of this information was derived from the 1977 monograph and systematically enriched (when material was available) by an account of more recent or contemporary research on the distribution, genetics and population genetics of SI. The second phase, by far the most challenging, consisted of the description and review of the very impressive and abundant research results that have accumulated, since the early 1980s, on the cellular and molecular biology of pollen-pistil interactions. The nature of this research, although the ultimate objectives (understanding and control of pollen-pistil barriers) remained unchanged, was radically modified by the new ability of man to perform analyses of life at the molecular level and to exploit all the techniques of modern biology to study the structure, action, function, mutation, reconstruction and evolution of the genes and mechanisms that participate in the recognition and rejection of incompatible pollen grains and pollen tubes.

A major difficulty in the preparation of a book of this type, which covers the interactions of many different disciplines, deals with the classification and distribution of information. It is necessary to compromise between repetition and overlap, essential if each chapter is to be considered separately, and cross-referencing, required for the unity of the work. Furthermore, I have not been able to avoid the creation of an opening chapter devoted to a general presentation of pollen-pistil barriers; it is not intended to summarize the book but to provide the perspective necessary for direct access to the matter covered by the other chapters. Again, I had to choose between cross-referencing and the duplication of information.

▓ The Role of Professor Linskens

The person to blame is Professor H.F. Linskens. Not only did he suggest the first edition of this book approximately 25 years ago, he proposed this "*bis repetita placent*" episode of pollen-pistil tribulations. However, he meant well. I must say that Professor Linskens tried to make up for his impulses through his constant support and the regular transmission of a very impressive amount of information. I not only owe him the text of articles by Correns (1913), Sutton (1918), Brieger (1930) and Sears (1937) but also articles discussing the latest state of the art (to the end 1999) of pollen-part mutations, the dimer hypothesis, pollen-tube-directed differential ovule development and the biochemistry of haplotype dominance. Through him, from several of his six working addresses in Europe and America, I received extensive collections of abstracts from recent papers that dealt with all possible aspects of pollen-pistil interactions. None of my mistakes, misinterpretations or omissions can be attributed to him, but he must be credited for his support and the suggestions that enabled me to improve my work.

▓ Help from the Scientific Community at Large and from Colleagues at Louvain-la-Neuve

Other people, in addition to Professor Linskens, gave advice and information. Dr. V.E. Franklin-Tong kindly provided explanations of the functional significance of the interaction between stigmatic S-proteins and SPB, a pollen protein, in the field poppy (Jordan et al. 1999, but unpublished at the time). She informed me, on this occasion, of the discovery, by a Birmingham group (N.N. Jordan and co-workers), of the first evidence that programmed cell death was involved in the rejection phase of SI. Professor B.A. McClure kindly commented on the research results at his laboratory and provided unpublished information and an illustration (Fig. 5.3) regarding work dealing with unilateral inter-species incompatibility. Professor Wehling described the results of his attempts to test the Bm2 probe of *Phalaris coerulescens*, supplied by Professor Langridge, as a candidate for the S-gene in rye. Professor M. Cresti contributed the unpublished micrographs of incompatible pollen tubes, which are now Fig. 3.1 and Fig. 5.1.

I thank all of them wholeheartedly, and I also express my deep gratitude to the scientists and the publishers who so kindly contributed photographs, drawings or tables from early or recent work and authorized me to use them in my book. Their very kind cooperation is acknowledged in the captions of figures and tables.

Finally, I want to thank the University of Louvain and the Unit for Physiological Biochemistry (FYSA) for the hospitality and the help that was generously and kindly given to me. My gratitude goes, in particular, to Professor A. Goffeau, former director of FYSA, and to Professor M. Boutry, who replaced him after his retirement. I appreciate the instructive discussions I had with them on the expression, cloning, sequencing and reconstruction of genes from yeast and higher plants.

Mr. Philip Menier competently advised and participated in the treatment and harmonization of 1200 references from several bibliographies, which Mrs. C. Durand reconstituted in part and which, at a later stage, Mrs. M. Rochat processed, checked and finalized with admirable patience. They cannot be blamed, of course, for flaws in the raw material provided to them. Mrs. A.M. Faber skillfully prepared the assembly of microphotographs and the computer drawing of diagrams.

D. DE NETTANCOURT
Emeritus Professor, Physiological Biochemistry Unit,
Catholic University of Louvain

Contents

CHAPTER 1 The Basic Features of Self-Incompatibility

▦ 1.1 A Definition

The majority of early workers (for instance, Herbert, Scott and Munro, cited in Darwin 1880), and many authors in the twentieth century (Sutton 1918; East and Mangelssdorf 1925; Riley 1936; Sears 1937), defined self-sterility as the inability of fertile plants to reproduce after selfing. Brieger (1930), who wrote the first book ever published on self- and cross-sterility, proposed that the phenomenon of self-rejection in flowering plants be referred to as "complete" or "incomplete" parasterility. As was pointed out by Stout (1917) and recognized by East (1940), self-incompatibility (SI) is the best name for a situation that involves the participation of both the pollen and pistil. As such, it is basically different from male or female sterility, where the phenotypic expression of the sterility genes is independent of the genotypic constitution in the mating partner before fertilization.

All definitions of SI implicitly or explicitly emphasize the function of the phenomenon as an outbreeding mechanism, but the literature is unclear as to whether or not the use of the term must be restricted to pre-fertilization processes or can be extended to describe all events that prevent fertile hermaphrodites from setting seeds after selfing.

Whereas, for instance, Brewbaker (1958) and Arasu (1968) defined SI as "the inability of a plant producing fertile gametes to set seeds when self-pollinated", other authors referred to SI as:

▦ "A social disease" (Dickinson 1994)
▦ "The hindrance to fertilization" (Lewis 1949a)
▦ "The impossibility of the pollen to fertilize the egg" (East and Park 1917)
▦ "The failure, following mating or pollination, of a male gamete and a female gamete to achieve fertilization where each of them is capable of uniting with other gametes of the breeding after similar mating or pollination" (Mather 1943).

Consulted on the matter, Lundqvist (1964, 1965) clearly stated that the term "incompatibility" should not include zygote lethality. With exceptions, such as that of *Borago officinalis*, for which Crowe (1971) demonstrated a post-fertilization system of self-rejection, it seems that most systems of SI are indeed pre-zygotic. Thus, Lundqvist was right in requesting that no confusion be made between SI and the diversity of accidents occurring after inbreeding during seed formation. Recessive embryonic lethality may, in some ways, be compared with SI. For expression, it effectively depends on a contribution from both the male and female partners, but the phenomenon (with few exceptions, such as the previously mentioned case of *B. officinalis*) does not seem to contribute extensively to the establishment and permanence of allogamy in natural populations. Its function, especially in cases where only a small number of unlinked recessive lethals operate, appears to be far more flexible and is essentially concerned with the maintenance of a minimum level of heterozygosity at the cost of a high mobilization and waste of ovules.

The need to distinguish between pre- and post-fertilization barriers to selfing is probably academic, but it seems advisable, in view of these basic differences in function, mechanism and occurrence, to adopt the attitude of Lundqvist and to define SI in higher plants as "the inability of a fertile hermaphrodite seed-plant to produce zygotes after self-pollination".

▩ 1.2 Nature of the SI Reaction

All SI systems are based on the inherited capacity of the flower to reject its own pollen. Such a characteristic is remarkable and unique in a biological world where discrimination mechanisms usually lead, on the contrary, to the acceptation of "self" and the rejection of unlike elements.

The various mechanisms through which incompatibility reactions are thought to occur are discussed in detail in Chapters 2 and 3, but it is necessary to define here, very briefly, the two processes that have been alternatively considered to constitute the basis of the rejection system: the stimulation of unlike genotypes and the inhibition of like elements.

In the first case, designated by Bateman (1952) as the *complementary system*, SI is hypothesized to be an absence of stimulation by the pistil after pollen growth; it can be equated to a mere absence of the substances necessary for pollen-tube penetration. After self-pollination, the pollen and/or the pistil fail to produce the component necessary to enable the pollen to germinate on the stigma or to grow through the style and within the ovary. In the second hypothesis, SI is defined as *oppositional* and is visualized as an active process that inhibits the growth of the pollen tube in the pistil. After combination or interaction, the incompatibility components in the pollen and pistil may, in this case, be considered to produce a substance having the property to interfere with the normal metabolism of the pollen grain or the pollen tube.

These two concepts were extensively discussed in the past, and numerous arguments (for instance, the ability of the pollen from self-incompatible species to germinate on artificial media and through the pistil of foreign species, and the inhibitory effects of high radiation doses on SI) have been advanced as evidence that the prevention of the germination or growth of incompatible pollen results from an active rejection process. Indeed, current knowledge of the mechanisms actually involved both in the recognition of an incompatible relation by the pollen and pistil and in the subsequent arrest of pollen-tube penetration through the pistil demonstrates the correctness of the oppositional hypothesis. A multigenic model (Mulcahy and Mulcahy 1983) based on the complementary hypothesis and heterotic interactions between pistil and outbred pollen has been clearly shown to be ill-founded (Lawrence et al. 1985). Of course, as shall be seen several times in this monograph, this does not necessarily imply that the recognition process involves the participation of one and the same gene in the pollen and pistil, nor does it imply the interaction of identical gene products. There is, on the contrary, growing evidence that this is not so. The pollen and pistil recognition genes appear to be separate (even in systems where SI is clearly inherited as a single Mendelian factor) but closely linked, with little or no crossover between them.

The oppositional hypothesis, as drafted by Bateman, restricts the consequences of the incompatibility reaction to the rejection of pollen, pollen tubes and sperm nuclei by the pistil. We now know (Sage et al. 1999; Sect. 2.3.7.3) that certain cases of SI probably result from an effect of incompatible pollen tubes on ovule development. There may be a need, as concluded by Sage and co-workers, to include a wider range of pollen–pistil interactions in current concepts of SI.

▦ 1.3 Classification of SI Systems

The man-made classification of SI systems is based on a correspondence between certain incompatibility features and a number of cyto-morphological attributes that concern:
- The time of gene action
- The association with floral polymorphism
- The site of expression
- The involvement of polyallelic series and the number of genetic loci.

1.3.1 The Time of Gene Action in the Pistil

In terms of the female part of the flower, the time of gene action (i.e., the stage at which the plant determines its incompatibility phenotype) coincides with the opening of the flower. Pistils at the receptive stage for cross-pollination are, as a rule, completely determined in terms of their ability to reject incompatible pollen. This means, in the case of an ordinary diploid plant and a recognition system inherited as a single Mendelian factor, that the pair of alleles responsible for the somatic determination of the incompatibility phenotype in the receptive tissue of the pistil establish (via an independent action of each allele or through allelic interactions or dominance relationships) the necessary basis for the rejection of self pollen.

Before and after this stage, and with variations that have been measured by means of bud pollination or senescent flowers, the pistils of self-incompatible plants may accept a certain amount of self pollen and produce illegitimate seeds. This capacity to tolerate selfing is variable and cannot be used, at the moment, as a convenient criterion for distinguishing among different systems of incompatibility.

1.3.2 The Time of Gene Action in the Stamen

In contrast with the situation in the pistil, the stage at which the male gametophyte receives the information necessary for the determination of its incompatibility phenotype can greatly vary among different families of plants and is one of the essential features upon which the classification of SI systems is founded. On this basis, most types of SI can be subdivided into two distinct groups:

▓ *Sporophytic incompatibility*, in which the incompatibility phenotype in the pollen is determined by the genotype of the pollen-producing plant

▓ *Gametophytic incompatibility*, where the genotype of the individual microspore determines the phenotype of the pollen

In other words, in sporophytic systems, the time of gene action in the stamen appears to be pre-meiotic (or, at the latest, meiotic) before individualization in the tetrads. However, as shall be seen below (Sects. 1.3.2.2, 3.1.3.2, 3.2.4.1, 3.2.4.2), there is evidence that, at least in some sporophytic systems, the S phenotype of the pollen is established at a later stage by a transfer of SI substances from the tapetum of the anther to the pollen exine or via the mixing of gametophytic S-products on the pollen coat. In gametophytic systems, the SI phenotype of the microspores is determined after the first metaphase of meiosis.

1.3.2.1 Determination of the Pollen Phenotype in Gametophytic Systems

The exact timing of events in the case of gametophytic incompatibility is unknown, but Lewis (1949b) and Pandey (1958, 1960, 1970a) held the view that gene action in gametophytic systems takes place immediately after meiosis and before microspore formation (Fig. 1.1). Their conclusions essentially stem from radiation work that shows that, when applied to post-meiotic stages, mutagens cease to be efficient for producing detectable mutations of SI genes. One must not forget, however, that mutations induced after the S-phase in mononucleated microspores will not be distributed to both the generative and the vegetative nuclei. Unless they occur si-

GAMETOPHYTIC SYSTEM   SPOROPHYTIC SYSTEM

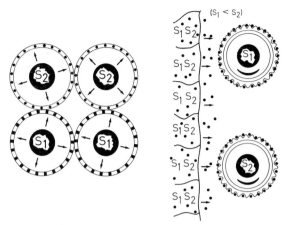

Fig. 1.1. Origin of the genetic information that determines the incompatibility phenotype of pollen grains in sporophytic and gametophytic systems. Final evidence for a contribution of the anther tapetum to pollen S-specificity cannot be obtained before the identification of S-determinants in the pollen and the demonstration of their migration from the tapetum. (de Nettancourt 1979)

multaneously in two different sites of the microspores, the mutations either will not be expressed in the mutated microspore or will not be transmitted to the next generation. Yet, observations by Mackenzie et al. (1967) and Heslop-Harrison (1967) indicate that the time of restoration of protein synthesis at the end of meiosis corresponds to the stage at which Lewis and Pandey believed the incompatibility phenotype is determined.

Brewbaker (1957, 1959) considered that the time of action of incompatibility genes during pollen formation in gametophytic systems was related to the second mitotic division and consequently occurred in the microspores of species with binucleate pollen and in the pollen tubes of species with trinucleate pollen. The rationale of Brewbaker is essentially based on the strong relationship between the stage at which the second mitotic division (pollen or pollen tube) and the site of the incompatibility reaction (style or stigma). The growth of pollen tubes depends on certain concentrations of a pollen growth factor, which is exhausted at the time of the second mitosis and can no longer be produced, utilized or transferred to the pollen in the case of an incompatible mating (Brewbaker and Majunder 1961).

Of course, the techniques available at the beginning of the year 2000 enable us to time gene action during microsporogenesis, pollen maturation and pollen growth. Dodds et al. (1993), working with *Nicotiana alata*, and Clark and Sims (1994), working on *Petunia hybrida*, have shown how it is possible to search immature anthers, microspores or pollen at different developmental stages (ranging from late meiosis to mature pollen) for the expression of specific genes (in this case, a stylar ribonuclease gene that controls the rejection of incompatible pollen tubes in the Solanaceae). The problem, as shall be seen in other sections of this volume, is that it is not yet known (with the recent exception of the *Brassica* system; Schopfer et al. 1999; Sect. 3.2.4.4) what the product of the incompatibility gene in the pollen is or, in the specific case of the Solanaceae, if the incompatibility determinant active in the pollen is related or partly homologous to the stylar ribonucleases.

1.3.2.2 Determination of the Pollen Phenotype in Sporophytic Systems

In sporophytic systems, with a few exceptions [such as those of tristylic species, of *Theobroma cacao* (Cope 1962) and possibly of the Cruciferae (Lewis et al. 1988), which are presented below], the determination of the incompatibility phenotype of the pollen depends exclusively on the genotype of the pollen-producing plant. This means that, in the case of a sporophytic system governed by a single locus S with two different allelic states (S1 dominant and S2 recessive), the heterozygous plant S1S2 will produce two classes of pollen grain, S1 and S2, which will both express S1 as a phenotype. The example will be extended to several others in Chapter 2.

Riley (1936) was the first to explain how the cytoplasm of pollen mother cells could disseminate incompatibility substances from the sporophyte to the gametophyte in *Capsella grandiflora*. Pandey (1960, 1970a) accepted this view and postulated that differences between gametophytic and sporophytic systems resulted from a small variation in the time of gene action; this variation would occur early in sporophytic incompatibility, at the onset of meiosis or at the pre-meiotic stages. Taking into consideration the complex network of relationships between genetic control, floral morphology, pollen cytology and the site of pollen inhibition (this section and Sects. 1.3.3, 1.3.4), Pandey even suggested that the time of action for incompatibility genes was likely to be earliest, and presumably pre-meiotic, in species where sporophytic incompatibility is associated with floral polymorphism.

Such conclusions were not shared by Heslop-Harrison (1968) and his co-workers (Heslop-Harrison et al. 1973), who considered that the synthesis of incompatibility substances in sporophytic systems does not depend on the meiocyte or its primordium but is the function of the tapetum (Fig. 1.1). The hypothesis was originally based on the observation that several constituents synthesized by the tapetal cells are transferred to the microspores during the late phase of pollen maturation (Heslop-Harrison 1968). It was reinforced by the finding (Heslop-Harrison et al. 1974) that, in *Iberis*, the material transferred from the tapetum to cavities of the pollen exine is responsible for the SI response on the stigmatic papillae a few hours after self-pollination. At approximately· the same time, Dickinson and Lewis (1973a, 1973b, 1975) discovered that extracts of the tapetum on the stigma surface of *Raphanus* elicit the stigma reaction in the complete absence of any sporocyte or pollen grain.

The concepts of Pandey (1960) and of Heslop-Harrison (1968) will be discussed in Chapter 3 together with additional data and new evidence regarding the mixing of the products of different S-alleles in pollen coats (Doughty et al. 1998; Schopfer et al. 1999). However, it is clear that the tapetum hypothesis can only be verified through the identification, in tapetal fractions, of the pollen component (as seen in Sect. 3.2.4.4 for SCR in the cabbage family) which participates in the pollen-pistil recognition reaction in sporophytic systems.

1.3.3 The Association with Floral Polymorphism

Several species with sporophytic incompatibility add a number of associated differences to the mechanism of pollen rejection; these differences reinforce the outbreeding potential of the self-incompatible plant and contribute directly to the prevention of self-pollination. Incompatibility systems characterized by such features are said to be heterostylic or heteromorphic.

The existence of different forms of flowers among individuals belonging to a single population appears to have been first noted by Clusius in 1583 (cited in Ganders 1979, who provides a detailed outline of early research). The first analyses were made by Hildebrand and Darwin, who later used the term "heterostyled" to designate a species composed of plants having different style lengths and anther levels. As can be seen from Fig. 1.2, such

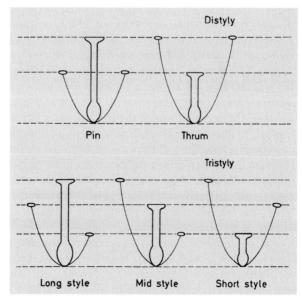

Fig. 1.2. Symbolic representation of distyly and tristyly. In each system, the compatible pollinations only involve anthers and styles at the same level. (de Nettancourt 1977)

species may be distylic, with two types of floral architecture (short style plus high anther and long style plus short anther) segregated in the population, or tristylic, i. e., composed of three distinct groups characterized by long-, mid- and short-styled flowers, each bearing anthers at two different heights (which do not correspond to the level of the stigma).

The main attributes of distyly and tristyly are briefly described below. Specific information on the morphology, genetics and evolution of heterostyly is provided in Chapters 2, 3 and 4. Detailed reviews of these and other matters have been prepared by Ganders (1979), Barrett (1992), who assembled and edited a multi-authored book on the evolution and function of heterostyly, Barrett and Cruzan (1994), and Richards (1997).

1.3.3.1 The Distylic Condition

Working with *Primula sinensis* and *Oxalis rosea*, Hildebrand (cited in Darwin 1880) was probably the first to establish a relation between distyly and SI. However, his earlier data merely suggested that crosses between plants with different morphologies (long×short or short×long) yielded more seeds than after self-pollination or after crosses with identical plants (long×long or short×short).

The observations by Hildebrand were contemporary or precursors to studies (Darwin 1862, 1868, 1876, 1880) carried out on other distylic species (such as *Fagopyrum*, *Pulmonaria*, *Linum*, *Hottonia* and *Polygonum*), where it was emphasized by Darwin that the number of plants with long styles and low anthers (later designated as the pin phenotype) and the number of plants with short styles and high anthers (thrum phenotype) were present in approximately equal numbers for any given population. All pollinations between flowers with identical morphologies (pin×pin or thrum×thrum) were usually sterile, whereas intercrosses between plants with different phenotypes (pin×thrum or thrum×pin) were fertile and gave rise to a progeny that was again segregated in a 1:1 ratio for short styles/high anthers and long styles/low anthers. In other words, Darwin and Hildebrand had not only established the genetic basis of the pin and thrum condition, they had also discovered, by means of self-pollination tests they performed manually, that the main barrier to selfing did not result from the architecture of the flowers but from the incapacity of self pollen to promote seed setting.

The work of Bateson and Gregory (1905), Althausen (1908), Dahlgreen (1916, 1922), Gregory (1915), Eghis (1925), and Garber and Quisenberry (1927) has shown that the pin–thrum alternative and the incompatibility barrier between pin and pin and between thrum and thrum are governed by a single gene complex that segregates as a simple Mendelian factor, with one dominant allele, S, only found in thrum plants and theoretically not obtainable under homozygous conditions, and one recessive allele, s, present in the homozygous state in all pin plants. Cases of reversion, where the pin phenotype is expressed by the dominant S allele and thrum plants are homozygous recessive, are known.

Due to the studies of Ernst (1932, 1936), we also know that, as could be expected, differences in cell size distinguish the pin phenotype from that of thrum. From the observation of abnormal heterostylic phenotypes in *Primula viscosa*, *P. hortensis* and *P. viscosa*×*P. hortensis* hybrids, Ernst concluded that the S-complex of morphological characters in *Primula* could be separated, presumably by crossing over, into at least three recombination units, which control stylar length (G), anther height (A) and the incompatibility of the pollen. It will be seen in Section 2.1.1.1 that Lewis and Jones (1992) and Kurian and Richards (1997) were able, through an examination of Ernst data, to order G, P and A and to add new recombination units to the structure proposed for the S-supergene more than 40 years ago (Ernst 1957).

Sharma and Boyes (1961) made somewhat similar observations for buckwheat, where they induced, by radiation treatment, the partial breakdown of the heteromorphic system and recovered thrum plants with branches bearing either pin or homostyled flowers. Obviously, the fact that floral polymorphism and incompatibility can be dissociated from one another is strong evidence that SI in buckwheat is basically independent from distyly. Lewis and Jones (1992) consider that the incompatibility character occurred first in heteromorphic systems and was followed by gradual changes in pollen size and style length. On the contrary, Lloyd and Webb (1992a) are of the opinion that the initial step in the emergence of heterostyly was stigma height polymorphism (Sect. 4.3.4 for a review of these and other hypotheses on the origin of heterostyly). Dulberger (1992), in her analysis of floral polymorphisms and their significance in the heterostylous syndrome, proposed an integrative concept of the morphological, structural, developmental and physiological functions governed by the S-supergene.

1.3.3.2 Tristyly

Darwin (1880) assembled the bibliography of his time on this more complex type of heterostyly and analyzed the progenies arising from legitimate matings between plants with different flower morphologies in *Lythrum* and *Oxalis*. As in the case of distyly, Darwin found evidence that fertile matings were only possible between plants with different morphologies and concluded, from hand-pollination data, that fertility (compatibility) or sterility (incompatibility) after any given pollination probably resulted from tiny differences between pollen grains and between pistils. The populations studied consisted of three different phenotypic groups of plants with either short, mid-length or long styles. In any given group, the flower bears anthers at two different levels, which never correspond to the height of the style in the flower but coincide with the stylar lengths of the two other groups. As in distyly, compatible pollination occurs only between stigmas and pollen at the same level, and all pollinations between stigmas and pollen at different levels are incompatible (Fig. 1.2).

Since every flower produces two different sets of anthers, it follows that, although the determination of the pollen is sporophytic, each flower (and, therefore, each plant) yields two kinds of pollen that are genotypically identical but phenotypically different. Evidently, the nature, activity and/or specificity of the S product in the pollen is highly sensitive to the internal environment. This sensitivity led Lewis (1949 a) to suggest that a very close connection exists between the associated character, anther height and growth reaction of the pollen. In homomorphic systems there is also ground to suppose that a given S-allele may express different products in different tissues or under different circumstances (Clarke et al. 1985; Chen and Nasrallah 1990; Trick and Heizmann 1992). However, the importance of the local environment for the incompatibility phenotype of the pollen in tristyly must not be exaggerated, because the genetic control typical of tristyly can arise without the morphological differentiation for stylar length and anther height (Riley 1936; Sects. 2.1.2.4, 4.3.4.2).

1.3.4 The Site of Gene Expression

Depending on the system involved, the genes governing SI express themselves at the level of the stigma, the style or the ovary. In heteromorphic incompatibility, rejection zones in different parts of the pistil may vary from family to family, species to species and, within the plant, from morph to morph. In such material, the ability of the flower to reject incompatible pollen or pollen tubes is often incomplete and is supplemented by a sequence of barriers from stigma to style and style to ovary. In homomorphic plants, on the contrary, the site of gene expression (stigma, style or ovary) constitutes a more reliable criterion for the classification of incompatibility mechanisms and can be used as an inference point for deducing other essential particularities of the rejection system in the species considered. Reciprocally, the presence of certain floral attributes in a self-incompatible species may permit one to make important conclusions regarding the site of gene expression.

1.3.4.1 Stigmatic Inhibition

An important characteristic of stigmatic SI is that it occurs in species where the pollen is trinucleate and the stigma dry. The relationship between pollen cytology and pollen behavior was discovered by Brewbaker (1957, 1959), who showed that the 17 genera known to display stigmatic incompatibility at the time were trinucleate, whereas 33 of the 36 genera recorded to exhibit pollen-tube inhibition in the style or the ovary produced binucleate pollen grains. Pandey (1970 a) attributed the relationship to the possibility that the second mitotic division is directly associated with the determination of the incompatibility phenotype in the pollen. Another explanation could reside in differences in the localization (exine-held in stigmatic SI and intine-held in stylar or ovarian incompatibility) of the pollen substances involved in recognition (Heslop-Harrison et al. 1974; Chap. 3). Indeed, Stephenson et al. (1997) demonstrated that the male determinant of stigmatic SI in *Brassica oleracea* is located in the pollen coating; accordingly, certain models of the mechanism of stylar incompatibility in the Solanaceae (Dodds et al. 1996; Kao and McCubbin 1996; Chap. 3) foresee that the pollen determinant (able to recognize the ribonucleases dispatched by an incompatible style) could be sited inside the pollen tube and not on its surface. In a few genera, such as *Beta*, *Helianthus*, *Boungainvillea* and *Fagopyrum*, trinucleate pollen is not associated with stigmatic incompatibility but is related to stylar incompatibility.

The occurrence of stigmatic incompatibility in species with dry, papillate stigmas has been extensively documented by Heslop-Harrison (1975), Heslop-Harrison et al. (1975) and Heslop-Harrison and Shivanna (1977). Heslop-Harrison and Shivanna (1977) observed that the differences between the "wet" and "dry" classes of stigma intergrade. They also explained how stigmatic incompatibility could "only function with a dry stigma, where single stigmatic cells tend to encounter single pollen grains, or at most very small numbers". In confirmation, Dickinson (1995) reported that the focusing of the SI response of the dry stigma is such that a single papilla is able, simultaneously, to accept cross-pollen and reject self-pollen. According to Roberts et al. (1980), the basic response in *Brassica* after the recognition reaction between the stigmatic pellicle and proteins from the pollen coating is the inhibition of water supply, from which all other manifestations of incompatibility result. It will be seen in Chapter 3 that the regulation of water is an important starting point of the complex rejection process in stigmatic sporophytic SI.

Stigmatic incompatibility is often typical of sporophytic systems, but the relationship is by no means absolute. Several species in many different genera, such as *Phalaris*, *Secale*, *Tradescantia*, *Papaver* and *Oenothera*, display stigmatic gametophytic SI.

1.3.4.2 Stylar Inhibition

The SI reaction occurs between the pollen tube and style in SI species where the stigma presents a "distinct surface secretion, with a free fluid surface" (Heslop-Harrison and Shivanna 1977). Such wet stigmas, generally found in species with binucleate pollen, are moistened by exudates produced by the style; they presumably contain promoters of pollen-tube growth (Franklin et al. 1995).

In the case of the Solanaceae, where the arrest (or, more often, the slowing down) of incompatible tubes results from an attack by specific ribonucleases, Lush and Clarke (1997) have noted that the literature is particularly unclear and contradictory with regard to the occurrence of callose plugs, the destruction of the tube apex and the distance incompatible tubes finally travel (Sects. 3.5.2.1, 3.5.7.2, 5.1.3). Indeed, the behavior of incompatible pollen tubes (which can lead to the swelling and bursting of tube tips in the upper portion of the style) varies among systems, species and plants. McClure et al. (1993) have shown that the specific S-ribonucleases of *N. alata* are not only present in styles but also in stigmas at concentrations that are not the same in different clonal populations. Variations can also occur among flowers of the same plant and among incompatible tubes in the same style. One reason for such variability is that the rejection reaction in several SI stylar systems involves interactions from many different genes or results from a general inhibition of protein synthesis in pollen tubes; this inhibition appears to be highly sensitive to the internal and external environment of the flower and to pollination conditions. The fact that incompatible and compatible tubes do not influence one another led Linskens to suggest, as early as 1965, that the incompatibility complex has its place of action on or in the pollen tube, not outside (in the style).

1.3.4.3 Ovarian Inhibition

It has been the mistake of several authors, and of the present one in previous reviews, to underestimate the importance and prevalence of ovarian SI in flowering plants. Reports published in the 1980s and the 1990s (Kenrick et al. 1986; Seavey and Bawa 1986; Sage et al. 1994; Sedgley 1994) demonstrate that ovarian self-incompatibility (OSI) operates in many woody and herbaceous species.

The arrest of self-tubes has been observed in different regions of the ovary, on the placenta adjacent to the micropyle, in the micropyle itself and within the embryo sac. In several species with OSI, the styles are hollow and, according to Brewbaker (1957), do not provide the pollen tubes with the contact necessary for growth inhibition to occur. Examples of species which combine hollow styles and ovarian responses are to be found in *Narcissus, Hemerocallis, Lillium, Gasteria, Ribes, Anona* and *Freesia* (Bateman 1954; Arasu 1968; Pandey 1970 a), among others. In certain instances, such as in *Gasteria verrucosa* (Sears 1937) and in *B. officinalis* (Crowe 1971), the reaction occurs after the first division of the endosperm (*Gasteria*) or is clearly post-zygotic (*Borago*). A well-known example of OSI taking place at the onset of gametic fusion has been reported in *T. cacao* by Cope (1962), who observed that the release of sperm nuclei was not followed by syngamy for some ovules. The detailed descriptions of OSI in *T. cacao* were provided by Knight and Roger (1953, 1955), Cope (1958, 1962) and Bouharmont (1960).

Sage and co-workers (1999) showed that the SI reaction in *Narcissus triandrus* does not primarily lead to differential pollen-tube growth but to an effect from incompatible tubes after ovule development (Sect. 2.3.7.3). Evidence (Sect. 2.3.7) that suggests that OSI is gametophytic in several species has also been obtained, but the genetic bases remain to be established. It is considered, in the case of *T. cacao*, that both gametophytic and sporophytic control operate (Cope 1962; Sect. 2.4.1).

1.3.5 The Number of Genetic Loci and the Involvement of Polyallelic Series

As pointed out by Nasrallah (1997), there are three phases in the SI reaction of *Brassica*: recognition (or signal perception), signal transduction and rejection. In stigmatic SI, these phases are often associated with certain processes (activation of kinase receptors and phosphorylation of essential pollen proteins) that have not been identified in vivo in the case of stylar SI, where it is customary to refer to only two steps in the SI reaction: recognition and rejection.

1.3.5.1 The Genetic Basis of Recognition

Depending on the system or the plant family concerned, the recognition of pollen–pistil incompatibility may be governed by one or several genetic loci.

1.3.5.1.1 Control by a Single but Complex Locus. In many SI species, the recognition phase is governed by one genetic factor, transmitted as a single Mendelian character, designated as the "S" gene. This terminology is improper because, even in systems where segregation ratios reveal monofactorial control, there appears to be no such thing as a single gene responsible for the pollen–pistil recognition phase. Different genes in the pollen and pistil (and presumably, but not necessarily, different gene products) are responsible for the identification of self pollen. In monofactorial SI, these genes are closely integrated within a same linkage group (or cluster of genes), apparently protected against disruption through the prevention of crossover, which is inherited as one determinant. This is the case in *B. oleracea* (SSI), where many different genes (McCubbin and Kao 1999; Chap. 3) reside at the S locus. Two of these (SRK and SLG) are active in the pistil, and the third one, SCR, between SRK and SLG, presumably encodes the pollen determinant, possibly a ligand for SRK (Stephenson et al., 1997; Nasrallah 1997; Schopfer et al. 1999; Sect. 3.2.4.4). Through a study of deletion effects in *Pyrus serotina* (GSI), Sassa et al. (1997) have demonstrated that the S-locus region contains two or more linked genes, including the S-ribonuclease gene and an unidentified pollen determinant.

The finding, predicted 35 years ago by Linskens (1965), that different linked genes in "monofactorial" GSI and SSI determine the incompatibility phenotypes of the pollen and the pistil casts doubt on the validity of the "dimer hypothesis" of Lewis (1965), as will be seen in Chapter 3. According to this hypothesis, the rejection of "self" results from the recognition of identical gene products from the pollen and pistil and the subsequent formation of dimer repressors. It is not impossible, however, in view of the stability of each gene within the linkage groups, that different pollen and pistil genes, which possibly originated from duplications, maintain identical or partly identical sequences (in particular, see Sects. 3.2.4, 3.5.8).

1.3.5.1.2 Recognition by Two Unlinked Loci in the Grasses. The recognition mechanism in the stigmatic GSI system of the grasses is governed by two unlinked multiallelic genes: S and Z (Lunqvist 1956, 1962a; Hayman 1956; Chap. 2). The S-gene encodes a protein erroneously suspected, in the past, to express similarity to thioredoxins. It is possible (Sect. 2.3.2; Chap. 3) that S and Z contribute the components of a kinase receptor. The efficiency of the system (i.e., the probability that a cross between two plants taken at random in a population yield seed) is increased by the absence of linkage between S and Z.

1.3.5.1.3 Recognition by Two or More Loci in Several Other Families. SI is also governed by unlinked loci in certain families with homomorphic or heteromorphic SSI. Two unlinked di-allelic loci operate in *Capsella* (Riley 1932, 1936), *Cardamine* (Correns 1912) and *Lythrum* (Barlow 1913).

Up to four polyallelic genes participate in the determination of GSI in *Eruca* (Verma et al. 1977), *Beta* and *Ranunculus* (Lundqvist et al. 1973; Larsen 1977, 1978, 1986; Lundqvist 1990a, 1990c), and *Lillium* (Lundqvist 1991). These genes are unlinked or incompletely linked and complementary in their effects. It is now generally considered (Lundqvist 1990a; Sect. 4.3.2) that they did not originate from duplications of an ancestral monogenic mechanism but, on the contrary, represent the more primitive system.

1.3.5.2 Polyallelism at the Incompatibility Loci

Recognition mechanisms can be classified into two groups:
- Those with polyallelic series for each or some of the genes participating in the recognition phase. Such series occur in many species with homomorphic SSI and GSI. They also have been detected (Sect. 2.1.3) in three heteromorphic species: *Narcissus tazetta* (Dulberger 1964), *Anchusa hybrida* (Dulberger 1970) and *A. officinalis* (Philip and Schou 1981). Interactions between different alleles of the same gene can in principle be established only in diploid (or polyploid) tissues – i.e., only in sporophytic systems – or only in the pistil in the case of gametophytic systems (however, see Sect. 3.2.4.2). Regardless of any relationship of dominance that may be expressed between alleles in diploid cells, each allele in a polyallelic series is designated by a capital letter (S in a monofactorial system, or S and Z when two unlinked loci are involved) followed by a serial number (S1, S37...; S1–Z4, S5–Z3...).
- Those with two alleles (one dominant, one recessive) per gene; these are very infrequent in homomorphic SSI (Sect. 2.1.2.1) and generally operate in heterostyly.

Lawrence (1996) re-visited earlier data regarding the number of S alleles in different SI species with polyallelic series. An important polymorphism was confirmed in clover populations (approximately 100 alleles for *Trifo-*

lium repens and 200 for *T. pratense*). In the populations representing the other species examined, the number of different S alleles recorded were much lower and ranged between 5 and 50.

1.3.5.3 How Many Genes Are Involved in the Rejection Process?

The answer depends on the system involved, but rejection, particularly when cascade reactions are involved, is certainly complex and leads to the modification or inactivation of many of the gene products normally required during the progamic phase (Linskens 1986).

1.3.5.3.1 Stigmatic SSI. According to Roberts et al. (1980), the basic response in *Brassica* after the recognition reaction, is the failure of pollen hydration, from which all other manifestations of SI result. Ikeda et al. (1997) consider that this inhibition of water supply, which occurs immediately after a kinase receptor has received a diffusible signal from the pollen coat, could be the consequence of the intervention of an activated aquaporin-related gene (Sects. 2.5.1, 3.2.5.2). Function losses affecting this gene prevent rejection and the expression of SI.

Of course, other downstream effectors are also involved in the rejection phase of SI; as noted by Nasrallah (1977), they may inflict several other types of damage in addition to the interruption of water uptake. She lists several examples of the Brassica SI response, such as the disruption of adhesion to the stigma, an important accumulation of calcium and, perhaps, modifications to the organization of the papillar cytoskeleton (nevertheless, see Dearnaley et al. 1999, who failed to observe any structural reorganization within papillar cells after compatible or incompatible pollination in *Brassica*) and exocytosis. Presumably, many of the changes result directly or indirectly from the phosphorylation/dephosphorylation of different gene products that are essential to the normal metabolism of pollen grains. Through the use of antisense DNA, Stone and co-workers (1999) clearly demonstrated the role of the ARC1 protein as a putative downstream effector for SRK in self-incompatible *Brassica* (Sect. 3.2.5.5.).

1.3.5.3.2 Stigmatic GSI. In rye, there is evidence that the products of the S and Z genes and of non-S-specific loci from three different chromosome arms participate in a transduction cascade that ultimately leads to the rejection of SI pollen (Wehling et al. 1995). In the field poppy, the SI signaling pathways involve changes in the phosphorylation of at least two pollen proteins (Rudd et al. 1997). The complexity of the situation is well illustrated by the finding by the Birmingham group that Ca^{2+}-dependent and Ca^{2+}-independent kinases are involved.

1.3.5.3.3 Stylar GSI. It is difficult, in this case, to dissociate recognition from rejection, because the S-specific ribonucleases apparently participate in the two phases. However, the group of Clarke at Melbourne (Kunz et al.

1996) has shown that the S-ribonucleases of *N. alata* are phosphorylated in vitro by Ca^{2+}-dependent protein kinases from pollen tubes. Therefore, it is possible, as shall be seen in Chap. 3, that a kinase receptor and protein phosphorylation/dephosphorylation play roles in gametophytic stylar SI. In principle, the involvement of such a mechanism in the rejection of incompatible pollen, unless it participates in the processing or activation of stylar S-ribonucleases, should not be required, because the effects of S-ribonucleases alone should be sufficient to prevent the synthesis of all the proteins normally produced in pollen tubes. There are indications, however, that this may not be the case (Sect. 3.5.7.2).

■ 1.4 Recapitulation on the Classification of SI Systems

In a study of almost 1000 SI and self-compatible (SC) species in about 900 genera from 250 families, Heslop-Harrison and Shivanna (1977) reviewed the relationships between pollen type, stigma type, the site of pollen rejection and the SI system involved (Table 1.1). The pollen barrier usually operates on or immediately below the stigma surface (in species characterized by a dry stigma) and within the style or (more rarely) the ovary in species with a wet stigma. Plants with wet stigmas tend to have bi-nucleate pollen, but bi-nucleate pollen also occurs with dry stigmas. SSI systems are often associated with dry papillate stigmas.

Of all the associations that appear in Table 1.1, the most constant and reliable concerns the physiological state (wet or dry) of the stigma and the

Table 1.1. Usual features of different self-incompatibility (SI) systems. (de Nettancourt 1997)

Homomorphic SI				
Tri-nucleate pollen on dry stigma	Stigmatic	Sporophytic	Di-allelic and poly-allelic loci Tapetum probably determines S-phenotype of pollen	One or more genes
Bi-nucleate pollen on wet stigma	Stylar	Gametophytic	Poly-allelic loci	
Heteromorphic SI				
Dry stigma	Stigmatic	Sporophytic	Di-allelic loci	One or more genes
Wet stigma	Stylar		Tapetum probably does not determine the S-phenotype of pollen	
	Ovarian			

site of the reaction (stigmatic or stylar). As pointed out by Heslop-Harrison (1975, 1979), dry stigmas allow a direct interaction between single pollen grains and single papillae (which is hardly possible in the common fluid medium, where pollen grains germinate when the stigma is wet).

▪ 1.5 The Distribution of SI Systems in the Angiosperms

1.5.1 Incidence of SI in the Families of Flowering Plants

East (1940) experimentally tested a total of 800 species from 44 orders of monocotyledonae and dicotyledonae or classified them from literature data. Assuming that such material represented a random sample, East calculated that the number of SI species among flowering plants is at least 3000. Darlington and Mather (1949) raised the number by estimating that half of the species in angiosperms display SI. Brewbaker (1959) confirmed this figure when he estimated that SI was known for at least 71 families in more than 250 of the 600 genera he analyzed (Table 1.2).

Homomorphic SI is present in most branches of the family tree, but with relationships to criteria (such as those of pollen cytology and the condition of the stigma) that are specific for the different families of flowering plants. Accordingly, with a few major exceptions, such as the Poaceae (Chap. 4), the system tends to be the same within any given family (Table 1.2). Thus, SI in the Solanaceae, Scrophulariaceae and Rosaceae is usually gametophytic, stylar, monofactorial and polyallelic; Poaceae (Graminaceae) generally display bi-factorial gametophytic control; Cruciferae and Compositae are sporophytic stigmatic and polyallelic, with a possible complementary gametophytic mechanism (Lewis et al. 1988; Lewis 1994). Lundqvist (1990a) notes that GSI is far more widespread than SSI, which is known from only six families of angiosperms.

In his review of the biology of heterostyly, Ganders (1979) indicates that the phenomenon is known in 24 families of flowering plants, including monocots and dicots belonging to 18 different orders. The Rubiaceae contain more heterostylous genera than all other families combined, but heterostyly, like homomorphic SI, is not restricted to a closely related group of families and must have evolved several times. Ganders also notes that heterostyly is usually associated with perenniality [with some exceptions, such as *Amsinckia* spp. and *Fagopyrum esculentum* (Ganders 1979) or *Crypyantha* and *Linum* (Dulberger 1974)]. Most of the heterostylous taxa recorded in Table 1.2 are distylous. Tristyly is only known in three families (Lythraceae, Oxalidaceae, Pontederiaceae) and is suspected in the Connaraceae.

Homomorphic SI occurs in many of the families that display heterostylic species (Table 1.2). The implications of our current knowledge of the distribution of SI with regard to the origin and evolution of incompatibility systems are discussed in Chapter 4.

Table 1.2. Distribution of homomorphic and heteromorphic self-incompatibility (SI) among the angiosperms

Order	Family	SI system recorded Homomorphic	Heteromorphic
Lilliales	Iridaceae	+	
	Amaryllidaceae	+	
	Liliaceae	+	
	Pontederiaceae		+
Commelinales	Commelinaceae	+	
	Bromeliaceae	+	
Poales	Poaceae	+	
Zingiberales	Zingiberaceae	+	
Orchidales	Orchidaceae	+	
Nymphaeales	Nymphaeaceae	+	+
Theales	Theaceae	+	+
	Clusiaceae		+
Malvales	Sterculiaceae	+	
	Bombaceae	+	
	Malvaceae	+	
Ericales	Ericaceae	+	
Primulales	Primulaceae	+	+
Cistales	Passifloraceae	+	
	Cisteaceae	+	
Capparidales	Resedaceae	+	
	Brassicaceae	+	
Cucurbitales	Begoniaceae	+	
Saxifragales	Saxifragaceae	+	+
Connarales	Connaraceae		+
Rosales	Rosaceae	+	
Fabales	Fabaceae	+	+
Geraniales	Tropacolaceae	+	
	Geraniaceae	+	
	Oxalidaceae		+
	Linaceae		+
Scrophulariales	Solanaceae	+	
	Nolanaceae	+	
	Scrophulariaceae	+	
	Bignoniaceae	+	
	Lentibulariaceae	+	
	Acanthaceae	+	+
	Plantaginaceae	+	
Santalales	Olacaceae		+
Olmeales	Oleaceae	+	+

Table 1.2 (continued)

Order	Family	SI system recorded Homomorphic	Heteromorphic
Gentianales	Loganiaceae	+	+
	Gentianaceae		+
	Menyanthaceae		+
	Apocynaceae	+	+
	Asclepadiaceae	+	
Lamniales	Verbenaceae	+	
	Labiatae	+	
Polemoniales	Polemoniaceae	+	
	Convolvulaceae	+	
	Boraginaceae	+	+
Campanulales	Campanulaceae	+	
	Goodeniaceae	+	
Asterales	Compositae	+	
Rubiales	Rubiaceae	+	+
	Caprifoliaceae	+	
Myrtales	Myrtaceae	+	
	Onagraceae	+	
Umbellales	Cornaceae	+	
Ranales	Ranunculaceae	+	
Papaverales	Papaveraceae	+	
	Fumariaceae	+	
Caryophyllales	Plumbaginaceae		+
Polygonales	Polygonaceae	+	+
Hamamelidales	Hamamelidaceae	+	
Urticales	Ulmaceae	+	
	Moraceae	+	
Fagales	Fagaceae	+	+
Betulales	Betulaceae	+	

Compiled by de Nettancourt (1977) from Fig. 2 in Brewbaker 1959 and with additions taken from Table 1 in Ganders (1979). Modifications have been made to take into account certain changes regarding ordinal and family levels found in Takhtajan (1969). For further information on SI distributions, see also East (1940) and Vuillemier (1967)

Table 1.3. Self-incompatibility systems recorded in some of the genera that play an important role in plant-breeding sciences. The list of authors in the third column does not always include the names of the scientists having identified the system but simply provides a basic reference from which more information may be obtained. (de Nettancourt 1977)

Genus	Self-incompatibility system recorded	Reference
Beta	Homomorphic, polyfactorial, gametophytic	Larsen (1977)
Brassica	Homomorphic, S-genes in haplotype, sporophytic[a]	Nasrallah and Nasrallah (1993)
Coffea	Homomorphic, monofactorial, gametophytic[b]	Devreux et al. (1959)
Dactylis	Homomorphic, bi-factorial, sporophytic	Lundqvist (1965)
Fagopyrum	Heteromorphic, monofactorial, sporophytic	Dahlgreen (1922)
Festuca	Homomorphic, bi-factorial, gametophytic	Lundqvist (1961)
Helianthus	Homomorphic, sporophytic	Kinman (1963)
Linum	Heteromorphic, monofactorial, sporophytic	Murray (1986)
Lotus	Homomorphic, one to several loci, gametophytic	Lundqvist (1993)
Lycopersicum	Homomorphic, monofactorial, gametophytic	Lamm (1950)
Malus	Homomorphic, monofactorial, gametophytic	Korban (1986)
Medicago	Homomorphic, monofactorial, gametophytic	Duvick (1966)
Nicotiana	Homomorphic, monofactorial, gametophytic	East and Mangelsdorf (1925)
Passiflora (passion fruit)	Homomorphic, bi-factorial, sporophytic	Do Rego et al. (1999)
Phalaris	Homomorphic, bi-factorial, gametophytic	Hayman (1956)
Prunus (sweet cherry)	Homomorphic, monofactorial, gametophytic	Crane and Brown (1937)
Prunus (almond)	Homomorphic, monofactorial, gametophytic	Tao et al. (1997)
Prunus (apricot)	Homomorphic, monofactorial, gametophytic	Burgos et al. (1998)
Pyrethrum	Homomorphic, sporophytic	Brewer (1968)
Pyrus	Homomorphic, monofactorial, gametophytic	Sassa et al. (1993)
Raphanus	Homomorphic, monofactorial, sporophytic[a]	Sampson (1957)

Table 1.3 (continued)

Genus	Self-incompatibility system recorded	Reference
Secale	Homomorphic, bi-factorial, gametophytic	Lundqvist (1956)
Solanum	Homomorphic, monofactorial, gametophytic	Pushkarnath (1942)
Solanum sp.	Homomorphic, bi-factorial, partly sporophytic	Pandey (1962a)
Solanum sp.	Homomorphic, bi-factorial	Abdalla and Hermsen (1971)
Theobroma	Homomorphic, multi-factorial, sporo-gametophytic	Cope (1962)
Trifolium	Homomorphic, monofactorial, gametophytic	Duvick (1966)

[a] With contribution from a gene acting gametophytically (Lewis et al. 1988)
[b] For further references on current research, see Lashermes et al. (1996)

It has been shown by East (1940) that, with the one exception of the Spatiflorae, where SI does not occur, cleistogamy and SI follow exactly the same pattern of distribution. SI was found to be more frequent in herbaceous plants (14 orders out of 25) than in woody plants (2 out of 18).

1.5.2 Distribution of SI among Species Important for Agriculture

SI in cultivated plants occurs less frequently than in wild species and often appears to be weakened by pseudocompatibility, i.e., the malfunctioning of the self-rejection mechanism in certain genetic backgrounds or under certain environmental conditions. Mather (1953) and Rowlands (1964) came to the conclusion that, during early domestication, pseudocompatibility in cultivated crops had been selected as a character conferring immediate fitness with regard to the environment and the production criteria imposed by the breeder (absence of cross-pollinating insects, wind shields, emphasis on maximum seed yield and, possibly, imposed inbreeding by deliberate self-pollination or very severe restrictions in the number of S genotypes). A somewhat similar situation, at least in terms of restrictions in the number of S alleles, probably also occurred in botanical gardens where only few genotypes were used repeatedly for the establishment and maintenance of collections [for instance, see Thompson and Taylor (1965) who concluded a survey on the proportion of identical S-alleles in different collections of cabbages]. Atwood (1947) observed that plant breeders dealing with forage crops performed inadvertent selection in two opposite directions. They tended to select for self-compatibility in all cases where they were searching for homozygosity in various phenotypic traits; at the same

Table 1.4. Discovery of self-incompatibility. (de Nettancourt 1977)

Discoverer	Species
Kölreuter	*Verbascum phoeniceum*
Herbert	*Zephyranthes carinata* and *Hippeastrum aulicum*
Bidwell	*Amaryllis belladona*
Bernet	*Cistus* sp.
Rawson	*Gladiolus gandavensis*
Mowbray	Several species of *Passiflora*
Munro	Several species of *Passiflora*
Gärtner	*Dianthus* sp., *Lobelia* sp., *Verbascum* sp.
Scott and Munro	*Oncidium* sp.
Lecoq	*Oncidium* sp.
Riviere	*Oncidium* sp.
Fritz Müller	*Oncidium* sp.
Darwin	*Eschscholtzia, Abutilon, Senecio, Reseda*

Description of distylic incompatibility

Hildebrand (1863)	*Primula sinensis*
Darwin (1877)	*Fagopyrum, Pulmonaria, Linum, Hottonia, Polygonum*

Description of tristylic incompatibility

Darwin (1877)	*Lythrum, Oxalis*

Effects of the environment of self-incompatibility

Darwin (1877)	Pseudocompatibility in *Abutilon darwinii*

Cross-incompatibility within self-incompatible populations

Munro (1868)	*Passiflora*

Self-incompatibility as a prezygotic event

▦ Stylar inhibition

Scott (1865)	*Oncidium*
Jost (1907)	*Secale* and *Lilium*

▦ Stigmatic inhibition

Müller (1868)	*Eschscholzia*
Hildebrand (1896)	*Cordamine*

The genetic bases of self-incompatibility

▦ Distyly

Bateson and Gregory (1905)

▦ Tristyly

Barlow (1913, 1923)

Von Übisch (1921)

Table 1.4 (continued)

Discoverer	Species
Sporophytic homomorphic bi-factorial systems	
Correns (1913)	
Gametophytic monofactorial polyallelic systems	
Prell (1921)	
East and Mangelsdorf (1925)	
Lehmann (1926)	
Filzer (1926)	
The inheritance of self-incompatibility in self-incompatible species	
Baur (1911)	*Antirrhinum* hybrids (segregation in F$_2$)
Compton (1913)	*Reseda odorata* (dominance of self-incompatibility in F$_2$)

All references to the work carried out in the nineteenth century were found in East and Park (1917) and in French translations of Darwin's, *Origin of species,* and Darwin's, *Animals and plants under domestication*, which are listed under Darwin 1879, 1880 and 1905 in the bibliography

time, they took advantage of SI and cross-compatibility for mass-selection purposes or for creating synthetic cultivars.

SI is not uncommon in cultivated species and occurs frequently among the wild relatives which are used for the transfer of important genetic traits, such as disease resistance, hardiness or precocity. From Table 1.3, it can be noted that several essential crop plants, such as clovers, alfalfa, cabbage, kale, sunflowers, coffee, rye and many grasses, sugar beet, cherries, pears, plums, apricots, apples, and olives are, to different degrees, completely or partly self-incompatible. SI is, however, generally restricted to diploid species because, as will be seen in Chapter 4, polyploidy usually leads to the breakdown of SI or, as in alfalfa and birdsfoot trefoil, to a complex and often erratic inheritance of the reproductive barrier. Sedgley (1994) established the (long) list of woody horticultural species that exhibit reduced fruit or seed set following selfing.

■ 1.6 Chronology of Early Researches on SI

It is not possible to retrace in any detail the succession of early researches that followed the discovery of SI by Kölreuter in 1764. For this purpose, information should be sought in Darwin (1876), Correns (1912, 1913), East and Park (1917), Stout (1917, 1920) and East (1929, 1932, 1934, 1940), who extensively reviewed the observations made on cross-breeding mechanisms in flowering plants before and during their time. In Table 1.4, attempts have been made to classify (in chronological order) some of the important findings reported during the period between the date on which Kölreuter reported his observations of the genus *Verbascum* and the year 1925, when East and Mangelsdorf confirmed the work of Prell (1921) regarding the genetic basis of the homomorphic monofactorial gametophytic system of self-rejection. For a historical overview of the discovery of sexual reproduction in higher plants, consult Cresti and Linskens (1999).

CHAPTER **2** **The Genetics of Self-Incompatibility**

The purpose of this chapter is to review the genetic bases of self-incompatibility (SI) systems and to show how one or several individual genes or groups of linked genes (1) control the breeding behavior of the plant that carries them and (2) govern the genetic structure of the population where they segregate. No other phenomenon in nature appears to provide such a clear example of the modalities through which an important recognition device integrated within a complex genetic network for essential reproductive functions is inherited, reconstituted in each generation and set to maintain outbreeding and heterozygosity.

2.1 Sporophytic Heteromorphic Systems

2.1.1 Distyly

The pin–thrum complex of characters (Table 2.1) is usually governed at a single chromosomal site with two allelic states, S and s, which control the thrum (Ss) and pin (ss) situations, respectively. Any given population con-

Table 2.1. Early description of the S supergene and its functions in distylic species. (de Nettancourt 1977, from Lewis 1949a and Sharma and Boyes 1961)

Gene	Function	Pin plants		Thrum plants	
		Genotype	Phenotype	Genotype	Phenotype
Is	Stylar incompatibility	is is	Rejection of pin pollen	Is is	Rejection of thrum pollen
Ip	Pollen incompatibility	ip ip	Rejection by pin style	Ip ip	Rejection by thrum style
G	Stylar length	g g	Long style	G g	Short style
P	Pollen size	p p	Small	P p	Large
A	Anther height	a a	Low	A a	High

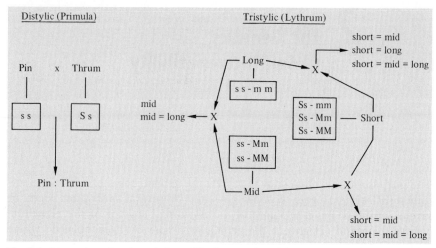

Fig. 2.1. Segregations for *pin* and *thrum* (distylic system) and *long, mid* and *short* (tristyly) after compatible mating. (de Nettancourt 1972)

sists of two classes of self-incompatible individuals that are reciprocally cross-compatible between classes and cross-incompatible within each class (Fig. 2.1). If there are no selective disadvantages operating against one of the two phenotypes, the progenies of cross-compatible plants will again consist of two phenotypic classes, pin and thrum, which are present in equal proportions (isoplethy). As noted by Lewis (1949a), the mechanism is similar to the XY system of sex determination operating in animals and plants and has similar consequences for outbreeding.

With the exceptions of three species in the Guttifereae (Ornduff 1975), the Plumbaginaceae (Baker 1954, 1966) and the Lythraceae (Lewis and Rao 1971; Lewis 1975; Lewis and Jones 1992) where the pin phenotype is determined by the Ss genotype, the pin (ss)/thrum (Ss) alternative is typical of all distylic species analyzed to date (Lewis 1949a; Vuilleumier 1967; Lewis and Jones 1992).

2.1.1.1 A Supergene ...

The S locus in distylic species hosts linked genes or linked groups of genes that control the breeding behavior of the plants. This integrated association of genes is referred to as a "supergene", but the designation of "haplotype" (Sect. 2.2.2.1), adopted in recent years for the complex of linked genes governing homomorphic SI in *Brassica*, would probably do just as well. The supergene in distylic species governs not only the pollen–pistil recognition reaction but several other associated characters (Table 2.1), such as anther position and stylar level, which result from changes in cell size and cell number or lead (as in *Jepsonia*) to dimorphism in the sculpturing of the pollen wall (Ornduff 1970). These characteristics reinforce the strength of the pollen–pistil barrier established by the SI genes.

Lewis and Jones (1992) reviewed the evolution of the "supergene" hypothesis, initially raised by Gregory (1915), drafted by Pellew (1928), Mather and De Winton (1941) and Mather (1950), and elaborated in depth by Ernst (1936, 1957), Lewis (1954) and Dowrick (1956). The total supergene in *Primula* is considered to consist of six "genes": *G* (style length), *S* (stigmatic surface), *Is* (stylar incompatibility), *Ip* (pollen incompatibility), *P* (pollen size) and *A* (anther height). Each of these genes is di-allelic. The six alleles are usually all recessive in Pin plants and all dominant in Thrum plants, but cases of reversion of dominance are known in distylic species where Pin is dominant and Thrum recessive (see above), and in tristylic *Oxalis* deviants (see below). These exceptions, particularly those observed in *Oxalis*, led Lewis and Jones (1992) to suspect the existence of a switch gene that would regulate the transcription of the entire supergene and the dominant status of the Thrum allele.

2.1.1.2 ... Within Which Recombination Occurs

The nature of these six genes is unknown. For some of them, physical existence is demonstrated by the occurrence of crossing over. The recombination of alleles that results from it may cause what are called "homomorphic variants", i.e., plants that have anthers and styles at the same level because they combine the stylar features of one morph with the anther features of the other morph (for an example of populations of *Primula vulgaris* containing high proportions of such variants, see Crosby 1940). The distances between the genes are such that the recombinant plants often share the style length and compatibility group of one morph or the stamen length and compatibility group of the other morph (Ganders 1979). This is why homomorphic variants are often self-compatible and self-pollinating.

However, not all homomorphic variants are recombinational in origin and, because of the non-symmetrical distribution of recombinants or the rates at which they occur, it is considered (Richards 1997) that several homostylous plants in *Primula* sp., *Primula* hybrids, *Linum* or *Anchusa* result from mutations or the actions of modifiers. Richards distinguishes primary homostyles, representing the initial stage in the original evolution towards distyly, from secondary homostyles, which result from recombination (or other mechanisms) during the heterostylic phase.

The recombination data available indicate that the whole supergene measures approximately 1 crossing-over unit and that the genes responsible for the size of the female gynaecium (G) and male androecium (A) and the pollen size (P) are in the order G, P, A for the material studied (Fig. 2.2). Lewis and Jones (1992) noted that, in controlled breeding experiments, there is no evidence suggesting that pollen incompatibility (Ip) is separated from pollen size (P) or that stylar incompatibility (Is) is separated from the stigmatic papillae character (S). Nevertheless, they are confident that several, if not all, of the six associated characteristics are governed by genes having a distinguishable sequence, even if they transcribe only minute amounts of RNA.

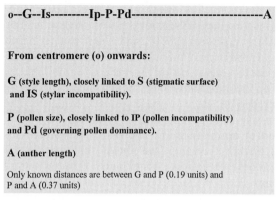

o--G--Is---------Ip-P-Pd----------------------------A

From centromere (o) onwards:

G (style length), closely linked to S (stigmatic surface)
and IS (stylar incompatibility).

P (pollen size), closely linked to IP (pollen incompatibility)
and Pd (governing pollen dominance).

A (anther length)

Only known distances are between G and P (0.19 units) and
P and A (0.37 units)

Fig. 2.2. Hypothetical structure of the S supergene. Adapted from Lewis (1949a), Sharma and Boyes (1961), de Nettancourt (1977), Lewis and Jones (1992) and incorporating data from Kurian and Richards (1997)

Recently, Kurian and Richards (1997) discovered a new recombinant product of the S supergene of *Primula*; they attributed it to crossing over between the gene governing pollen size (P) and a gene controlling dominance for pollen size (Fig. 2.2). Their observations not only suggest the presence of a recessive lethal closely linked to P on the Thrum chromosome but also confirm the independence of pollen size from male incompatibility (Ip). Furthermore, Kurian and Richards (1997) substantiated the earlier proposals that style length, stigma papilla length and style cell length are developmentally correlated and governed by at least two loci with additive effects; the genes for male and female incompatibility were shown to be recombinable.

2.1.1.3 The Supergene is Controlled by Modifier Genes

There is considerable evidence that the S locus is subject, in distylic species (as in all other SI systems), to modifications by a number of genes not linked to the S complex. A classic example is *P. sinensis*, where two independent recessives affect both the SI character and the floral architecture (De Winton and Haldane 1933). The dependence of the S locus, in this system and in others, is not limited to the action of such major genes but is probably also integrated in a complex network of polygenes that immediately respond to a rupture of balance provoked by inbreeding or genetic engineering, as suggested 50 years ago (Mather and De Winton 1941; Sects. 4.2.5, 4.3.4.2, 4.4.4.3).

2.1.2 Tristyly

The inheritance of tristyly has been established in *Lythrum* by Von Übisch (1921) and Barlow (1913, 1923) and confirmed, after a controversy with

East (1932), by Fisher and Mather (1943). The genetic control is carried by two loci, *M* and *S* (with linkage in *Oxalis valviviensis* and independence in *Lythrum salicaria*, as shown by Fisher and Martin 1948), each with a dominant and a recessive allele. The dominant S allele is epistatic with M and characterizes all plants with short styles. Long-styled plants have the recessive alleles of both genes, while individuals with mid-styles lack the dominant S allele but are either heterozygous or homozygous dominant at the M locus. In *L. salicaria*, where the loci are unlinked, inheritance is tetrasomic, with 10% double reduction at both loci (Fyfe 1953).

All possible genotypes of the three forms, together with the separation of phenotypes resulting from cross-compatible matings, are shown in Fig. 2.1. Table 2.2 provides the expectations for cross-compatibility and cross-incompatibility relationships prepared by Riley (1932, 1936) for the similar genetic system he thought was operating in *Capsella grandiflora*, without the morphological differentiation. The removal of M, S or m from the pop-

Table 2.2. Sporophytic incompatibility by two loci with epistatic relations or independence between loci (Correns 1912 and Riley 1932, 1936, presented and discussed by de Nettancourt 1977). In the case of independence (Correns, working with *Cardamine pratense*), all individuals are self-incompatible, and any two plants with one of two identical dominant alleles in common or homozygous recessive at the two loci are cross-incompatible. When the two loci are epistatic (Riley, with *Capsella grandiflora*), the population is composed of three cross-compatible groups of plants that are self-incompatible and cross-incompatible within their own group

Epistatic interactions					
I			II		III
Male TtSS	TtSs	Ttss	ttSS	ttSs	ttss
Female					
TtSS —	—	—	+	+	+
TtSs —	—	—	+	+	+
Ttss —	—	—	+	+	+
ttSS +	+	+	—	—	+
ttSs +	+	+	—	—	+
ttss +	+	+	+	+	—

Independence between loci			
I	II	III	IV
Male BbGg	Bbgg	bbGg	bbgg
Female			
BbGg —	—	—	+
Bbgg —	—	+	+
bbGg —	+	—	+
bbgg +	+	+	—

ulation leads to the elimination of the mid-styled, short-styled or long-styled class, respectively, while the disappearance of s causes the losses of the three classes and the extinction of the population. As in the case of distyly, all reciprocal crosses yield the same genotypic and phenotypic classes in all instances, and differences in cell size are associated with variations in floral morphologies.

2.1.2.1 Homomorphic Variants and Supergenes in Tristyly

The occurrence of homostyles in tristylic species has been reviewed by Lewis and Jones (1992) and by Richards (1997). Lewis and Jones classify all variants as semi-homostyles, because the change only affects one of the two anther levels in the flower. They have been found in *L. salicaria* by Stout (1925) and Esser (1953), in *O. dillenii* by Ornduff (1972) and in *Eichhornia paniculata* by Barrett (1988). Lewis and Jones note that the various characteristics in tristyly can be dissociated in the same way as in distyly and, therefore, are assumed to be under the control of separate genes. Accordingly, there is, in their opinion, "nothing against and something in support of the supergene concept of both S and M in tristyly".

2.1.2.2 One Genotype, Two Phenotypes

An exceptional feature of tristyly is the fact that, for any given plant, the phenotype of the pollen is related to the level of the anther from which it has been produced. In other words, although a plant yields the same distribution of pollen genotypes at all anther levels, the phenotype of the pollen depends on the location of the anther within the flower. Lewis (1949a) noted that such a relationship indicates a close physiological connection between anther height and the growth reaction of the pollen. Ganders (1979) considers that, "...in morphological and biochemical aspects, tristyly probably represents the most complex breeding system known".

2.1.2.3 Dominance Change in *O. articula*

In contrast with all other tristylic species in the genus *Oxalis*, the short-styled plants in *O. articulata* are homozygous recessive at the S locus but may have any genotype at the M locus (Von Übisch 1921). As summarized by Ganders (1979), "...if plants are non-short, a dominant allele at the M locus makes them mid-styled and the homozygous recessive at the M locus is long styled". In this change of dominance involving several linked genes, Lewis and Jones (1992) saw evidence for a master switch able to turn dominance on and off (Sect. 2.1.1.1).

2.1.2.4 Breeding Behavior Can be Independent of Floral Heteromorphism

The rare situation of breeding behavior independent of floral heteromorphism has been reported in *Narcissus triandrus* (Bateman 1952; Barrett et al. 1997; Sage et al. 1999) and in *Narcissus* species with stigma-height dimorphism (Dulberger 1964 and Baker and Barrett's unpublished data cited by Sage et al. 1999). It is characterized by SI (which Sage and co-workers have shown to be basically different from a simple manifestation of inbreeding depression) and the fertility of both intermorph and intramorph pollinations.

2.1.3 Multi-Allelic Series in Species with Incomplete Heterostyly?

Dulberger (1992) provides examples of species in which distyly is incomplete (there is no reciprocal placement of stigmas and anthers in the morphs) and where the mating system is not based on di-allelic incompatibility. One of these examples is *Anchusa officinalis*, for which Schou and Philip (1984) postulated a single locus with two alleles for the morphological traits and two loci, possibly multi-allelic, that appear to involve gametophytic and sporophytic control. Dulberger also refers to the heterostyly of *Narcissus* [for which Bateman (1952), working with *N. triandrus*, proposed a multi-allelic system] and to the special case of *N. tazetta* (Dulberger 1964), where incompatibility, not linked to distyly, could be polyallelic.

■ 2.2 Sporophytic Homomorphic Stigmatic Control

2.2.1 Two Di-Allelic Loci

In Table 2.2, the example of sporophytic incompatibility governed by two di-allelic loci with epistatic relations may only occur in heteromorphic SI (Sect. 2.1.1.2). As seen in Section 2.1.2, it was prepared by Riley (1932, 1936) to explain segregation data in homomorphic *C. grandiflora*, which did not fit a model with two independent di-allelic loci established by Correns (1912) for *Cardamine pratense*. Bateman (1954, 1955) has shown that the figures presented by Riley and by Correns could also be explained by a single genetic locus with polyallelic series. Furthermore, in the case of *C. grandiflora*, it was considered unlikely that the inheritance of the SI character should be basically different from that of the other SI species identified in the Crucifers. Bateman was probably right, but the ultimate demonstration will come from a co-segregation analysis of several different incompatibility proteins and incompatibility alleles. Meanwhile, in the field and in the greenhouse, it is not an easy task to distinguish the various segregation ratios that can be expected from different hypotheses dealing with

(1) gametophytic or sporophytic control, (2) one or several loci, (3) di-allelic or polyallelic series, and (4) unknown interactions between alleles (for an appreciation of the difficulties, see Lawrence 1975; Lawrence et al. 1978 and Lewis et al. 1988).

2.2.2 A Single Locus with Polyallelic Series, Dominance and Competitive Interaction: the *Brassica* Type

Homomorphic sporophytic incompatibility with polyallelic control at a single locus (S) was first discovered in *Parthenium argentatum* (Gerstel 1950), *Crepis foetida* (Hugues and Babcock 1950), *Cosmos bipinnatus* (Crowe 1954) and in several species of the Cruciferae (Bateman 1955). Despite differences in relationships among alleles, the features characterizing polyallelic series in this SI system do not vary much among species or among families. The most important features can be summarized as follows:

▦ *In natural populations*, a relatively large number of different alleles segregate at the S locus. Bateman (1954) established, as an underestimate, that 22 different alleles were present in a natural population of *Iberis amara* that was composed of only 47 self-incompatible plants. Sampson (1967) identified nine different S alleles in 45 *Raphanus raphanistrum* plants representing five different wild populations. In his study of the number of incompatibility alleles in clover and other species, Lawrence (1996) reported that 49 alleles are known in *Brassica oleracea* (Ockendon 1985) and 35 alleles are known in populations of *Sinapis arvensis* (Stevens and Kay 1989). From the small overlap of identical S-alleles between different populations of *B. campestris*, Nou et al. (1993a) inferred that more than 100 different alleles segregate in the species throughout the world. Kowyama et al. (1994) identified a total of 49 different S-alleles in 224 plants from six populations of diploid *Ipomea trifida*. In contrast to the situation in the glutamine synthase I (GSI) system of *Papaver rhoeas* (Lawrence et al. 1993; O'Donnell et al. 1993), there appear to be large variations in allele composition among different populations in *Ipomea*. However, as noted by Franklin et al. (1995), the exact amount of overlap among populations cannot be calculated for a polyallelic sporophytic system. The factors that affect the numbers, frequencies and distributions of SI alleles in all systems with polyallelic series are presented below in Sections 2.7 and 2.8.

▦ Despite the effects of dominance (see below), the SI plant in a population is usually heterozygous and, therefore, carries *two different S alleles*.

▦ These alleles may express relationships of *dominance, independence, interaction or mutual weakening* in the anther or in the pistil (Sampson 1957 for *Raphanus*; Thompson 1957 for *B. oleracea*; Richards and Thurling 1973a for *B. campestris*; Fig. 2.3 for *I. amara* and *C. bipinnatus*). Lundqvist (1990a) suggests that dominance relationships are common in

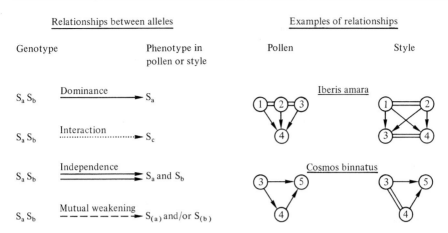

Fig. 2.3. Relationships between S alleles in the sporophytic monofactorial multi-allelic system of *Brassica*, and examples of relationships in *Iberis amara* (Bateman 1954) and *Cosmos bipinnatus* (Crowe 1954). For a facilitated understanding of the effects of such relationships on mating performance, consider the simpler, monofactorial, multi-allelic GSI system in Figs. 6 and 7, where all alleles express independence. (de Nettancourt 1977)

sporophytic systems because they lower the number of S specificities and increase, through the reduced degree of cross-incompatibility, the efficiency of the system. When dominance occurs among alleles, the hierarchy of relationships among alleles is not necessarily the same in stigma and pollen. In such a case, certain combinations of alleles lead to the breakdown of SI and to the expression of self-compatibility (SC).

▪ *Dominance relationships* in natural populations of *B. campestris* were observed by Hatakeyama et al. (1998a) in 249 of 276 possible pair-wise combinations among 24 S-alleles isolated from two natural populations in Turkey and Japan (Nou et al. 1993a, 1993b, and references provided by Hatakeyama and co-workers on early research on the subject). Their conclusions are that co-dominance in *Brassica* species is more frequent than dominance/recessiveness relationships and that dominance occurs more often in the pollen than in the stigma. Alleles are not necessarily dominant to the same alleles in pollen and stigma. Independent weakening was found between 20 pairs of different alleles in the pollen but between only two in the stigma. These features are generally those of other mono-factorial stigmatic SI species within the Brassicaceae, such as *Lesquerella densipila* (Sampson 1958), *R. raphanistrum* (Sampson 1957, 1967) or *S. arvensis* (Ford and Kay 1985), a species for which Stevens and Kay (1989) estimated, in South Wales populations, the number, frequency and dominance relationships. Comparable data were also observed outside the Brassicacea in taxa such as *Cosmos bipinnatus* (Crowe 1954) and *I. trifida* (a close relative, in the Convolvulaceae, of the sweet potato). However, in *Ipomea*, the pattern of dominance was less sensitive than in *Brassica* to different factors such as genetic background, flower age and environmental conditions and was, for 28 S-alleles tested, strongly linear in stigma and pollen (Kowyama et al. 1994). Uyenoyama (1999) has shown that pollen dominance reflects interactions among haplotypes at the S locus itself. Pollen–part dominance strongly affects the probability that a newly introduced S haplotype will be incorporated into populations; it may influence the relative numbers and frequencies of class-I and class-II haplotypes.

▓ When *unequal frequencies* are observed, the most common S alleles are the most recessive ones (Kowyama et al. 1994).

▓ *Alleles acting independently* from one another in the anther and/or the pistil usually occupy the same position in the scale of dominance. Lundqvist (1990b) related the case of a one-locus SSI system in *Cerastium arvense* (Caryophyllaceae) with no dominance relationships in the pollen and the independent action of the S alleles (Sect. 2.4.3).

▓ *In cultivars*, S-allele diversity is low, and dominance usually high. Ockendon (1974, 1975) made a detailed analysis of the distribution of S alleles and the breeding structure of open-pollinated cultivars of Brussels sprouts. He found 19 different alleles in 488 plants representing 16 cultivars. The number of S alleles per cultivar varied from 4 to 13. In every cultivar, the S-allele frequencies were unequal, and relatively unselected cultivars tended to host more alleles than highly selected ones. Some alleles were identified in all cultivars tested, while the rare S alleles (occurring at low frequencies in the populations where they were detected) could be found in only few cultivars. These rare alleles were usually dominant, and the most common ones were recessive. As a general rule, the rare dominant alleles conferred a higher degree of SI. The ratio of dominant to recessive alleles was considered by Ockendon to regulate a dynamic balance between outbreeding (favored by dominant alleles) and inbreeding (promoted by the recessive alleles). Ockendon emphasized the practical value of rare dominant alleles for the production of hybrid seeds. In cabbage (*B. oleraceae* variant capitata), a total of 31 S alleles was found in 11 cultivars (Ockendon 1985). Five of these apparently occur only in cabbage. A study of five cultivars of Spring cabbage exhibited only 12 S alleles in all, with six to ten alleles in four older cultivars and only three in the more recent and highly selected cultivars.

2.2.2.1 The *Brassica* Haplotypes

Several linked transcriptional units are situated at the S locus of *B. oleracea*, one of the most studied species with sporophytic polyallelic SI. Two of these genes, the SLG (S-locus glycoprotein) and the SRK (S-locus receptor kinase) genes are expressed in the stigmatic papillar cells, where they are thought to cooperate, possibly through the variable regions of their sequences, and establish the incompatibility phenotype of the pistil (Sects. 3.2.3–3.2.5; Takasaki et al. 1999). However, as seen below in Sections 3.2.3.2 and 3.2.5.4, certain SI phenotypes express very low levels of SLG glycoproteins (Gaude et al. 1995), while SC mutants or SC amphidiploids display normal SLG levels (Gaude et al. 1993). Furthermore, recent evidence (Luu et al. 1999) suggests that SLG has other functions and plays a major role in pollen adhesion (Sects. 2.2.2.3, 3.2.5).

The SLG and SRK genes are polyallelic, and each particular pair of SLG/SRK alleles at the S locus of any given plant has been referred to by Boyes and Nasrallah (1993) as a *haplotype*. Two different classes of haplotypes have been identified; for now, they only specify the determinants of the stigma.

2.2.2.1.1 Class-I Haplotypes. Class-I haplotypes occupy a high position in the dominance hierarchy of the pollen and are associated with a strong SI reaction. Within each haplotype, a specific portion of the variable region of the DNA sequence of the SRK allele is homologous to a portion of the variable region in the SLG sequence. The percentage of homology between the two genes can be very high in certain haplotypes (Stein et al. 1991; Watanabe et al. 1994) and not in others (Kusaba et al. 1997). However, no differences at all were found between the SLG sequences from the same haplotype of *B. campetris* in Turkey and those in Japan (Matsushita et al. 1996), and striking similarities were observed between SLG sequences of *B. oleracea* and those of *B. campestris* (Kusaba et al. 1997). Therefore, it appears that SLG divergence occurs very slowly in the genus *Brassica*; it predated speciation many millions years ago (Dwyer et al. 1991; Nasrallah and Nasrallah 1993; Hinata et al. 1995; Boyes and Nasrallah 1995; Chap. 4).

2.2.2.1.2 Class-II Haplotypes. Class-II haplotypes are recessive in pollen and are considered to be structurally different. The first evidence of the functional distinctness of class II was reported by Tantikanjana et al. (1993), who found that the SLG in the S2 haplotype of *B. oleracea* produces two transcripts that differ at their 3' ends. One transcript encodes the expected secreted glycoprotein, whereas the second exon encodes a putative membrane-anchored glycoprotein not present in class-I haplotypes, which was suspected to play a role in the leakiness of class-II haplotypes. Hatakeyama et al. (1998c) reported the results of RNA gel-blot analyses of the SLG transcripts of the pollen-recessive SLG29, SLG40 and SLG44 haplotypes of *B. rapa*. They found that the recessive nature of one (SLG29) of the alleles studied could be attributed to the absence of the transcript for a membrane-anchored form of SLG (not encoded by exon 2). However, Hatakeyama and co-workers suggest that the unusual structure of SLG29 is perhaps not the only determinant of its pollen-recessive nature. They also note, from published data, that SLG genes from class-I haplotypes exhibit pairwise sequence identity ranging from 78% to 98%. The values (in excess of 95%) calculated for class-II SLG29, SLG40 and SLG44 could imply a relatively low rate of divergence among recessive alleles and, as suggested by Schierup et al. (1997; cited by Hatakeyama and co-workers), a greater loss of alleles through genetic drift and a shorter life span. On the subject of pollen S-recessiveness, Cabrillac et al. (1999) reported the presence in *B. oleraceae* of two different SLG genes (SLGA and SLGB) at the S-locus; these form part of the pollen-recessive class-II S15 haplotype. Both genes are interrupted by a single intron. As in the class-II S2 haplotype studied by Tantikanjana et al. (1993), SLGA encodes both soluble and membrane-anchored forms of SLG (mSLG), whereas SLGB encodes only soluble proteins. The important point raised by Cabrillac and coworkers – in addition to the possibility that mSLG, like the soluble, truncated SRK (eSRK) discovered by Giranton et al. (1995), modulates the SI response – is that certain haplotypes carry only one of the two genes. Either SLGA and SLGB are redundant, or they are not really required for the SI response.

2.2.2.2 Extension of the Haplotype Concept to Other Genes, Other Families and Other Systems

The complexity of *Brassica* haplotypes extends beyond that of SRK and SLG and essentially results from the additional presence of highly diverged and specific sequences known, or suspected, to be expressed in the stigma (Boyes et al. 1997; Nasrallah 1997; Susuki et al. 1999). It is now necessary to include some of these sequences in the description of each *Brassica* haplotype and, of course, to introduce SCR (S-locus cysteine-rich protein), assumed by Schopfer and co-workers (1999) to participate as a pollen determinant in the recognition reaction. SCR, considered to provide the ligand for the SRK receptor, is related to the group of pollen-coat proteins (PCPs) discovered by Stephenson and co-workers (1997; Sect. 3.2.4.3) and suspected to play a role in the SI reaction (Sect. 3.2.4.4).

    B. oleraceae and *B. campestris* are not the only self-incompatible species to which the new terminology will have to be applied during the coming years to take into account the complexity of the S-locus. In other homomorphic plants, one can expect, from the work carried out in *Ipomea* (Kowyama et al. 1995, 1996) and in several species with stigmatic GSI (Chap. 3), that receptor kinase genes and related linked loci are often involved, as in *Brassica,* in the control of SI. Even in the more distant case of stylar GSI, Sassa et al. (1997) have suggested, on the strength of their finding that the pollen determinant in the Rosaceae is not allelic to the stylar ribonuclease gene, that the notion of the haplotype be used to refer to the specific associations of the alleles of different pollen and pistil S genes. The observation also applies, of course, to heteromorphic self-incompatible plants – particularly distylic species, where the S locus has been known, since Pellew (1928), to host a closely integrated complex of SI genes. It is interesting to note, in this connection, that the presence in distylic *Primula* of two distinct genes for pollen and pistil incompatibility at the S locus had been predicted more than 50 years ago by Ernst (Sect. 2.1.1.1). This was long before Linskens' antigen–antibody model (1965) for different pollen and pistil determinants in homomorphic systems (Sect. 3.5.3.3).

2.2.2.3 S-Locus-Related (SLR) Genes in *Brassica*

Three genes (SLR1, SLR2 and SLR3), unlinked to the S locus, have been detected in *B. oleracea* and found to display strong homology with SLG/ SRK and other members of the S-multigene family (Sect. 3.2.3.7). These genes appear to be diverged products (with deletions and rearrangements) of the duplication of the SLG or SRK genes. Their functions are not known and probably concern the general physiology of the flower rather than the recognition phase of SI.

SLR1, which encodes the protein referred to as NS (non-specific) by Isogai et al. (1988), was first described by Lalonde et al. (1989). It is expressed in the stigma papillae, and its product, secreted with SLG, accumulates in the papillae cell walls (Umbach et al. 1990). The SLR1 DNA is, in part, homologous (70%) with class-I SLG. Franklin et al. (1996) found that the SLR1 protein exhibits minimal allelic variation and is not required for the proper mechanism of SI or SC in *Brassica*. This result was confirmed by Luu et al. (1997a, 1997b) who did not dismiss, however, through a statistical analysis of adhesion variables and the study of transgenic plants modified for SLR expression, the possibility of a participation of SLR1 in pollen adhesion. Further work (Luu et al. 1999) involving transgenic suppression and pre-treatments of wild-type stigmas with anti-SLRl and anti-SLG antibodies or pollen-coat protein extracts (Sect. 3.2.5.1) showed that both SLR1 and SLG play a significant role in pollen adhesion. Divergences among SLR1 sequences in the Brassicaceae are at least 43–47%.

SLR2 is linked to SLR1 and was analyzed by Scutt et al. (1990) and Boyes et al. (1991). Different amino-acid sequences of SLR2 exhibit little variation (less than 1%). The gene shares approximately 90% homology with the S2 haplotype.

SLR3 was identified by Cock et al. (1995); it is expressed in petals, sepals, vegetative apices, stigmas and anthers, and is linked to two or three closely related genes. SLR3 appears to be derived from an SRK-like gene through a series of deletion events and is predicted by Cock and co-workers to encode a secreted glycoprotein lacking both the trans-membrane and kinase domains.

2.2.2.4 Many Genes in the S-Linkage Group

In addition to the genes coding for SRK, SLG, SCR, SLGA, SLGB, SLA and PCP, several other genes, possibly more distantly related to the expression of SI, have been identified within the S-linkage group. This complementary list, established and enriched by Susuki et al. (1999), includes SLL1 (Yu et al. 1996), SLL2 (Yu et al. 1996; Boyes et al. 1997), ClpP (Conner et al. 1998) and ten new genes in the 76-kb SLG/SRK region of the S9 haplotype of *B. campestris* (Susuki et al. 1999). Among the new genes discovered by Susuki and co-workers, one must particularly note SP11, which is tightly linked to the S locus, downstream of the SRK9 gene. The SP11 gene produces a cysteine-rich protein possibly related or corresponding to the small PCP protein, which Stephenson et al. (1997) and Schopfer et al. (1999) associated with the manifestation of SI (Sects. 3.2.4.3, 3.2.4.4).

2.2.3 A Single Sporophytic Stigmatic Locus with Multiple Alleles but without Dominance and Competitive Interaction

This system, found by Lundqvist (1990b, 1994a) in the Caryophylaceae, is presented below, in Section 2.4.3. As emphasized by Lundqvist (1994a), its main features (association with trinucleate pollen, dry stigma, reaction on stigma, persistency in polyploids) appear to be quite typical. What was unexpected was the absence of dominance and of competitive interaction, which seems to result from a dual nature (sporophytic and gametophytic) of the system. The origins of the gametophytic dimension of the system could be a premature formation of the cell wall in the dyad and the isolation of microspores before transcription or the distribution of S-transcripts. This gametophytic interference creates a link with the GSI system that operates in other Caryophyllales and in the closely related order of the Ranunculales.

2.2.4 Three or Four Polyallelic Loci in *Eruca sativa*

SI in the oleiferous crucifer *E. sativa* is, as expected, sporophytic and stigmatic (Verma et al. 1977). What is more surprising is that Verma et al. (1977) and Lewis (1977), who re-investigated an earlier conclusion of Narsingdas (see Singh 1958) that SI in this species was gametophytic, not only discovered a sporophytic control but also found that at least three polyallelic loci (possibly four) were in command of the mechanism. These loci are complementary, i.e., relationships among the different loci are interactive rather than simply additive. With the exclusion of complete reversal of dominance relationships from pollen to pistil (which leads to SC), all possible combinations of dominance and co-dominance among alleles in the pollen or in the stigma of *E. sativa* could be inferred from the outcome of reciprocal crosses between self-incompatible plants. On this occasion, Lewis (1977) showed how the percentage of reciprocal and non-reciprocal incompatible matings among different genotypes in a di-allelic matrix can be used to discriminate among different SI systems.

▪ 2.3 Gametophytic Homomorphic S Systems with Polyallelic Series

Very different plant species display gametophytic SI. The fact that the genetic control appears to be monofactorial in some and is clearly bi-factorial or multigenic in others suggests, on the basis of phylogenetic evidence, that SI in the ancestors of these species was governed by several genes that became partially eroded by mutations or rendered invisible through homozygosity at some of the loci involved (Lundqvist 1975; Lawrence et al. 1978). An earlier school of thought (Whitehouse 1950; Brewbaker 1957, 1959; Pandey 1958, 1960; Crowe 1964) had reached the opposite conclusion (Sects. 4.3.2.1, 4.3.2.2).

2.3.1 One-Locus Stigmatic Control: the Case of the Style-Less Field Poppy

The SI system of the field poppy (*P. rhoeas*) was shown to be gametophytic by Lawrence (1975) and monofactorial by Lawrence et al. (1978). Its population genetics and molecular biology have been intensively studied by integrated groups of researchers at the School of Biological Sciences at the University of Birmingham (Chaps. 2–4).

The SI reaction could involve, as in *Brassica*, the activation of a receptor, (initiated by a "ligand-receptor"-like matching of identical or complementary S-specific pollen and pistil determinants) followed by specific phosphorylation events (Rudd et al. 1996, 1997; Sect. 3.3.4.2). Although SI in *Papaver* is considered to be monofactorial, it is possible (in this system as in many others) that several tightly linked genes cooperate in the establishment of the pollen–stigma barrier. However, an important difference with the *Brassica* recognition model is that in *Papaver* and other species with stigmatic GSI, the putative receptor is contributed by the pollen or pollen-tube surface and is activated by a ligand provided by the stigmatic cells (Chap. 3).

The stigmatic nature of the SI barrier in *P. rhoeas* does not seem to be the direct consequence of the absence of a style. Sykes, cited by Lawrence et al. (1978), found that in the related genus *Meconopsis*, one species, *M. horridula*, displays a short style but stops incompatible pollen at the stigma surface.

2.3.1.1 Genetics of SI Polymorphism in *P. rhoeas* and Other Species with Polyallelic, Monofactorial SI

In a one-locus gametophytic polyallelic system, dominance relationships and interactions between alleles do not occur. The genetic features are simple and common to all monofactorial GSI systems (Figs. 2.4, 2.5). They imply that any pollen grain may germinate and accomplish fertilization

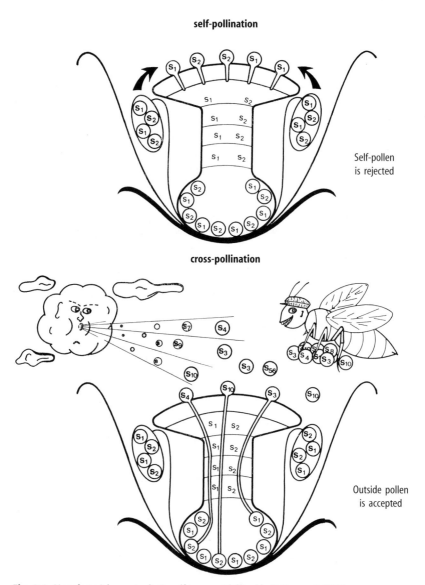

Fig. 2.4. Monofactorial gametophytic self-incompatibility. (de Nettancourt 1972)

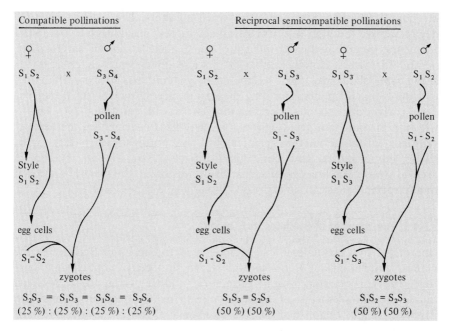

Fig. 2.5. Monofactorial gametophytic self-incompatibility: segregation of S genotypes in progenies derived from compatible and semi-compatible pollinations. Note the differences in S genotypes among progenies arising from reciprocal crosses between semi-compatible plants. (de Nettancourt 1977)

provided the S allele it carries is not present in the diploid tissue of the female organs. Only three types of pollination can occur:

- An illegitimate and incompatible pollination on selfing and between plants with identical S genotypes.
- A semi-compatible pollination between plants with one S allele in common.
- A fully compatible pollination between individuals with no S allele in common.

After cross-pollination, the number of different compatibility classes in the progeny of any given cross will be the product of the number of alleles in the pistillate parent and the number of alleles in the staminate parent that are not present in the pistillate partner (2×0 on incompatible mating, 2×1 after a semi-compatible pollination and 2×2 in the case of complete cross compatibility). Reciprocal crosses between semi-compatible plants yield different results.

Under conditions of open pollination, and assuming complete panmixis, the number of different compatibility classes composing the progeny of any given self-incompatible plant is $2 \times (n-2)$, where n represents the number of different S alleles segregating in the population. The different S alleles in the population should be present in equal numbers. The main reasons for deviations are the high selective advantage of new alleles with a low initial frequency, genetic drift and linkage of the S allele to one or several non-neutral characters. However, other factors and interferences, such as population size, plant size, the overlapping of generations, species differences and the extent of pollen and seed dispersal, are also involved; these render a correct analysis of the population genetics difficult (Sect. 2.8). This is why the Birmingham group (Brooks et al. 1997a) developed a mathematical theory of the population genetics of SI polymorphism when SI is multi-allelic, monofactorial and gametophytic (Sect. 2.8.6). The theory can be used to investigate a range of scenarios and provides the general approach to any variable that can be expressed linearly as a time series.

2.3.1.2 The Number of S Alleles in *P. rhoeas*

One of the first estimates of the number and frequency of S alleles in the natural populations of *P. rhoeas* was issued in 1981 (Campbell and Lawrence 1981). Thereafter, from an estimation of allele overlap (53%) between the allelic complements of Spanish and British populations, Lane and Lawrence (1993) established that the number of alleles in the species is unlikely to be much greater than 66. This low number was tentatively attributed to a dynamic restraint, i.e., an attenuation of the strength of frequency-dependent selection as the number of alleles increases, even if the population is very large. The second explanation considered by Lane and Lawrence was molecular restraint to allelic diversity in *P. rhoeas* (Sect. 2.8.1.2). However, the estimation made by Lane and Lawrence of the number of alleles in *P. rhoeas* is not lower that the numbers Lawrence (1996) calculated (*Trifolium pratense* and *T. repens* excepted) for several other SI species (Sect. 2.3.4.1).

2.3.2 Two Loci-Stigmatic Control in the Grasses

Stigmatic gametophytic incompatibility by two loci was discovered in different species of grasses by Lundqvist (*Secale cereale*: 1954, 1956; *Festuca pratensis*: 1955, 1961; *Hordeum bulbosum*: 1962b, 1964; *Dactylis aschersoniana*: 1965), Hayman (*Phalaris coerulescens*: 1956) and Murray (*Briza media*: 1974) and for *Lolium perenne* by Cornish et al. (1979). It seems to characterize the great majority of self-incompatible species belonging to the Graminaceae. The two loci involved (S and Z) are independent, play an active role in the recognition phase between pollen and stigma, and are each controlled by multiple alleles. *Identity between pollen and pistil at either of the two loci alone gives no incompatibility. Each specific pair of S and Z genes leads to one unique specificity and identity between pollen and pistil in one such specificity is sufficient to lead to incompatibility* (Lundqvist 1965).

At one time, work by Li et al. (1994, 1995, 1996) suggested that the S-gene product carried two sections: a variable one determining specificity and a conserved terminus with catalytic activity and strong similarities to thioredoxins. Recent observations in *Phalaris* (Langridge et al. 1999) and rye (Wehling, personal communication 1999) revealed, however, that the probe (Bm2) utilized as the S gene translates into thioredoxin-like protein only and does not contain an allelic domain. The occurrence of recombination between the S locus and the thioredoxin sequence provides evidence that two different genes are involved (Sect. 3.4.2).

At the time the S-gene of *Phalaris* was thought to include a thioredoxin region, McCubbin and Kao (1996) tentatively suggested that the S locus modulates a kinase activity of the Z gene. In rye, genes that may participate with the S and Z loci to form a signal-transduction cascade within the pollen grain have been identified on three different chromosomal arms by Wehling et al. (1995). Gertz and Wricke (1989) found that the Z locus is linked to a β-galactosidase gene in rye.

2.3.2.1 Breeding Efficiency of the S–Z System

In such a system, breeding efficiency (the probability that any two plants are cross-compatible) is much higher than in the monofactorial mechanism, because the number of different specificities within a given population will correspond, in the absence of any interference by internal or external factors, to the product of the number of alleles segregating at each of the two loci. Hence, the total number of different specificities in a population of *F. pratensis* with six alleles at the S locus and 14 at the Z locus is 84 (Lundqvist 1964, 1969). Of course, as far as fitness is concerned, the drawback, if several independent loci participate (see below), is that cross-compatibility among sibs will increase. Nevertheless, it is likely, if S and Z loci have a duplicative origin (Lundqvist 1964, 1990a; Pandey 1977, 1980; Chap. 4), that the breakage of linkage between S and Z accelerated the establishment of the bi-factorial recognition system in the Graminaceae and greatly contributed to the expansion of this family.

Cross-compatibility relationships among heterozygous individuals at the S and Z loci are illustrated in Fig. 2.6. Examples of relationships (expressed as percentages of compatible pollen) among plants with identical alleles in common and homozygous at either the S or the Z loci have been provided by Lundqvist (1965).

2.3.2.2 The Size of Polyallelic Series

The conclusion of Lundqvist (1965) is that the incompatibility loci of grasses have allelic series of large sizes; he notes rare exceptions, such as that of *D. aschersoniana*, where only few different S and Z alleles were found in a small and isolated population. Lundqvist was able to estimate that 11 S alleles and 12 Z alleles were segregating in a population of mea-

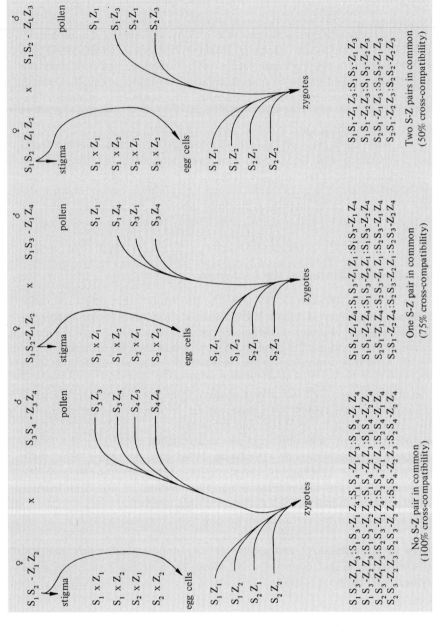

Fig. 2.6. Cross-compatibility relationships in grasses among individuals heterozygous at the S and Z loci. (de Nettancourt 1977)

dow fescue sampled at five different localities. He did so using a method
that involved (1) crossing of the population to be tested to a double homozygous line and (2) back-crossing of the progenies to other double homozygotes having either the S or Z allele present in the original tester line
(Lundqvist 1964, 1968). The data are surprising, because they were collected from a strain known as "Svalöf late", established through selfing and
continuous sib crossings; this strain normally should have segregated for
only two different S alleles and two different Z-alleles. The high number of
alleles recorded could be attributed to the rapid establishment of stray pollen or to the capacity of incompatibility loci to mutate in inbred backgrounds (Sect. 4.2.7).

More recently, the number and frequency of the S and Z alleles were calculated for *Lolium perenne* in natural populations (Fearon et al. 1994) and
in a cultivar (Devey et al. 1994). Crossing procedures were used to extract
one S and one Z allele at random from each of a number of plants. In the
natural population, 17 different S alleles and 17 different Z alleles were
found in a sample of 38 and 39 plants, respectively. The frequencies of
these alleles were unequal; some of the alleles occurred 12 and nine times
each. In the experimental cultivar, 17 S and 13 Z alleles were detected in
samples of 40 and 43 plants examined for estimates of 26 S and 18 Z alleles in the population. Because the cultivar had initially been grown from
only five plants having a maximum of ten different alleles at each locus, it
appears that the increase in the number of S and Z pollens must result
from either pollen contamination or high mutation rates. As emphasized
by Devey et al. (1994), who present arguments in favor of the hypothesis of
stray pollen contamination, the issue is important. The plant breeder needs
to know if the added variability results from the homemade generation of
a few alleles (Sect. 4.2.7.2) or implies large-scale invasion by entire (pollen-
borne) genomes. DNA testing or restriction fragment-length polymorphism analyses could help.

2.3.3 Four-Loci Stigmatic Gametophytic Control in the Ranunculaceae, the Chenopodiaceae and the Liliaceae

The control of stigmatic GSI by four loci has been found in the Ranunculaceae (Osterbye 1977; Lundqvist 1990a, 1990c) and the Chenopodiaceae
(Lundqvist et al. 1973; Larsen 1977), which belong to the same lineage in
the phylogeny of the orders Ranunculales and Caryophyllales (Takhtajan
1969, cited by Lundqvist 1991), and in the monocots within the Lilliaceae
(Lundqvist 1991). As with the grasses, pollen–pistil recognition occurs
through the complementary action of independent genes. So far, few cases
of tetra-factorial stigmatic GSI have been identified. It must be remembered, however, that their analysis is difficult and that there are often several ways to interpret segregation ratios when they are distorted by linkage
relationships among different S genes or by lethal or sterility factors. Ex-

amples of possible confusions have been provided by the study (Bateman 1954, 1955) of Riley's data (1932, 1936) with *Capsella* and the re-analysis (Lewis et al. 1988) of the monofactorial SSI system in the Cruciferae and Compositae. In addition, some of the genes involved in multigenic control of SI will not be detectable without the help of the molecular biologist if they are represented by a single allele in the populations under study.

This last possibility led a number of scientists to consider that the complementary GSI stigmatic system did not evolve from duplications of a primitive S locus but is itself of ancient origin, as suggested by its presence in both the dicots and the monocots. Many or all of the contemporary mono-factorial systems may have originated from homozygosity, inactivation or integration (in a same linkage group) of several of the unlinked or loosely linked genes that initially operated in a multigenic, complementary mechanism. Conflicting arguments and references regarding such an eventuality can be found in Hayman (1956), Osterbye (1975), Pandey (1980), Larsen (1986), Lundqvist (1964, 1990a, 1990b, 1991) and Chapter 4.

No information is available regarding the *molecular biology* of multigenic complementary SI. Lundqvist (1990a) noted that partial S-allele identity between pollen and pistil in the four-gene system does not reduce the strength of the SI reaction. Such an "all-or-none" process, typical of a complementary mechanism, suggests, to quote Lundqvist "that the system operates biochemically by the product from cooperating loci and not by the sum of additive effects from independent loci". Lundqvist (1990a) and Wehling et al. (1995) agree that, to be functional, a recognition mechanism involving the specific participation of many different gene products probably requires simplicity.

It is possible that some of the specific incompatibility genes active in tetra-factorial SI systems are related to the unspecific genes (on chromosomes 3R, 5R and 6R in a species like rye) that, in the grasses, contribute to a signal-transduction cascade triggered by pollen and pistil S and Z products (Wehling et al. 1995; Sect. 2.5.1; Chap. 3).

2.3.3.1 Few Alleles per Locus in Tetra-Factorial Stigmatic GSI

It is the opinion of Lundqvist (1990a, 1990c) that the molecular requirements for complementary interactions among incompatibility genes possibly limit the size of polyallelic series at each of the loci concerned. There should be, in other words, a negative correlation between the number of loci and the number of alleles at each locus. At the same time, the level of cross-compatibility is so high in a gametophytic system governed by four independent or loosely linked genes that new alleles have little selective advantage. Indeed, the data available (Lundqvist 1990a, 1990c) show that polymorphism at these SI loci is restricted. Larsen (1978) found that there are few different SI alleles in the sugar beet, the sea beet and the forage beet, and most of them are common to the three sub-species. Osterbye (1986) observed that one of the four SI loci in *Ranunculus acris* was represented by only two allelles in her material, and different accessions from several European countries share identical alleles. In the genus *Ranunculus*, progeny studies also revealed little polymorphism of the SI genes (Lundqvist 1990c).

2.3.3.2 Linkage between the Four SI genes

Larsen (1978) reported significant deviations of the segregation ratios expected if the four SI genes (Sa, Sb, Sc and Sd) that operate in *Beta vulgaris* segregated freely. In fact, the genes are linked, and their order is either Sd–Sa–Sb–Sc or Sd–Sb–Sa–Sc. Indications of linkage among the four SI loci have also been obtained in *R. acris*, *R. bulbosus* and *R. polyanthemos* by Lundqvist (1990a). No data on a possible linkage between the SI genes in *Lillium martagon* are available, because this species, the only known monocot characterized by what appears to be a four-locus control of SI, has a very long life cycle (Lundqvist 1991), which prevents rapid progeny testing.

2.3.3.3 SI is Maintained in Tetraploid *R. repens*

Lundqvist (1994b) used *R. repens*, known as the creeping buttercup, to find out why tetraploids maintain the SI character of diploid relatives that express multifactorial complementary SI. The starting material consisted of 11 original plants that gave rise to inbred, hybrid and back-cross families. To explain the data obtained from different intercrosses (I1×I1, F1×F1, P×BC) between the parents (P), the hybrids (F1), the back-crosses (BC) and the inbreds (I1), Lundqvist considered different combinations of genomic structures (alloploidy, autotetraploidy), cytological behavior (chromosome assortment, random chromatid assortment), mode of inheritance (disomic, tetrasomic) and system dependence (independence, interdependence). The tentative conclusion reached by Lundqvist was that SI in *R. repens* involves the disomic inheritance of a team of at least four loci with complementary cooperation. A comparable situation was also observed in the related tetraploid species *Caltha palustris* (Lundqvist 1992) and in the genus *Cerastium* (Lundqvist 1990b).

2.3.4 Monofactorial Stylar GSI with Polyallelic Series

Monofactorial stylar GSI with polyallelic series, once the star system for most "incompatibilists", is known to operate in the Solanaceae, the Leguminosae, the Rosaceae and the Scrophulariaceae. Discovered by East and Mangelsdorf (1925) in *Nicotiana sanderae*, it is frequently referred to as the *Nicotiana* type of SI.

For the SI species of at least eight genera (*Nicotiana, Petunia, Solanum, Lycopersicum, Prunus, Malus, Pyrus, Anthirrhinum*), it is now known that the stylar products of the S locus in polyallelic monofactorial stylar GSI are S-specific ribonucleases (Sect. 3.5). The complementary DNA of the S gene that codes for the stylar S-ribonuclease was sequenced for the first time, in *N. alata*, by the team of Adrienne Clarke in Melbourne (Anderson et al. 1986). McClure et al. (1989, 1990) provided evidence that the products of S alleles were ribonucleases and that S-allele-specific degradation of pollen RNA occurs in vivo after incompatible pollination. In transformation experiments, Lee et al. (1994, in *Petunia*) and Murfett et al. (1994, with transgenic interspecific *Nicotiana* hybrids) demonstrated that S proteins were necessary and sufficient for the rejection of self pollen by the style. At the same time, Huang et al. (1994) established, through loss- and gain-of-function experiments, the final proof of the involvement of ribonucleases in the rejection process.

The pollen component in stylar GSI has not been identified. Sassa et al. (1997) observed that the complete deletion of the S-ribonuclease gene in *Pyrus* leads to the non-functioning of the S-allele in the style but does not affect the S-phenotype of the pollen tubes. Their conclusion – that the S-locus region may contain several linked genes, including the S-ribonuclease gene and an unidentified pollen S-gene – consolidates the hypothesis that the S-products in pollen tubes and in styles are encoded by different genes (Sect. 3.5.8).

2.3.4.1 The Size of Polyallelic Series in Stylar Monofactorial GSI

Lawrence (1996), who re-visited data reported for clover (stylar monofactorial GSI) by Atwood (1942, 1944) and Williams (1947, 1951), estimated that the number of different S alleles is approximately 100 in *T. repens* (breeders stocks and commercial varieties), 200 in *T. pratense* (natural populations) and only 17 in *T. hybridum* (commercial varieties).

In an attempt to identify the factors that could be involved in the establishment of such large polyallelic series, Lawrence compared the estimates obtained from the clover populations to those available for nine well-investigated SI species of flowering plants representing the one-locus gametophytic system, the two-locus gametophytic system and the one-locus sporophytic system. The results emphasized the exceptionally large size of the *T. repens* and *T. pratense* series and the relatively low numbers of alleles (ranging from 5 to 50) estimated in the nine species chosen for comparison purposes. The conclusion does not apply to the values (31 for S and 31 for Z) obtained for the two-locus gametophytic system of *L. perenne* (a species among the nine studied), because the number of different pollen SI phenotypes is, in this case, very high ($31 \times 31 = 961$). It also may not apply to *B. campestris*, for which Nou et al. (1993a) have concluded, from the analysis of relatively small samples of plants, that more than 100 S alleles segregate in the species.

To account for the very large numbers of alleles in the two clover species, Lawrence (1996) considered three explanations: inflation of the esti-

mates of the numbers of alleles through undetected errors in classification, possible differences in the molecular basis of SI in the genus *Trifolium*, and sub-structuring of *T. repens* and *T. pratense* populations into large numbers of semi-isolated neighborhoods. It is this last explanation that Lawrence favors, essentially because frequency-dependent allele selection is stronger in small and isolated colonies than in a large, panmictic population.

2.3.4.2 The Structure of the S Locus in Stylar Monofactorial GSI

One of the main reasons for the popularity of this system in research laboratories during the late 1950s and early 1960s was its capacity to promote the self-screening of pollen grains that expressed a loss or change of S specificity. The attempt to find new, mutagen-induced S specificities was disappointing (de Nettancourt et al. 1971). However, a complete range of different "specificity-loss" mutations that suggested a tri-partite structure (one specificity gene expressed in pollen and pistil, and two promoter-like genes for regulation of the specificity part in the pollen and pistil) for the S-locus of many species was obtained (Fig. 2.7). With few exceptions [see,

STRUCTURE OF THE S. LOCUS

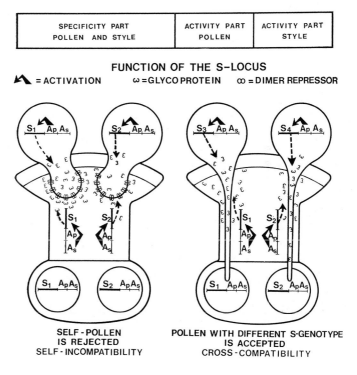

Fig. 2.7. Structure and function of the S locus, as envisaged by Lewis (1960) from the hypothesis that pollen and pistil S phenotypes are specified by a same determinant. (de Nettancourt 1972)

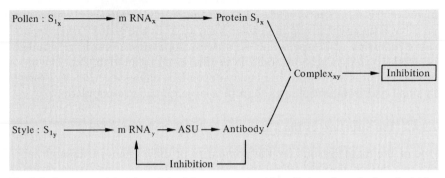

Fig. 2.8. Structure and function of the S locus, as envisaged by Linskens from the hypothesis that different S-specific proteins are produced in the pollen and pistil. The model foresees the activity in the style of an antibody synthesizing unit (*ASU*) that produces an antibody (*y*) regulating its own production (compatibility) or combining with an x protein in the pollen tube to form an inhibiting xy incompatibility complex. (Linskens 1965, Fig. 1, with kind permission from Kluwer Academic Publishers)

in Fig. 2.8, the suggestion of Linskens (1965) that the pollen and pistil specificity determinants code for different RNA molecules] the model of Lewis, because of its attractive simplicity and because it fitted all available data, was widely accepted.

However, the product of the S gene was completely unknown at that time. As seen in Sections 1.3.5.1 and 3.5.8.2, the deletion studies of Sassa et al. (1997) and the fact that only traces (at best) are found in mature pollen grains of the pistil S proteins (which play an active role in recognition) now cast doubt on the validity of the dimer hypothesis.

2.3.4.3 Identification of S-Bearing Chromosomes

Within the Solanaceae, the S-bearing chromosome has been identified in four species from four different genera: chromosome 1 of *N. alata* (Pandey 1967; Labroche et al. 1983), chromosome 1 of *Lycopersicum peruvianum* (Bernatzky and Tanksley 1986), chromosome 1 of *Solanum tuberosum* (Gebhardt et al. 1991) and chromosome 3 of *Petunia hybrida* (Ten Hoopen et al. 1998). In all four cases, the S locus was sited in a region syntenic for the Solanaceae, hosting peroxidase isozyme loci and possibly close to the centromere (Ten Hoopen et al. 1998).

2.3.5 Bifactorial Stylar GSI with Epistatic Relationships

Bifactorial stylar GSI with epistatic relationships has been discovered by Pandey (1957) in a Mexican solanaceous species, *Physalis ixocarpa*, and is characterized by two independent series of incompatibility alleles. Epistasis between the alleles of the two loci in pollen and style may render the pollen incompatible even when only one of the alleles present in the pollen is also present in the style. The number of specificities produced is, therefore, much lower than the product of the number of alleles at each of the two loci and, as a consequence, the system does not lead to the high level of cross-compatibility that characterizes the self-incompatible grasses. Pandey considers that this two-locus mechanism evolved from a duplication of the S locus. Because pollen bearing two (different) S alleles is usually self-compatible (Sect. 4.2.6.2 on competitive interaction between two different S alleles in a same pollen grain), Pandey suggested that the duplication of the S locus in *Physalis* was followed by selection against competitive interaction in digenic hetero-allelic pollen.

Other cases of bifactorial incompatibility in the Solanaceae have been reported by Pandey (1962a) and by Abdalla and Hermsen (1971). For *S. pinnatisectum*, Pandey concluded that two unlinked loci, S (with polyallelic series) and R (with two mutant alleles, RF and RIC) governed the incompatibility behavior of the species. There is dominance between the S alleles, and these are epistatic over the mutant R alleles when these are in the heterozygous condition. On the basis of his model of primary and secondary specificities within the S-gene complex and of reports by Pushkarnath (1953) and Crowe (1955) that certain mutant genes can shift the determination of the incompatibility alleles to the sporophytic stage, Pandey postulates that the mutant alleles Ric and RF act sporophytically on the incompatibility substances in the pollen. However, the S gene continues to express itself gametophytically in the pollen grain and to ensure the qualitative determination of the incompatibility substance.

In the Solanaceae, which is an interesting family for studies on the secondary evolution of SI, Abdalla and Hermsen found (in *S. phureja* and *S. stenotomum*) a situation that closely resembles that described by Pandey for *S. pinnatisectum* and *S. stenotomum*. They discovered two loci (S and R), which they assumed to have evolved from chromosome doubling in an ancestral species with monofactorial incompatibility. Here also, the S locus is epistatic to the R locus, but dominance relationships are not apparent between S or R alleles. One R allele (Rgi) has the property (when in homozygous condition in the pistil) of preventing all fertilizations and can be equated to a female sterility recessive. Abdalla and Hermsen tested their model of the genetic features of SI in *S. phureja* through progeny analyses of a cross alleged to be S1S2 RaRgi X S3S RaRgi. Their expectation that the offspring would form five intra-incompatible inter-compatible groups with a frequency ratio of 4:3:3:3:3 [with female fertility (RgiRgi) in the group represented by 4] was, on the whole, satisfied. However, as in the case of Pandey's experiments, a number of irregularities could be observed; the authors attributed these, in part, to environmental conditions.

2.3.6 Three or Four S Loci in a Complementary System of *Lotus tenuis*

SI in the legume *L. tenuis*, to which Bubar (1958) tentatively attributed a monofactorial gametophytic SI system, has been re-investigated by Lundqvist (1993), who suggests the presence of at least three (probably four) S loci operating in a gametophytic complementary system. This finding illustrates the complexity of the breeding behavior of *L. tenuis*, for which Bullen (1960) and de Nettancourt and Grant (1963) observed the mentor effects of foreign pollen (from self-compatible *L. filicaulis*) on self-fertilization. As noted by Lundqvist, the presence of a multigenic complementary system in a family usually characterized by conventional monofactorial GSI may represent the end result of a process giving rise to the evolution of a one-locus mechanism from a primitive multigenic system (Sect. 4.3.2.2). *L. tenuis* and related species (such as *Vicia faba*, tetraploid *Medicago sativa* and polyploid *T. medium*), which Lundqvist (1993) cites as examples of unexpected breeding behavior in the Fabaceae (Sect. 4.3.1.3), may express certain properties of this primitive system.

2.3.7 Ovarian Gametophytic SI

Few studies (Seavey and Bawa 1986; Sage et al. 1994; in the case of *Gasteria*, Willemse 1999) have been made of the genetics of ovarian self-incompatibility (OSI). In *Acacia retinodes*, Knox and Kenrick (1983) and Kenrick et al. (1986) reported that a gametophytic system of SI prevented the growth of incompatible tubes at the level of the nucellus (the wall of the mega-sporangium that encloses the female gametophyte). For *Thrytomene calycina*, Beardsell (1991) suggested that placental incompatibility (an arrest of the incompatible pollen tube in the portion of tissue of the ovary to which the ovules are attached) also resulted from GSI. Willemse and Franssen-Verheijen (1988) observed that the incompatible pollen tube grows slowly in *Gasteria*, with low amounts of callose in the tube wall and higher numbers of callose plugs than the compatible tube. The SI system of *Gasteria* is governed by two or more loci (Brandham and Owens 1977; Naaborg and Willemse 1992). Willemse (1999) described the progamic phase of *Gasteria* from pollination to penetration in the micropyle. Signal substances that promote recognition, activation and orientation of pollen tubes are present in the pollen coat, along the pollen-tube pathway and in the micropyle. After an incompatible pollination, the incompatibility rejection reaction is finalized in the micropyle, but the symptoms of (gametophytic) incompatibility are expressed throughout the entire progamic phase.

2.3.7.1 Post-Zygotic OSI?

While OSI is primarily a pre-syngamic barrier (Klekowski 1988), several scientists wondered if late OSI could not exert a delayed action immediately before, during or just after double fertilization. Sage et al. (1994) have defined criteria that can sometimes be used for distinguishing between inbreeding depression, such as it often occurs in progenies resulting from obligate selfing, and the delayed action of an OSI system, which inhibits the development of zygotes, embryos or endosperm. These criteria include:

- The timing of abortion, which takes place at a single stage of development when SI is involved (see also Seawey and Bawa 1986 who refer, as criteria of identification for inbreeding depression, to the blockage of embryos at different developmental stages and to the expression of depression in the seedling stage).
- Variability in seed-setting among individuals of a population, which is usually indicative of inbreeding effects.
- Evidence that the genetic basis of abortion does not depend on progeny genotype (any indication of S-allele segregation among sibs or through the analysis of parent/offspring incompatibility relations is suggestive of SI).
- Embryo rescue in tissue culture, as the effects of lethal genes could be less amenable to rescue than those of SI genes.
- The effects of mutagens, which should increase self-fertility in cases where sterility results from the action of mutable SI genes (Klekowski 1988, cited by Sage et al. 1994).

From a review of examples, Sage et al. (1994) conclude that it is often impossible, despite these criteria, to distinguish inbreeding depression from late ovarian incompatibility. One case she refers to was observed in *L. corniculatus*, where Dobrofsky and Grant (1979) found that, in certain cases, ovule and embryo abortion on self-pollination occur at a "continuum" of stages after syngamy and, in other instances, result from the failure of the pollen tube to enter the ovule.

Sage suggests that the terminology "non-random mating" be used each time it is established that the sorting process takes place before syngamy. Sorting events of unknown timing or known to occur before and after syngamy could be referred to as "pistillate sorting" (Bertin et al. 1989). She also notes that there is a strong argument against inbreeding depression resulting from recessive embryonic lethals in species with complete or almost complete self-sterility, because very large numbers of lethals (at least 15) would be required in order to reach a high level of sterility (Waser and Price 1991).

2.3.7.2 Cyclic, Post-Zygotic, Polygenic SI

Cyclic, post-zygotic, polygenic SI, which falls outside the definition of SI by Lundqvist (Sect. 1.1), has been described by Crowe (1971) for the species *Borago officinalis*. It is facultative and versatile. The barrier, as estimated by mixed pollinations designed for measuring the amount of ovules mobilized by incompatible pollen, occurs after fertilization and interferes with embryo development. Its efficiency depends on the degree of homozygosity of the plant subjected to self-pollination, inbred plants usually being far more self-incompatible than heterozygous individuals. SI is thus expressed as a quantitative character with wide and continuous variation between generations and between families; it cannot be attributed to a small number of genes. Crowe (1971) considers that polygenic incompatibility in *Borago* acts as a safety valve against excessive homozygosity. According to Crowe, the selection of specific modifiers that increase the effectiveness of certain sterility genes at the expense of others may have been accompanied by an aggregation of incompatibility (sterility) loci into a complex S supergene. This would have led to the transformation of cyclic polygenic incompatibility into the apparently more rigid and non-facultative mechanism of self-rejection observed by East (1940) in certain populations of *B. officinalis*. However, in opposition to the opinion of Bateman (1952), who considers that polygenic incompatibility is only a step towards obligate outbreeding, Crowe notes that the great flexibility of a polygenic system of the *Borago* type favors its maintenance during evolution. Indeed, it is probable that post-zygotic polygenic restrictions to inbreeding operate in certain species tabulated as regular self-pollinators (for an appraisal of genetic diversity in autogamous plants, see Allard 1965; Allard et al. 1968).

2.3.7.3 Incompatible Pollen Tubes that Prevent Ovule Development

It has been suggested in the past (Sears 1937; Cope 1962; Pimienta and Polite 1983; Sage and Williams 1991, cited in Sage et al. 1999) that the incompatible tubes of certain species with OSI fail to provide the information appropriate for normal integumentary growth and normal ovule development. Thereafter, the demonstration was made (Sage et al. 1999) of differential ovule development following compatible and incompatible pollination in a species (*N. triandus*) where there appears to be no differences in the growth rate of compatible and incompatible pollen tubes. This heteromorphic species is characterized by self-sterility that clearly occurs in the ovary (Bateman 1954) and is uncoupled to floral morphology, with fertile cross-pollinations between and within morphs (Sect. 2.1.2.4). Earlier studies (Bateman 1954; Dulberger 1964; Barrett et al. 1997) attributed the low seed-setting on selfing to either SI or early inbreeding depression.

What Sage and co-workers did was show that ovule development within an ovary is asynchronous and that there are no significant differences in the behaviors of self- and cross-pollens during the first 6 days of growth, although double fertilization was markedly higher after cross-pollination. Prior to pollen-tube penetration, a distinctly greater number of ovules ceased to develop after self-pollination. In other words, self-sterility (which corresponds to SI) in *N. triandrus* operates pre-zygotically and does not involve differential pollen-tube growth. The recognition of self pollen tubes presumably leads to the emission of messages that prevent or fail to stimulate normal ovule development and, thereby, reduce pollen-tube penetration and double fertilization. The finding is remarkable, because it shows that the ovule, and not only the male gametophyte, can be a target of SI and because it suggests long-distance messaging between tube tips and young ovules.

▪ 2.4 Sporophytic-Gametophytic Systems

The objectives of inheritance studies, once the genetic nature of a given character has been established, is to identify, through the interpretation of segregation ratios, the number and relationships of linkage groups, genes and alleles that participate in the transmission and expression of the character. The work is relatively simple when a single Mendelian trait is concerned. It can be extremely difficult in the case of a complex mechanism like that of SI, where the precise site and time of gene expression and gene action and the cohort of genetic loci involved (and often inter-connected) in the recognition and rejection phases are usually unknown. In most cases, however, the geneticist has been able to identify an SI system as gametophytic or sporophytic and to indicate the theoretical combination of genes, alleles and interactions that best fit the observed segregation figures. However, in a number of instances reviewed elsewhere in this chapter (Sect. 2.3.5, an account of Pandey's observation in *S. pinnatisectum*), the genetic data do not satisfy the requirements of either the gametophytic or the sporophytic system and suggest that the phenotype of the pollen grains is determined during micro-sporogenesis by both its own genotype and that of the mother plant.

2.4.1 Three Genes Participate in the Ovarian Gametophytic–Sporophytic System of *Theobroma cacao*

In the ovarian gametophytic–sporophytic system of *Theobroma cacao*, one is dealing with a situation where sporophytic and gametophytic factors not only determine the pollen phenotype but also appear to condition the incompatibility reaction of the ovules (which, in most species with ovarian barriers, participate directly in the process of self-rejection). The opposition is not, as in other SI systems, between a gametophyte and a sporophyte but is directly between gametophytes. The detailed genetic situation has been analyzed by Cope (1958), Knight and Rogers (1953, 1955), and Bouharmont (1960) and has been amended and summarized by Cope (1962) and Bartley and Cope (1973). It involves three genetic loci, S, A and B, which regulate syngamy on self- and cross-pollination. The S locus displays polyallelic series having independence and dominance relationships, but only the gametes containing a dominant or independent allele are activated against fusion with similar gametes. This determination of the potentiality for participating in the incompatibility reaction is thus accomplished in two steps: a pre-meiotic one during which, when dominance is involved, only one allele imprints its specificity to the cell, and a second one (post-meiotic) requiring the presence of the imprinting allele for further specification of the gamete phenotype. The entire process cannot take place, however, unless the dominant alleles of two accessory loci, A and B, are present to provide non-specific precursors to which the S alleles supposedly impart their specificities. In other words, only three types of self-incompatible trees ought to be found in nature:

▪ Individuals where two S alleles have a different dominance status. On selfing, 25% of the ovules of such individuals will be non-fused.
▪ Plants with two different but independent S alleles, which will cause approximately 50% non-fusion in the selfed ovaries.
▪ A group with complete SI (100% non-fusion), which characterizes individuals having identical S alleles.

These predictions were verified (Knight and Rogers 1953, 1955; Cope 1958; Bouharmont 1960) and confirmed by the fact that, in some cases, self-compatible clones (which the hypothesis postulates to be homozygous at either the A, B or S loci) give rise to partly or fully self-incompatible progenies when crossed to other self-compatible clones (Cope 1962). Hence, it is possible to substantiate a prediction by Lundqvist et al. (1973) that crosses between different self-compatible forms may lead to the reconstruction of a functional SI system (Sect. 5.5). The implications for agriculture could be important.

2.4.2 Sporophytic Stigmatic SI Revisited in the Cruciferae and the Compositae

Di-allele crosses rarely yield clear and unequivocal data. Unexplained deviations seldom but regularly occur in the distribution of plus and minus signs that form the basis for the classification of individuals from a progeny into intra-incompatible and inter-compatible groups. These exceptions predominantly characterize relationships between sibs rather than the outcomes of self-pollination. As speculated by de Nettancourt (1999) and in Chapter 4, it is possible that some of the plus signs reported in the literature (for instance, the data from de Nettancourt et al. 1971 for stylar GSI in *L. peruvianum*) after di-allele crosses within an intra-incompatible group result from point mutations affecting different parts of the specificity (variable) region of the S gene in different siblings. Certain pollen–pistil combinations of initially identical alleles may, in such cases, turn out to be compatible. The transformation process may be slow, with transitory phases during which a mutated allele expresses two distinct specificities (the original one and the new one; Matton, personal communication 1997).

However, this unverified hypothesis certainly does not explain the regular and relatively frequent occurrence of plus signs in the wrong columns for 2–10% of the di-allele crosses conducted in species with sporophytic SI (Lewis 1994). Such exceptions, summarized by Lewis et al. (1988) and Zuberi and Lewis (1988), consist of compatible (positive) pollinations in groups that should be 100% cross-incompatible. They are not typical of self-pollinations but of crosses between plants with identical S phenotypes. Their occurrence had first been reported in the little-known Ph.D. thesis of M.B. Hughes (cited by Lewis 1994), which not only presented the complete description of the sporophytic system for the first time but also clearly referred to the positive and negative anomalies and to the fact that cross-compatibility between sibs rarely but regularly occurs within intra-incompatible groups. The interpretation of D. Lewis, S.C. Verma and M.I. Zuberi is that SI, in most species of the Cruciferae and of the Compositae, is governed by two genes: S (which is sporophytic and multi-allelic) and G (which is gametophytic and probably exists only in two forms, G1 and G2). Its discoverers consider G to be an S-gene ancestor that has been maintained in certain dicots for the value of its cooperation with S. In the species fully tested (*R. sativus* and *B. campestris*), the G and S genes were found to be on the same chromosome, with linkage values ranging from 10% to 30% (approximately 20% in *B. campestris,* where G and S are linked to a pollen lethal, and approximately 25% in *R. sativus*). The values from published data for the Brassicaceae and the Compositae were estimated as 9.7% in *B. oleracea*, 19.1% in *Crepis foetida* and 27.7% in *E. sativa.*

According to Lewis and his co-workers, the G gene can exist in the homozygous state in natural populations (and even more in cultivars). The gene is then silent and remains hidden. In heterozygous plants (G1G2) that also carry S alleles high in the dominance series, the incompatibility reaction will not take place unless the G and S alleles present in the pollen and the style are able to match. If they do not match, a plus sign will substitute for a minus sign in the column where the outcome of the pollination is recorded. G is not required for proper functioning of the SI system in plants with S alleles that are low in the dominance hierarchy. The mechanism through which G acts on S is not known, but Lewis (1994) suggests that "G produces a two-ended molecule which fits into a groove in the S glycoprotein, with the other end projecting to change the specific recognition of the S glycoprotein. In S alleles which are not affected by G the specific end is imbedded in the groove and the other end, which is neutral, leaves the S glycoprotein unaltered in its recognition".

These results and their interpretation are complex and, as noted by Franklin et al. (1995), they should be independently confirmed as soon as possible. The most puzzling feature of the mechanism proposed is not the restriction of G action to S alleles with strong dominance; it is indeed known that these dominant alleles have serologically different S glycoproteins (Nasrallah and Wallace 1967 a, cited by Lewis 1994). The surprise comes from the fact that the G gene (or equivalent genes) is widespread among more than a dozen of different species from two and perhaps three different families (Sect. 2.1.4.3). Gametophytic/sporophytic incompatibility in these families seems to have a high adaptive value, and Lewis (1994) considers the possibility that the G gene is universal.

2.4.3 A One-Locus Sporophytic System with Traces of Gametophytic Pollen Control in the Caryophyllaceae

With the one-locus sporophytic system with traces of gametophytic pollen control of the Caryophyllaceae, we are not dealing with the involvement of two different genes, one acting sporophytically and the other gametophytically. Instead, there is a single sporophytic gene that, due to a shift in the developmental stage of transcription and translation, is suspected to be in a state of transition between gametophytic and sporophytic control of the pollen. This property was suggested by Lundqvist to explain the breeding behavior of two species of the Caryophillaceae: *C. arvense* (Lundqvist 1990 b) and *Stellaria holostea* (Lundqvist 1994 a).

As mentioned above (Sect. 2.2.3) and below (Sect. 4.3.3), the SI system in these two species is clearly sporophytic for many of its features (trinucleate pollen, dry stigmas, stigmatic reaction and persistence in polyploids). However, the S alleles tend to act independently, without any dominance.

Through an analysis of the pollination performances of population plants, their F1s and back-crosses, Lundqvist found that S heterozygotes produced both S substances in the pollen and the pistil in a co-dominant way. Any of these substances was sufficient to produce incompatibility. However, seed setting was such that it appeared that a fraction of the pollen population reacted as if they carried "slow" alleles (producing insufficient amounts of S substances) and "quick" alleles (meeting a barrier to the penetration of their S products within the tetrad of microspores). The hypothesis of Lundqvist (1990 b, 1994 a) was that "quick" alleles were likely to enter early and vigorously in transcription, and "slow" alleles were assumed to be late and weak in transcription. He then showed how, for a proportion of alleles, cell-wall formation in the tetrad could lead to the substitution of sporophytic control by gametophytic imprinting. The process would enable certain cross-compatible pollinations to occur between plants that are normally cross-incompatible.

Lewis (1994) considered that the observations of Lundqvist closely paralleled the basis of the G gene in the Cruciferae and the Compositae. Obviously, the two mechanisms, which stimulate cross-compatibility without endangering SI, may contribute usefully to the replacement of a mechanism of dominance relationships. However, Lundqvist (1990b) observed that the situation in *Cerastium* and *Stellaria*, where the S alleles at a single locus are in a state of transition between gametophytic and sporophytic control, is quite different from the two-loci G–S system presented by Lewis and his co-workers (Sect. 2.4.2).

▪ 2.5 Genes Involved in the Rejection Phase of SI

Throughout the four first sections of this chapter, an attempt has been made to describe the genes that control the recognition of pollen and pistil in different SI systems. The presentation is necessarily vague when, as in the case of pollen determinants, no or very little information is available on the function of these genes. It is sometimes misleading, because some of the genes identified in Section 1.1 may also intervene directly in the rejection process (as the ribonuclease gene of stylar GSI probably does). It may also be inaccurate because, particularly when not polyallelic, certain genes considered to participate in stigmatic SI possibly restrict their intervention to the transduction pathway established after recognition or to the end points of the rejection phase and do not directly contribute to recognition.

2.5.1 In Stigmatic SI Systems

In self-incompatible species of the *Brassica* type, the products (or the formation of products) of many pollen genes are initiated, affected or repressed at different steps of the SRK pathway as a consequence of phosphorylation or de-phosphorylation (Hiscock et al. 1995; Nasrallah 1997). The numerous post-pollination processes (adhesion, hydration, regulation of calcium concentration, organization of the pollen-tube cytoskeleton, exocytosis, tube emergence, tip growth, ingress into the papillar cell wall) suspected or known to be disrupted by SI have been listed and outlined by Nasrallah (1997).

An example of a genetic locus active in the SRK signaling pathway is an aquaporin-like gene (ALG) that, through the analysis of *Brassica* SC mutants, was shown to provide the trans-membrane water channel necessary for the SI response (Ikeda et al. 1997). A recessive mutation at an unlinked locus (the MOD locus) epistatic to S prevents the encoding of ALG and, by so doing, cuts off a specific supply of water and causes SC. Another example of the effects of the S locus at the onset of pollination is pollen–stigma adhesion, which results from a joint action of SLG and S-unlinked SLR1 (Luu et al. 1997a, 1999; Dumas et al. 1999; Sects. 3.2.2.1, 3.2.3.2). However, in kale, Luu et al. (1997 b) have demonstrated that pollen–stigma adhesion is independent of the self or cross status of the pollination. A third example is that of the ARC1 gene, which has been shown to display phosphorylation-dependent interaction with SRK (Sects. 3.2.3.7, 3.2.5.5).

Obviously, and independently of water supply or pollen adhesion, many other functions, such as cutinase activity or pollen-tube orientation, are upset in SI species of the Brassicacea at the start of tube emergence and signaling pathways (Sect. 3.2.5). In the presentation of their model for SI in rye, Wehling and co-workers (1995) suggested that several unknown loci on chromosomes 3R, 5R and 6R participate in a signal-transduction cascade within the pollen grain (Fig. 2.9).

It is probable that what occurs through the phosphorylation/de-phosphorylation of pollen proteins during the rejection phase in plants of the *Brassica* type also takes place, with adaptations, in *Papaver* and in the grasses. In *Papaver*, Rudd et al. (1996, 1997) have demonstrated in vitro that the SI signaling pathway involves increases in phosphorylation in at least two proteins (p26.1 and p68) that possibly mediate the SI response. It is possible that p68 activates genes responsible for the irreversible inhibition of pollen-tube growth. Working with *P. coerulescens*, Li and co-workers (1995) have shown that the product of a gene closely linked to the S locus has thioredoxin activity in the mature pollen grain. It is possible that

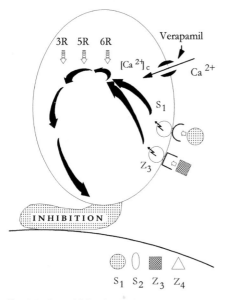

Fig. 2.9. A model for the SI mechanism in rye. The loci S and Z and a number of unknown loci on chromosomes 3R, 5R and 6R encode gene products that form a signal-transduction cascade in the pollen grain. The cascade is triggered by the binding of soluble S- and Z-allele-specific components to their receptor counterparts on the pollen grain. The stimulus is mediated by verapamil-sensitive influx of Ca^{2+} (Wehling et al. 1995). Verapamil is known to act as a Ca^{2+}-channel blocker in animal cells and to interfere with certain Ca^{2+}-dependent processes in plants. (Wehling et al. 1995)

this gene plays a role in the rejection phase of the SI reaction (Sect. 3.4.2). In *Secale*, the products of the S and Z genes and of non-specific loci on three different chromosome arms form a signal-transduction cascade that ultimately leads to the rejection of SI pollen (Wehling et al. 1995).

2.5.2 The Rejection Phase in Species with Stylar GSI

The involvement of rejection genes [other than the S-ribonuclease gene, which appears to control both phases (recognition and rejection) of the SI response in the one-locus GSI system] has not been demonstrated. There does not seem to be, at first sight, any need for an additional rejection gene, because through the systematic destruction of RNA (messenger and/ or ribosomal), ribonucleases prevent the formation of the proteins essential for proper function and development of pollen and pollen-tube structures in the incompatible tube (for a description of the organization of these structures and of their functional interactions see Pierson and Cresti 1992 and Li et al. 1997). It is not certain, however, that pollen tubes are unable to synthesize ribosomal RNA, particularly if the initial supply provided by the pollen has been degraded (Lush and Clarke 1997; Sect. 3.5.7.2). There-fore, it is possible that some of the enzymes (peroxidase isozymes, cyto-chrome oxidase, amylase, acid phosphatase, glycan hydrolases, etc.) that were identified in the 1960s and 1970s in self-pollinated styles and are sus-pected to play a specific role in stylar SI have a residual or secondary func-tion in the rejection phase of GSI. References to early research reports on these enzymes can be found in de Nettancourt (1977).

It may also be that the Ca^{2+}-dependent protein kinases from pollen tubes, which Kunz et al. (1996) observed to phosphorylate the stylar ribonucleases of *N. alata* in vitro, played a role in the recognition and rejection phases of SI. In particular, they may have led to the inhibition or alteration of the numer-ous signals and reactions required from the stigma, style or ovary for the orientation, pulsated growth and penetration (in ovule) of the pollen tube. The possibilities for intervention are particularly numerous: see, for in-stance, the analysis by Li et al. (1997) of functional interactions among the cytoskeleton, membranes and cell walls in pollen tubes of flowering plants.

2.5.3 SI in the Ovary

In the ovary also, one is dealing with complex relationships between the pollen and pistil. For instance, in self-incompatible *Gasteria verrucosa*, Willems (1999) has shown that signal substances are present in the pollen coat, the pollen-tube pathway and in the micropyle; these lead to the recog-nition of the incompatible pollen tube and to its rejection after penetration in the ovule at the time of the first endosperm mitosis. In this case, the fi-nal rejection process also appears to be the end point of a long and complex chain of events. Signaling obviously occurs in the two directions: towards and away from the incompatible pollen tube (Sect. 2.3.7.3).

■ 2.6 SI in Polyploids

The effects of polyploidy on the expression of SI are reviewed in Section 4.2.6, which deals with the breakdown of SI in diploid pollen. In summary, one can state that the tetraploid relatives (or artificially induced tetraploids) of diploid dicotyledons with monofactorial GSI usually display an SC phenotype. The SC character does not result from modifications to the pistil (the stigma in the Oenotheraceae or the style in most other dicots with monofactorial GSI) but from interactions (competitive effects?) among different S alleles in heterogenic di-haploid pollen. The loss of SI is not observed in families where the interaction of incompatibility alleles normally contributes to the normal expression of SI in haploid pollen, such as occurs in species with SSI or in many taxa (particularly among the monocots) with multigenic complementary GSI. SI is also retained in the autotetraploids of certain monocotyledonous families with a one-locus system (Lundqvist 1991) where other (complementary) incompatibility loci are possibly present in the homozygous condition.

■ 2.7 Equilibrium Frequencies of SI Alleles

Assuming randomized pollination and the complete absence of pressures from mutations, migrations and selective influences not implied by the incompatibility system itself, the equilibrium frequencies of SI alleles in infinite populations can be estimated from generalizations of the Hardy-Weinberg law for the majority of systems. As can be seen throughout this chapter, the determination procedures range from extreme simplicity (as in the distylic system) to high complexity requiring the use of computer simulation techniques for sorting out intricate relationships (such as those characterizing polyallelic series with different levels of dominance in pollen and pistil, or multigenic polyallelic systems combining gametophytic and sporophytic control). Information on the properties of populations with multi-locus homomorphic GSI can be found in Charlesworth (1979).

2.7.1 Two Alleles at One Locus in a Sporophytic System

This situation corresponds to the incompatibility mechanism in *Primula*, *Fagopyrum* and *Pulmonaria*. It represents the case of a single locus (S) with two alleles (S dominant and s recessive) and two phenotypes (pin and thrum) and can be summarized as follows:

Genotype	SS	Ss	ss
Phenotype	–	Thrum	Pin
Frequency	a	b	c

If $a=0$ (SS plants do not occur), frequencies b and c are in equilibrium when they can be reproduced in the next generation (i. e., when $b=c=0.5$). Regardless of the initial frequencies b and c, equilibrium will be reached, as predicted from the Hardy-Weinberg law (Finney 1952), in one generation, and will be maintained thereafter (in the absence of pressure forces).

2.7.2 Trimorphism

In this system, representative of SI in genera, such as *Lythrum* and *Oxalis*, one is dealing with three phenotypes (long, middle and short) and six different genotypes. As for the one-locus di-allelic heteromorphic system, the equilibrium equation (Fisher 1941) infers that the sum of the frequencies of all constituted genotypes is the same for each phenotypic class produced (isoplethy) and, consequently, the three style lengths are present in equal numbers. The frequencies of phenotypes and genotypes under conditions of equilibrium are the following (Finney 1952):

Genotype	MM–Ss	Mm–Ss	mm–Ss	MM–ss	Mm–ss	mm–ss
Phenotype	Short	Short	Short	Middle	Middle	Long
Frequency	0.02393	0.13077	0.17863	0.02393	0.30940	0.33333
	0.33333				0.33333	0.33333

In the calculation of genotype frequencies, Fisher (1944) and Fisher and Martin (1948) made allowance for linkage between the M and S loci in *O. valdiviensis* and for double reduction in *L. salicaria* (where inheritance is tetrasomic and M and S are unlinked).

2.7.3 One Polyallelic Locus in a Sporophytic System

By means of computer simulation, Imrie and co-workers (1972) calculated the equilibrium gene frequencies in infinite populations of the monofactorial species *Carthamus flavescens*. Based on segregation of three and six alleles with dominance in the pollen (S1<S2<S3<S4<S5<S6) and independence in the style, their estimates showed that:
▪ Depending on the initial frequencies of alleles and of genotypes, the number of generations to equilibrium ranged from 8 to 19 in the three-allele model and from 27 to 46 with six alleles in segregation.
▪ The equilibrium frequencies naturally varied with the number of alleles in the model but were not dependent on the initial frequencies.

■ Equilibrium frequencies were closely related to the level of dominance; the frequency of any allele increased as its degree of dominance decreased from S6 to S1 (for cases where this does not happen, see Sampson 1974).

2.7.4 Polyallelic Series in a Monofactorial Gametophytic System

The equilibrium frequencies for a monofactorial gametophytic system where all alleles are considered to have independent action in the pollen and pistil are approximated (in the absence of selection pressures, breeding restrictions, migrations, drifts and mutations) by $P_i = 1/n$, where n is the number of different alleles segregating in the population. As n has a lower limit of three (a necessary condition for the system to operate) and an upper limit of twice the number of individuals composing the population, it follows that the equilibrium frequencies of such alleles in a population of x individuals will necessarily lie between $1/(2x)$ and $1/3$. The total number of different S genotypes is given by $n(n-1)/2$, where n is the number of different alleles.

2.7.5 Two Polyallelic Gametophytic Loci

In this system, typical of the grasses, two loci, S and Z, are functionally integrated but inherited independently and display allelic series of different sizes. The frequencies of the individual alleles will be $1/ns$ at the S locus and $1/nz$ at the Z locus, where ns and nz represent the numbers of alleles segregating at each of the two loci. The frequency of any $S_i Z_i$ genotype will be equal to the product of these frequencies (weighted to allow for the non-occurrence of double homozygotes). It follows, because individuals can be homozygous at one or the other of the two loci, that the minimum number of alleles required to maintain the two-locus system is two at each locus.

Lundqvist (1962b) and Blom (cited in Lundqvist 1962b) elaborated and applied a formula for estimating the number x of different genotypes represented in a sample from the progeny of two heterozygotes with no alleles in common. With N genotypes (each occurring with probability $1/N$) and n plants in the sample, the mean of x will be NP, where $P = 1-(1-1/N)^n$. The variance of x will be $NP(1-P)-RN(N-1)$, where $R = (1-1/N)^{2n}-(1-2N)^n$.

Lundqvist (1962a) has considered the two levels of heterozygosity allowed at the S and Z loci (hom–het: plants homozygous at one locus, producing two specificities; het–het: plants heterozygous at both loci, producing four specificities) in his calculation of the probability that a specific S–Z gene pair, taken at random, will recur in the plants of the population at equilibrium. For the het–het and ho–het situations, he provided (Table in Lundvist 1962 a; Charlesworth 1979 for the relative frequencies of single versus double heterozygotes) the different formulas leading to the determination of pollen genotype numbers, the frequency of the individual genotype and the number of incompatibly reacting pistil genotypes for different numbers of the pollen genotypes tested.

The crossing procedure necessary to determine the SI genotypes of a sample of unrelated plants from a population with a two-locus gametophytic system has been proposed by Lundqvist (1962 a) and improved by Fearon et al. (1994). The modified scheme requires an extra round of preliminary crosses and four double homozygous testers instead of three. However, as emphasized by Fearon and co-workers, in the new procedure, these homozygotes do not have to be obtained from the population under study, and the entire sample can be investigated regardless of the allelic constitution of individual plants.

2.7.6 The Number of Possible Allelic Combinations in *Theobroma*

In *T. cacao*, the three types of SI phenotypes (25, 50 and 100% non-fused ovules on selfing) have been analyzed by Cope (1962), who calculated the number of possible allelic combinations for each of these classes when n S alleles segregate in the population and a fraction p of these are individual in action. The number of combinations calculated by Cope are the following:

- ▥ Class 1 (100% non-fusion): $n-1$
- ▥ Class 2 (50% non-fusion): $\frac{1}{2}np(np-1)$
- ▥ Class 3 (25% non-fusion): $\frac{1}{2}n^2(1-p^2)-\frac{1}{2}n(1-p)$

Copes notes that the third class generally will be the largest because its numerical expression contains terms in n^2, and the third and second classes will be equal only if p exceeds the limiting value of $\sqrt{2}/2$. Bartley and Cope (1973) worked out in detail the segregation patterns for the alleles of the A, B and C loci and the distribution of phenotypes after mating among self-incompatible and/or self-compatible individuals.

▥ 2.8 The Maintenance and Efficiency of Incompatibility Systems

The segregation patterns and equilibrium frequencies presented above have all been estimated under the assumption, made by Wright in 1939 in his paper regarding the distribution of self-sterility alleles in populations, that the populations considered are panmictic and that the parents for each new generation are selected at random from the previous generation. However, many events and fluctuations occur in nature, such as variation in population size or in plant size (that is to say in the amounts of seed and pollen produced by each plant), the arrival or generation of new alleles, overlap between generations, genetic drift, or linkage of SI loci to non neutral genes. These may lead to considerable differences in the frequencies of individual alleles, allelic losses, modifications of cross-compatibility levels between sibs and population extinction.

2.8.1 Population Sizes and Numbers of Incompatibility Alleles

2.8.1.1 Smallest Numbers of Alleles Required

Although an SI system can, in theory, be maintained in populations of only two individuals, allele loss due to random genetic drift or to selection will decrease the number of alleles to values below the minima allowed for the proper functioning of the system in small populations. In polyallelic systems, the number of generations leading to the loss of a given allele will increase as the number of alleles decreases, and the maintenance of n alleles in the population will be a function of the size (x) of the population. Imrie et al. (1972), who calculated the rate of loss of S alleles by genetic drift for *C. flavescens* (sporophytic monofactorial polyallelic system with dominance series in the pollen), showed that four was the maximum number of alleles that could be maintained in a population of 32 plants originally segregating for six alleles. Populations of 8 and 16 plants were found to be unable to maintain the critical number of three alleles and to become extinct (Table 6 in the 1977 edition). Similar trends had been estimated by Wright (1964) for the gametophytic system (which, for all mathematical purposes concerned, can be treated in the same manner).

The maintenance of two-locus GSI systems has been discussed and treated by Mayo and Hayman (1968), who also examined inbreeding effects and the time to extinction for S alleles linked to deleterious genes (Mayo and Hayman 1973). With many alleles at each SI locus, the time until extinction increases as the frequency of recombination decreases, and the performances are similar to those calculated for the one-locus system. However, with three alleles at each SI locus, the time to fixation is greater for the two-locus case. The minimum numbers of alleles per locus required for maintenance of the system are, as explained above (Sect. 2.8.1.1), three for monofactorial GSI and two for bifactorial GSI.

2.8.1.2 "Molecular Restraints to the Coding Capacity of the S Gene in *Papaver*"?

A total of 66 S alleles has been estimated for the field poppy (Lane and Lawrence 1993; O'Donnell et al. 1993). One explanation (for others, see Sect. 2.3.1.2) suggested by Lane and Lawrence to account for this relatively low number is that there are narrow limits to molecular diversity in the encoding regions of the S gene that participates directly in the pollen–stigma recognition reaction of *P. rhoeas*. Such severe limits, if they occur, could be ascribed to a variety of causes, including the small size or particular structure of recognition sequences, a possible protection against mutations, their lethal effects or an association with a very efficient repair system.

2.8.1.3 Consequences of High Numbers of Alleles at the SI Loci

Fearon et al. (1994) noted that the total number of genotypes expected with two-locus GSI is so large (for instance, 245,055 if nS and nZ each equal 31) that a population cannot be expected to contain more than a subset of the total number of possible genotypes. Under such conditions (which, of course, occur less frequently in the case of monofactorial GSI), practically all plants in the population are cross-compatible, and the strength of the frequency-dependent selection (which stimulates the introduction and diffusion of rare pollen originating from migration or mutation and genetic polymorphism) is strongly attenuated (Lawrence et al. 1994). The importance of the phenomenon (i.e., the decrease of cross-incompatibility and pollen incompatibility when the numbers of different SI alleles augments) has been demonstrated and measured by Fearon et al. (1994). They also showed how its main effects (distribution of only a subset of possible genotypes, and attenuation of the frequency-dependent selection of alleles) lead to unequal allele frequencies.

2.8.1.4 Linkage Effect as a Main Cause to Unequal S-Allele Frequencies in British Populations of *P. rhoeas*

Lawrence and Franklin-Tong (1994) found that the segregation ratios of SI alleles in three sets of full-sib families deviated substantially from Mendelian expectations. They attributed such distortions to the linkage of the S locus to a number of genes, including those governing seed dormancy and albinism. The extra effect of selection acted on the female gametophyte and, in one case, on the pollen. The small size of the populations studied and the relative facility with which the extra effect of selection was detected suggest that linkage to the S locus is particularly strong. The possibility that sampling effects also play a major role in the occurrence of unequal allele frequencies was dismissed by Lawrence et al. (1994) in populations where the number of S alleles was sufficiently high to maintain the number of alleles they were estimated to contain (Sect. 2.8.3.2).

2.8.2 The Selection of Rare Alleles and Replacement Processes

The fact that small populations of self-incompatible plants [like those described by Bateman (1947) in *Trifolium* and by Emerson (1938, 1939) in *Oenothera*] do not reach extinction, and maintain polyallelic series of large size, demonstrates that efficient forces counterbalance S losses due to genetic drift or selection. Nagylaki (1975) derived a sufficient condition for the increase of rare alleles by finding a lower bound for the strength of selection. For a system of n SI alleles, and neglecting mutation and random drift, the completely symmetric equilibrium is locally stable. Any allelic frequency less than

$$q = 1 + a - \sqrt{1 + a^2},$$

where $a = [2(n-1)]^{-1}$, will increase. For all n, $q > (2n)^{-1}$ but, if $n \geqslant 1$, $q \approx (2n)^{-1}$.

The replacement of S alleles may occur through migrations from adjacent populations, mutations within the population or, under certain circumstances, hard seed carryover. Wright (1939) developed the theory for calculating the mutation rate necessary for maintaining a gametophytic polyallelic monofactorial system in equilibrium. Assuming n′ alleles to be possible and n of them to be present at any moment in the population, Wright calculated that the chance that any given allele absent from the population will arise in the next generation is Nv, where N is the population size and v the mutation rate of the allele considered. Attributing an equal mutation rate to all S alleles, Wright finds that U, the mutation rate from one allele to all others, is given at equilibrium by:

$$U = (n' - 1)v = \frac{n(n - 3)(n' - 1)}{4N(n - 1)(n' - n)} f(1/2N)$$

Wright also showed how the equation for replacement can be written to express introduction by outbreeding rather than by mutation (substitution of v by mq_t, of U by $m - mq_t$ and of n′ by $n_t = 1/q_t$, where m is the proportion of migrating gametes and q_t is the frequency of any given allele in the species, taken as a whole).

2.8.3 Explanations to the Large Numbers of Alleles Found in *Oenothera*, *Trifolium*, *Carthamus* and *Lolium*

2.8.3.1 High Mutation Rates

In his work on allele replacement, Wright next proceeded, to calculate the number of S alleles maintained in a population with an indefinitely great number of possible alleles, and replacement rates (by migration or mutation) ranging to values as low as 10^{-8}. His calculations enabled him to find out that the 34 alleles estimated by Emerson from among a few hundred plants of *Oenothera* could not have been maintained in the population without a mutation rate at least equal to 10^{-3}.

2.8.3.2 Subdivisions of Populations

However, Wright (1939) also proposed, as an alternative explanation, that the *Oenothera* population surveyed by Emerson was in fact subdivided into many isolated groups that essentially bred within themselves and only exchanged 2% foreign pollen with other groups. The total number of different S alleles maintained in the entire population under such conditions of local inbreeding would increase proportional to the number of groups, and 40–50 alleles could be maintained in a population of 500 plants by a mutation rate of 10^{-5} per generation if cross-pollinations were restricted (in at least 98% of the cases) to plants immediately adjacent to one another. Since one would expect a larger diffusion range of pollen grains by insects and wind, it is possible that the concept of isolated groups, for which there is no evidence (Fisher 1961), should be considered in conjunction with the hypothesis of the regulation of mutagenesis by inbreeding factors (Sect. 4.2.7). The subdivision of populations has also been proposed by Lawrence (1996) as an explanation for the exceptionally high number of alleles in certain populations of clover. He also considered the possibility of sampling errors or of a particular molecular biology facilitating allelic diversity in the genus *Trifolium* (Sect. 2.3.4.1).

2.8.3.3 Migration and Hard Seed Carryover

In their simulation studies, Imrie and co-workers (1972) demonstrated that 10% migration increased the number of generations needed for the loss of an allele from a population of 32 plants (*C. flavescens*) from 5.60 to 9.93. The same authors analyzed the effects of hard seed carryover and found that a combination of 20% hard seed carryover and 10% migration would bring the number of generations to allele loss to 11.11. In addition to these effects, Imrie and co-workers also observed that migration and hard seed carryover caused the re-introduction of an allele one or more generations after it had been lost by genetic drift. In one computer run of 100 generations for a population of 32 plants, 24 alleles were lost; they were all re-introduced through a fixed migration rate of 10% in subsequent generations.

The effects of migration were also observed in a ryegrass cultivar founded from only five plants. Devey et al. (1994) concluded that the high number of S and Z alleles resulted from pollen immigration and not from high mutation rates at the SI loci. In this case, as in that of meadow fescue (Sect. 2.3.2.2) and small clover sub-populations (Sect. 2.8.3.2), one could perhaps consider the hypothesis that the generation of new S alleles is not a random process occurring regularly and frequently within the population but is a mechanism that is switched on in inbred plants.

2.8.4 The Efficiency of SI Mechanisms for Preventing Unions Between Near Relatives

Bateman (1952) defined the outbreeding efficiency of a given SI system as the ratio of general cross-compatibility to cross-compatibility between sibs. The definition is accurate and appropriate for annual species but needs to be modified (in the case of perennials) to including the proportion of matings between near relatives, which results from parent–offspring relationships.

As calculated by Bateman (1952), the outbreeding efficiency of the gametophytic system (case B, in the terminology of Bateman) and of the sporophytic systems with independence (case F) or dominance (case J) increases with increased numbers of alleles and fails to exceed unity only when the number of alleles is kept below five in case F (outbreeding efficiency of 0.67 for four alleles) or below three in case J (outbreeding efficiency of 1.00 for two alleles). With three alleles, the minimum number for a workable mechanism in case B, the outbreeding efficiency of the monofactorial GSI system equals 1.33. For very large numbers of alleles (in equal frequencies), outbreeding efficiency tends towards four in the gametophytic system and towards two in the two other cases considered by Bateman. An increase in the number of loci from one to two squares the frequencies of cross-compatibility and, consequently, squares the ratio of general-to-sib compatibility.

With the help of the method elaborated by Lundqvist (1954), Mayo and Hayman (1973) calculated the proportion of incompatible pollination in an indefinitely large population. For the case of equal numbers of gametophytic alleles (n) at all segregating loci, these proportions are given by:

$$\frac{2}{n} \text{ for a single locus} \quad \frac{4}{n(n+3)} \text{ for two loci} \quad \frac{8(n^3-1)}{n^3(n+1)^3-8} \text{ for three loci}$$

Thus, although it may appear that the three-locus system is clearly at an advantage in forcing out-crossing and reducing the chance of a plant being unable to set seed, the proportion of incompatible pollination remains low in all cases as soon as more than a few alleles are present in the population (Mayo and Hayman 1973).

2.8.4.1 A Comparison of Parent–Offspring Relationships in Heterostylic and Gametophytic Systems

The analyses of the efficiency of SI mechanisms for diminishing matings between near relatives carried out by Fisher (1949) apply to perennials and, consequently, cover the possibility of compatible unions between parents and offspring. After noting, like many authors before him, that neither sib matings nor parent–offspring matings are in any degree diminished by the mechanism of distyly, Fisher proceeded next to compare the tristylic mechanism in *O. valdiviensis* (linkage between the S and M loci) to that of *L. salicaria* (independence between loci) and to polyallelic systems. Fisher found that tristyly reduces the proportion of parent–offspring matings in the ratio of 82.596% and the proportion of sib matings in the ratio of 77.452% + (10.946)pq, where pq = 0 for absolute linkage between M and S and pq = 1/4 for independence. In contrast, a gametophytic system with large polyallelic series will provide no protection at all (progeny of a cross S1S2×S3S4) or only partial protection (progeny of a cross S1S2×S1S3) against parent–offspring matings, but will reduce (progeny of SIS2×S1S3) or 75% (progeny of S1S2×S3S4) the frequency of compatible sib matings to 50%. In other words, a polyallelic gametophytic system efficiently restricts the occurrence of homozygosity through sib matings in annuals but is of little use for decreasing the frequency of compatible parent–offspring intercrosses in perennials.

2.8.5 Effects of Pollen and Seed Dispersal, Overlapping Generations and Plant-Size Variations in Populations at Equilibrium

These effects were examined by Brooks and co-workers (1996) on the basis of field data from the literature. They used a simulated population of 3840 plants containing 16 alleles of equal frequencies. The time to steady state (~50 generations) did not change, but the steady-state variance was affected in practically all cases: an increase of 228% by variations in plant size and of 12% through a combined effect of limited pollen and seed dispersal, and a decrease of 30% with overlapping generations (resulting from seed dormancy). Together, the four factors caused a large increase in the average steady variance, and the authors observed that, even when a population is in a steady state, the variance for a particular generation can be considerably higher than the average value. This means that the frequencies of the S alleles of a population in a steady state can be very different and that unequal allele distributions observed in samples taken from natural populations may reflect stochastic effects operating under steady-state conditions (Campbell and Lawrence 1981; Lawrence and O'Donnell 1981; Lawrence et al. 1994; Sects. 2.8.1.3, 2.8.1.4).

2.8.6 A New Mathematical Approach to SI Polymorphism in a One-Locus Gametophytic System

The work referred to above was accompanied by the development of a new mathematical theory of SI polymorphism (Brooks et al. 1997a) derived from that of Wright (1939), which is based on an equation of the probability that a particular allele will occur in a new plant. The frequency distribution for this single established allele in a population submitted to the effects of different factors (i.e., overlapping generation and variation in plant

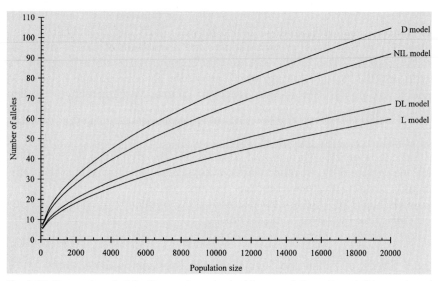

Fig. 2.10. The number of alleles that can be maintained in a population with an infinite number of possible alleles and a mutation rate of 10^{-6}. (Brooks et al. 1997b)

size) was derived as a normal distribution, with a variance easy to calculate. This distribution may be used to study alternative scenarios and facilitate our understanding of interfering factors. The time to reach steady state can be easily calculated.

2.8.7 The Concept of a Frequency-Equivalent Population

As a continuation of the analyses that permitted the estimation of the number of alleles that can be maintained for populations with overlapping generations (D model) or variations in plant size (L model), Brooks et al. (1997b) introduced the concept of a frequency-equivalent population. This is the panmictic distribution (NIL model) that has the same allele frequency distribution as a given population with overlapping generations and variation in plant size (DL model). For a given DL model, this alignment can be obtained by modifying the population size and the number of alleles of the NIL model until the frequency distribution has the same mean and variance as the DL model. The concept differs from the classical notion of effective population size in that variation in plant size makes the frequency-equivalent population larger than the actual population size, whereas the effective population size is smaller. Overlapping of generations has the reverse effect. Brooks and co-workers estimated the number of alleles that can be maintained in a population at equilibrium, with overlapping generations (D) increasing the number of alleles that can be maintained and variations in plant size (L) reducing it (Fig. 2.10).

CHAPTER 3

Cellular and Molecular Biology of Self-Incompatibility

Systematic efforts to identify and understand the cellular and molecular bases of self-incompatibility (SI) were initiated at the start of the second half of the twentieth century with the very first investigations regarding the involvement of glycoproteins in pollen–pistil relationships (Linskens 1954, 1955, 1958, 1960). From then on, Linskens (1961, 1962, 1964, 1975, 1986, 1988), who undertook an important part of the early work in the biochemistry of SI, regularly reviewed the state of the art and continued to contribute actively to research on the differential evolution of selfed, cross-pollinated and castrated buds and flowers and their RNA, amino acid (AA; Tupý 1961) and specific protein contents. New models of the recognition mechanisms, particularly those of Lewis (1965) and Ascher (1966) – both of which were inspired by earlier suggestions from Straub (1946, 1947) – and of Linskens (1965) began to flourish.

Essentially concentrated on homomorphic monofactorial systems and on the relatively simple model of Lewis (1965), the research performed in the late 1980s and the 1990s revealed a complexity and diversity of mechanisms far greater than expected. This is why, despite the remarkable arsenal of methods and techniques developed for the characterization, cloning, transformation and transfer of individual genes, there is still no indication of the identity of the substances that determine the incompatibility phenotypes of pollen grains in the different plant families analyzed to date, with the recent exception of the Brassicaceae (Schopfer et al. 1999). Several contributions to our understanding of the incompatibility phenomena in flowering plants, once viewed as almost final, may have to be reconsidered in the light of new knowledge on the nature, origin and properties of the specific products of SI genes. These uncertainties include the role of the tapetum in sporophytic SI (SSI), the dimerization of like male and female elements as the end point of recognition, the involvement of a thioredoxin S gene in the rejection mechanism of grasses and the linear evolution of SI from a single ancestral system.

▦ 3.1 Heteromorphic Incompatibility

3.1.1 A System of Its Own

3.1.1.1 The Research Approaches are Different

Barrett and Cruzan (1994) consider that the differences between heterostyly and homostyly led the students of each system to follow divergent research routes. Indeed, the economic importance of homomorphic SI has, to some extent, encouraged research on physiological and biochemical factors (affecting cross-pollination, fructification, seed production, hybrid vigor) likely to play a role in crop production. In addition, the unique properties of monofactorial polyallelic SI for pollen–pistil studies on cell–cell communication and for the analysis of allelic diversity, although not yet exploited in plant breeding and biotechnology, have been extremely profitable for our understanding of the cell biology of SI and the study of S-gene evolution at the molecular level. However, in the case of heteromorphic incompatibility, which raises little agricultural interest and where mating types can easily be identified, much attention has been given to the genetics and ecology of natural populations and, consequently, to the evolution, adaptive significance and breakdown of a very unique set of characteristics. In contrast, very little information is available on the physiology and chemistry of recognition and rejection in heteromorphic systems.

3.1.1.2 Several Rejection Sites

Rejection zones within the stigma, the style or the ovary have been found to vary from family to family, species to species and, in a species, from morph to morph. Some of the important characters associated with the differences (Sect. 3.1.2) may include pollen morphology, sculpturing of exine, cell size, stigma type (usually wet in thrum flowers and dry in pin flowers) and style type (hollow or solid). The relationships have been described by Dulberger (1992), together with examples of the taxa concerned, and have been shown to fall into two major categories:

▪ A category in which Dulberger includes "the majority of taxa with di- or trimorphism in pollen size, and where the morph-specific behavior seems to depend on a factor strongly linked with pollen size". This category (for an example of pollen-size variations among morphs, see Fig. 3.1) can be classified into three sub-groups corresponding to differences in the site of occurrence of different morphs and the structure of the style. These three subgroups host species with:

 – Two or three dissimilar sites of rejection (*Primula sinensis, Fagopyrum esculentum, Lythrum salicaria, L. junceum*)
 – Identical sites in both morphs, usually the stigma or the upper portion of the style (*Linum* and *Oxalis* species, *P. elatior*)
 – Flowers with hollow styles and a rejection site in the ovary for at least one of the morphs (*Pontederia* species, *Narcissus tazetta*)

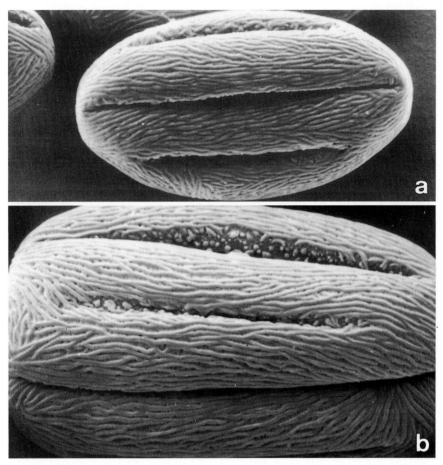

Fig. 3.1. Pollen of mid-style flower from *Lythrum salicaria*. **a** In short anther. **b** In long anther. ×2900. (Professor M. Cresti, University of Siena)

▪ A second category, typical of *Limonium meyeri* and *Plumbago capensis*, in which Dulberger includes the taxa "where the pollen grains display dimorphic exine sculpturing combined with dimorphic shape and/or structure of the stigmatic papillae".

3.1.1.3 Rejection Cascades

As in the case of homomorphic systems, in heteromorphic SI, the recognition of incompatible pollen or pollen grains by the pistil may be followed by the prevention or disturbance of a wide range of essential processes, such as pollen adhesion, hydration, germination, tube penetration and growth through the stigma, the style or the ovary (Barrett and Cruzan 1994; Sect. 3.2.5). In heteromorphic species, however, the site and nature of the rejection mechanism is often morph-specific, and several different barriers may operate in succession within a given flower (Shivanna et al. 1981, 1983).

3.1.1.4 Stigmatic Rejection May Occur on Wet Stigmas

Correlations between the state (wet or dry) of the stigma and the site of rejection, such as were statistically demonstrated for homomorphic systems by Heslop-Harrison and Shivanna (1977), are not always maintained in heterostylic species (Schou 1984; Dulberger 1987a, 1987b; Murray, cited in Lloyd and Webb 1992a, 1992b). Dulberger (1987a, 1987b) explained such exceptions on the basis of her observations in *Linum* (1987a) and of those of Schou (1984) in *Primula obconica*. In each case, she attributed the ability of the wet stigmas of heteromorphic species to reject incompatible pollen to basic physiological properties. She considers that these stigmas differ from typical gametophytic SI (GSI) stigmas (where pollen exudates circulate freely) in that they retain the secretory product on small groups of individual papillae. The rejection response is thus restricted to a very narrow space, with large areas available for many other interactions between papilla and (compatible) pollen.

3.1.1.5 The Rejection Sites Are Not Always Typical of a Sporophytic System

In spite of the capacity of wet stigmas in several heteromorphic systems to reject incompatible pollen, cases where rejection occurs in the style or the ovary are also known. This fundamental difference with homomorphic SSI may indicate (Stevens and Murray 1982) that the pollen incompatibility substances in heteromorphic SI do not lie in the pollen coat but within the grain and are probably not of tapetal origin (Sect. 3.1.1.6) or are hosted by the exine in an intermediate or inactive form.

These explanations and others (i.e., the ability of incompatible pollen to germinate and grow after the recognition reaction has taken place on the stigma) have been presented or reviewed by Richards (1986), Scribailo and Barrett (1991) and Barrett and Cruzan (1994). Instances of heteromorphic SI occurring in the ovary were reviewed by Sage et al. (1994), and differential ovule development following self- and cross-pollination was studied in depth in *Narcissus triandrus* (Sage et al. 1999; Sect. 2.3.7.3).

3.1.1.6 The Incompatibility Loci Are Usually Di-Allelic

The monopoly and persistence of di-allelism at the SI loci of heteromorphic species have been attributed (Lloyd and Webb 1992a, 1992b) to their close association with floral polymorphism, which protects them from invasion by new alleles, population extinction and evolutionary changes. As will be seen later, it is the opinion of Dulberger (1992) that, in heteromorphic SI, there is no equivalent to the incompatibility loci identified in homomorphic systems. Lloyd and Webb (1992b) consider that heteromorphic SI does not depend on the recognition of shared pollen and pistil specificities. However, cases (Sect. 2.1.3) where heterostylic systems appear to be governed by multiple alleles have been reported.

3.1.1.7 Role of the Internal Environment in the Specificity of Incompatibility Products

The surprising dependence of the specificity of the incompatibility genes participating in tristyly – which Lloyd and Webb (1992a) consider to be "the most dramatic demonstration of the genetic distinction between multi-allelic and di-allelic self-incompatibility systems" – was defined in Sect. 2.1.2.2. It certainly seems to demonstrate an intimate relationship between the differentiation of floral morphs (Fig. 3.2) and the qualitative expression of SI genes in tristyly. The mechanism involved is perhaps analogous to one of the processes [different splicing of the same transcription unit to

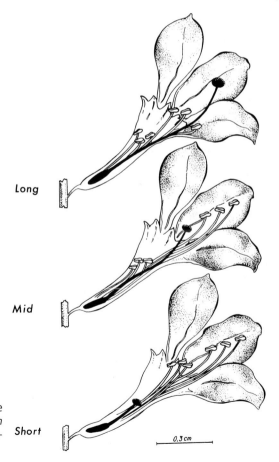

Long

Mid

Fig. 3.2. Drawings of short, middle and long flower morphs in *Lythrum junceum*. (Dulberger 1970; de Nettancourt 1977)

Short

0.3 cm

produce different messenger RNAs (mRNAs) or different post-translational modifications, such as glycosylation] proposed by Clarke et al. (1985) to explain how a given transcriptional unit could give rise, in GSI species, to different S products in the pollen and style.

3.1.2 Occurrence and Function of Stigma and Pollen Polymorphism

Several studies of stigmatic pollen loads reveal that the pollination of short-styled flower morphs may have been a "crucial factor, with little margin of error" (Lewis 1982) in the evolution of heterostyly (Ornduff 1970, 1971, 1975, 1979; Ganders 1974; Lewis 1982 and Dulberger 1992). However, the involvement of reciprocal herkogamy [defined in Barrett and Cruzan (1994) as "the reciprocal arrangement of stigma and anther heights in the floral morphs"] in the promotion of cross-pollinations between morphs has been firmly ascertained (Ganders 1979; Barrett 1992; Lloyd and Webb 1992a; Barrett and Cruzan 1994).

Other traits, particularly those giving rise to pollen and stigma polymorphism, are also typical of heteromorphy (Iversen 1940; Baker 1966; Vuilleumier 1967; Erdtman 1970; Ornduff 1970; Bokhari 1972; Dulberger 1974, 1975, 1987a, 1987b, 1992; Gosh and Shivanna 1980; Heslop-Harrison et al. 1981; Schou 1984; Barrett and Cruzan 1994; Wong et al. 1994a). Pollen and stigma differences (such as those shown in Figs. 3.3, 3.4 for *Carambola*, one of the rare heteromorphic SI species exploited commercially) essentially result from variations in cell size, the presence or absence of stigmatic outgrowths, osmotic pressure, the sculpturing of pollen exine and the amount of stigmatic exudate. In several heteromorphic families, certain of these variations have been described at the ultra-structural level [for instance, the work of Gosh and Shivanna (1980) and Dulberger (1987a) in *Linum*, Heslop-Harrison et al. (1981) and Dulberger (1987b) with *Primula*, Wong et al. (1994b) for *Averrhoa*; Sect. 3.1.2.1].

Species may be dimorphic for both the pollen and stigma (with or without association with heterostyly) or dimorphic for the pollen and monomorphic for the stigma. The different pollen phenotypes produced are referred to as A and B, while the two classes of stigma are called "cob" and "papillate". Baker (1953) found that, in species having correlated pollen and stigma dimorphism, A pollen is associated with cob stigmas, and B is associated with pollen with papillate stigmas. When heterostyly is involved, long flowers are characterized by cob stigma and A pollen; the short flower morphs bear papillate stigmas and shed B pollen.

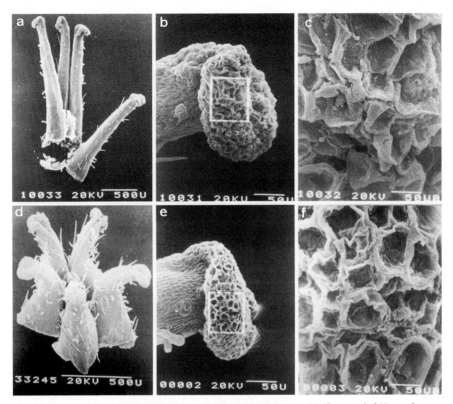

Fig. 3.3. Styles and stigmas from di-stylous *Averrhoa carambola*. **a–c** Pin flowers. **d–f** Thrum flowers. Note the vesicles in the stigma structure, the stylar bundles consisting of five styles and the greater lengths of the pin styles. The stigmatic head of the pin morph (and, presumably, the receptive surface) is larger than that of the thrum morph. (Wong et al. 1994b)

3.1.2.1 Analyses of Pollen Walls

The fine structures of types A and B pollen or of pin and thrum pollen have been described by several authors (for instance: the analysis of pollen dimorphism by Erdtman and Dunbar 1966; the description of types A and B in monostylic species of the Statitaceae and of pin and thrum phenotypes in *Ceratostigma* and *Plumbago* by Dulberger 1974, 1975; the study of different morphological and physico-chemical characters likely to play a role in pollen affixation by Mattsson 1983; the general review by Dulberger 1992; and the fluorescence and scanning electron microscope study of SI in *Averrhoa* by Wong et al. 1994a). Some 30 years ago, pin and thrum pollen grains in *Jepsonia parryi* were examined through scanning electron microscopy by Ornduff (1970), who found that the smaller size of pin pollen was associated with distinct sculpturing of the wall. The ektexine of pin and thrum is reticulate, but the lumina of the walls of thrum pollen are much larger and more verrucate than those of pin-pollen grains.

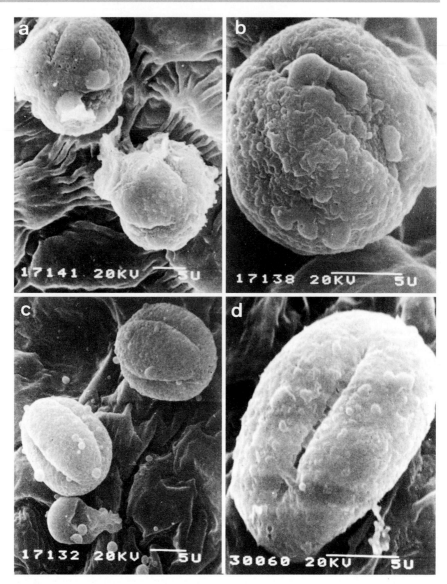

Fig. 3.4. Pollen grains from di-stylous *Averrhoa carambola*. **a, b** Pin morph. **c, d** Thrum morph. Note the three colpi of pollen grains and differences in pollen shape between the two morphs (round for pin pollen and oblong for thrum pollen; Wong et al. 1994 b)

3.1.2.2 The Role of Pollen and Stigma Dimorphism on Pollen Affixation and Pollen Metabolism

Detailed analyses of the possible effects of pollen and stigma polymorphism (such as that resulting from morphological, physical or chemical differences among morphs) on the mechanism of pollen rejection by incompatible pollen have been made or reviewed by Mattson (1983), Richards and Mitchell (1990), Dulberger (1992) and Barrett and Cruzan (1994). In some instances, these effects are strong and contribute to the establishment or reinforcement of the SI barrier. The most spectacular example, as seen in Section 3.1.1.7, is probably the specification of the incompatibility phenotypes in tristyly by the internal environment of flower morphs. Dulberger (1975, 1992) emphasized the developmental relationship between incompatibility and the morphological and structural traits of heterostyly. However, in addition to the fact that SI has been found to be expressed in the absence of the morphological differences (Sects. 2.1.1.2, 4.3.4.4) in homomorphic deviants, the effects of floral polymorphism on crossing behavior are not always constant or significant. A number of typical examples are provided below.

3.1.2.2.1 Effects of Differences in the Morphology of the Stigmatic Cuticle and in the Sculpturing of Pollen Exine.
Through a detailed examination of several dimorphic species in the Plumbaginaceae, Dulberger (1975) discovered that the stigmatic papillae differ in the way the cuticle is attached to the cellulose layer. In *P. capensis*, the stigmatic cuticle of the short-styled plants displays minute protuberances, whereas the cob and papillate stigmas in *Limonium* and *Armeria* are distinguishable by the thickness of the cuticle layer at the papillae apex. In some of the species studied, the architecture of papillae and pollen exine appears to prevent pollen affixation after the self-pollination of plants producing type-B pollen. In *Linum grandiflorum*, where Lewis (1943) found that pin pollen usually does not adhere to the pin papilla, Dulberger (1981) has shown how precise and diversified the differences between morphs can be. The exine of a short-styled flower (Fig. 3.5a) "has truncated processes of roughly uniform size and structure, each bearing a marginal ring of six to nine papillae. Rarely, a few smaller processes also occur" (Fig. 3.5a–c). In long-styled flowers, "the exine bears two distinct kinds of processes (Fig. 3.5d,e): large processes of about the same size as in pollen from short-styled flowers but with a marginal ring of six to eight spinules rather than papillae, and with a conspicuous central spinule, and smaller processes that terminate in a single spinule". In the sister species, *L. mucronatum* (Fig. 3.5f,g), Dulberger notes that "the processes of the pollen exine in short-style flowers are monomorphic and have four to six peripheral papillae (Fig. 3.5f) whereas the pollen from long-styled flowers has dimorphic processes (Fig. 3.5g). The large processes bear five to eight peripheral spinules and a prominent central one, while the small processes have only a central spinule". It is likely that these

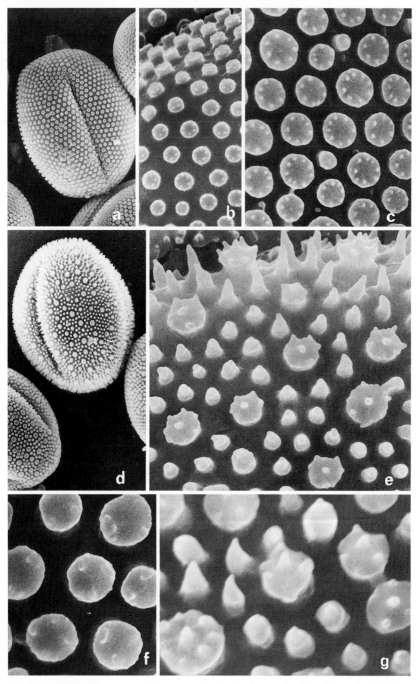

Fig. 3.5. Whole pollen grains and details of exine sculpturing in *Linum grandiflorum* (**a–e**) and *L. mucronatum* (**f, g**). **a–c, f** Short-styled morph. **d, e, g** Long-styled morph. **a, d** ×635. **b** ×3200. **c, e** ×5100. **f, g** ×10,500. (Dulberger 1981)

pointed processes play a role in the prevention of intimate contact between exine-borne material and the smooth surfaces of pin papillae (Dulberger 1999).

3.1.2.2.2 Function of the Pollen Exine in *Jepsonia*. The effects of exine sculpturing on pollen adhesion are not noticeable in all cases. As a follow up of his observations of pin and thrum pollen in *Jepsonia parryi* (Sect. 3.1.2.2.1), Ornduff (1970) compared the results of open pollination for each morph. A pin stigma was observed to host an average of 57.8 thrum-pollen grains and 203.7 pin-pollen grains. On the thrum stigmas, average numbers of 53.9 and 144.7 grains were calculated for thrum and pin pollen, respectively. From his knowledge of the production of microspores by pin flowers (200 000 average) and by thrum flowers (80 000) and of the differences between stigma areas (0.24 mm^2 for pin, 0.14 in the case of thrum), Ornduff reached the conclusion that the variations in morphology between pin and thrum pollen do not function to prevent illegitimate pollination. The possibility remains, nevertheless, that contact between the pollen and stigma is more intimate and instrumental to interaction in the case of compatible pollination. In another species of *Jepsonia* (*J. heterandra*), Ganders (1974) measured the effectiveness of heterostyly at promoting disassortative pollination. The results showed that intact flowers of both pin and thrum types receive an excess of pin pollen and that pins are subject to more self-pollination than thrums. Analysis of pollen loads from emasculated flowers revealed than, in inter-flower pollinations, thrums experience more disassortative pollinations than pins.

3.1.2.2.3 Availability of Exudate on the Stigma Surface. Barrett and Cruzan (1994) consider that insufficient moisture levels on the stigma may prevent the germination of illegitimate pollen. Their opinion is based on the observation of differences among morphs in the amount of stigmatic exudates in *Linum* (Gosh and Shivanna 1980; Dulberger 1987a), *Primula* (Heslop-Harrison et al. 1981; Schou 1984) and *Ponderia* (Scribailo and Barrett 1991) and on the finding by Shivanna et al. (1983) that different pollen types differ in their ability to take up water from the atmosphere.

3.1.2.2.4 Variations in Osmotic Pressures. An observation of somewhat analogous nature applies to the SI mechanism of *Linum grandiflorum*, for which Murray (1986) found that the differences in relative turgor pressure between pollen and styles observed by Lewis (1943) were only accessory to the failure of water absorption (pin pollen on pin pistil) or the bursting of pollen tubes (thrum pollen on thrum pistil).

3.1.3 The Molecular Biology of Heteromorphic SI

3.1.3.1 Identification of S-Recognition Factors

The efforts made to identify the products of SI genes in heteromorphic species have provided some indications of possible binding sites and substances that could be involved in recognition. In *P. obconica*, Golynskaya et al. (1976) reported that water-soluble proteins from pin and thrum pistils inhibit or delay the growth of pin and thrum pollen, respectively. Similar results were obtained by Shivanna et al. (1981) with *P. vulgaris*. Marked differences in water economy, which Shivanna et al. (1983) demonstrated to result from the transfer to thrum pollen from the sporophytic parent of a factor facilitating hydration, were found between pin and thrum pollen. Dulberger (1987b) found, for *Primula capensis*, that the thrum (wet) papillae produce a secretion and that the clefts between adjacent papillae are plugged with protein bodies. Neither the secretion nor the protein plugs can be observed in the larger pin papillae, which are characterized by a thick cuticle impermeable to Neutral Red solution.

In *Linum grandiflorum*, Coomassie-blue staining material is observed only on the surface of the thrum (wet) papillae stigma; esterases and acid phosphatases occur on the stigma of both morphs but with an activity significantly higher on the thrum stigma (Gosh and Shivanna 1980). In the leachates of pin and thrum pollen, Gosh and Shivanna (1983) detected different glycoproteins (of exine or intine origin) and lipids. As hexane washing of pin pollen was found to increase its adherence, it is possible that these lipids are responsible for the lack of adherence of pin pollen after pin–pin pollinations.

For both *P. obconica* (Stevens and Murray 1982) and *Linum grandiflorum* (Gosh and Shivanna 1980), lectin-binding sites have been identified in the stigma surface. Stevens and Murray (1982) found that the removal of the stigma, which bears a proteinaceous pellicle and binds the lectin concanavalin A (Nishio and Hinata 1980; Hinata and Nishio 1981; to react with *Brassica* S-proteins), destroys the incompatibility barrier. Carraro and co-workers (1996) showed that stylar apoplastic peroxidase could be involved in the heteromorphic incompatibility response of "thrum" morphs in *Primula acaulis*. Wong et al. (1994a) compared protein profiles (Table 3.1 and Fig. 3.6) in pin

Table 3.1. Levels of total soluble proteins (mean standard error: ±4.2) in mature pin and thrum floral buds of *Averrhoa carambola* (Wong et al. 1994a)

Stamen			Style[a]			Ovary			Tepal	
P	T1	T2	P	T1	T2	P	T1	T2	P	T1
37.5	25.9	20.4	79.2	50.8	53.8	59.2	28.8	53.5	36.9	25.4

P, pin flower clone B2; *T1*, thrum flower clone B10; *T2*, thrum flower clone B17
[a] Including stigmas

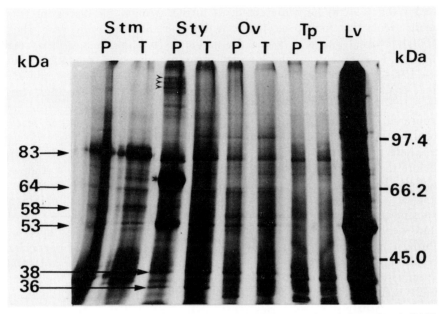

Fig. 3.6. One-dimensional Shine-Dalgarno sequence (SDS) polyacrylamide gel electrophoresis (PAGE) of proteins in various organs of pin and thrum flower buds of carambola at the mature stage 1 day before anthesis. Proteins from leaflets are included for comparison value. Polypeptides common to all parts are marked with *arrows* on the left, while the positions of molecular weight standard are indicated on the right. Polypeptides unique to the style of pin morph are marked by *arrowheads* on the left side of the lane. *Lv* Leaflet, *Ov* ovary, *P* pin morph, *Stm* stamen, *Sty* style, *T* thrum morph, *Tp* tepal. SDS is a particular sequence of nucleotides in an RNA molecule to which a ribosome will bind. PAGE is used for the separation of proteins and RNA molecules. (Wong et al. 1994a)

and thrum floral organs of the distylous starfruit tree (*Averrhoa carambola*, family Oxalidaceae). A 72-kDa polypeptide was detected exclusively in styles of the pin morph. Other polypeptides (45 kDa and 70 kDa) that bind concanavalin A are found in pollen and styles at pI values that differ for each morph.

3.1.3.2 Is There a Fundamental Difference between SI in Heteromorphic Species and SI in Sporophytic–Homomorphic Systems?

The almost complete absence of data regarding the molecular biology of heteromorphic SI (as reflected in Sect. 3.1.3.1) renders any answer to this question impossible. In addition, the apparent diversity of SI reactions in homomorphic species (Chap. 1 and below) should certainly complicate any systematic comparison between homomorphic and heterostylic SI. However, the opinion that heteromorphic SI is, by necessity, qualitatively different from sporophytic–homomorphic SI because it results (at least in tristyly) from differential pollen responses to particular pistil environments that do not occur in homomorphic species has often been expressed. Lloyd and Webb (1992 a) have presented, as further ground for discussion, the main events that took place in the Lythraceae and the Oxalidaceae when tristyly evolved into distyly and the mid-style form was lost. Pollen from the mid-level anthers of the long-styled morphs became compatible with the low stigmas of the short-styled morphs, while pollen from the mid-anther position of the short-styled morph was compatible with the high stigma of the long-styled morph. In other words, when the mid-style form disappeared, the incompatibility pollen phenotypes of mid-anther levels changed in opposite directions in long- and short-styled morphs.

These arguments and others (the early origin of floral polymorphism before the occurrence of the SI character, the relationship of the stigma–anther distance to the strength of intramorph incompatibility in tristylous and certain distylous species, the occurrence of SI at different inhibition sites that, within a population, are often specific for each morph) led Lloyd and Webb (1992 a) to suggest that "SI in heterostylous species operates through the failure of each type of pollen tube to grow under particular conditions, rather than from the recognition of shared pollen and style specificities". Lloyd and Webb do not, however, entirely dismiss the possibility that an immunological system of pollen–pistil recognition functions in certain distylic species. Nevertheless, they suggest a parallel between heterostylic SI and interspecific incongruity, which assimilates de facto heterostylic SI to the absence of a relationship between the pollen and pistil. It is difficult to envisage the occurrence of such a divergence (generally the end result of a speciation process followed by a very long period of reproductive isolation) in a panmictic population of inter-fertile groups of plants (Chap. 5).

Another illustration of the difference between SI in heteromorphic species and SI in sporophytic–homomorphic systems has been contributed by Stevens and Murray (1982), who demonstrated that the proteinaceous materials in the walls of *Primula* pollen could be washed out without any consequence for the pollen incompatibility phenotype. The observation implies that the tapetum does not participate in the determination of this incompatibility phenotype. Either SI substances or their precursors are formed in pollen mother cells (PMCs) before tetrad formation and are first stored inside the pollen grains, or they do not occur in heteromorphic plants. Whatever the case may be, the work of Stevens and Murray could explain why, in distyly and tristyly, the incompatibility phenotype of the pollen is expressed (for certain morphs) by the pollen tube in the style or the ovary and not on the stigma by the pollen grain itself (Sect. 3.1.2.2 above). However, such an observation does not necessarily contribute to our knowledge of differences between heteromorphy and sporophytic homomorphy, because there is no certainty that the male determinants [probably S-cysteine-rich protein (SCR)] in sporophytic homomorphic SI have a tapetal origin (Sect. 3.2.4.2).

Therefore, in the absence of new information on the molecular biology of species with heteromorphic SI, it is too early to draw any picture of the recognition and rejection mechanisms involved. However, when conclusions are finally formulated, it should be remembered that the pollen and stigma incompatibility sub-genes in *Primula* are real (Kurian and Richards 1977; Lewis and Jones 1992; Chaps. 2, 4), recombinable from each other and other sub-genes, and have independent functions and controls in the integrated expression of heteromorphic SI. Also, it must not be forgotten that homostylic plants, occurring through crossing-over (Lewis and Jones 1992) or otherwise, remain self-incompatible in the absence of the associated floral characters, and cases are known where heterostyly is not coupled to mating type (Sects. 2.1.2.4, 2.3.7.3).

▓ 3.2 Homomorphic Sporophytic Stigmatic SI: the *Brassica* Type

3.2.1 Morphology and Structure of Stigma and Pollen Surfaces

For the Crucifers and related families, a large part of the current knowledge regarding pollen and stigma surfaces can be traced back to research performed during the late 1960s and the 1970s (Heslop-Harrison 1968, 1975; Dickinson and Lewis 1973a, 1973b, 1975; Howlett et al. 1973; Mattson et al. 1974; Heslop-Harrison et al. 1975; Dickinson 1976; Roberts et al. 1980). More recent descriptions and reviews have been presented by Elleman and Dickinson (1986, 1990, 1994, 1996), Kishi-Nishizawa (1990) and Elleman et al. (1992). Scanning electron micrographs showing part of a longitudinal section of a *Brassica* stigma and incompatible pollen grains connected to papillar cells (Dzelzkalins et al. 1992) are assembled in Fig. 3.7.

3.2.1.1 The Stigma Surface

Plants with homomorphic stigmatic SI are usually characterized by dry stigmas that carry little or no secretion fluid at maturity (Heslop-Harrison et al. 1975). The stigmatic cuticle is covered by a thin, apparently uninterrupted, proteinaceous pellicle, which was first revealed by scanning electron microscopy (Mattson et al. 1974). It presents small protrusions and shows non-specific esterase activity. Mattson et al. (1974) found that this hydrophilic pellicle could function to initiate and maintain water movement through the cuticle and considered that it constitutes the primary recognition site of incompatible pollen. The main features of the pellicle, its extra-cellularity and its origin and development from cytoplasmic microbodies in young papillae have been described by Heslop-Harrison et al. (1975). Below the pellicle lies the cuticle, which is electron opaque and is crossed by micro-channels, which probably participate in the hydration of captured pollen grains (Elleman and Dickinson 1994). The cuticle presents numerous interruptions over the stigma (Mattson et al. 1974). Beneath it is

Fig. 3.7. Compatible and incompatible pollinations of *Brassica*. **a** Scanning electron micrograph of a longitudinally sectioned portion of a *Brassica* stigma showing the surface papillae (*P*) and the inner transmitting tissue (*T*) (×270). **b** A compatible cross-pollinated stigma, showing a hydrated pollen grain (*Po*) with its pollen tube (*Pt*) growing into the papillar cell (*P*) (×900). **c** A self-pollinated stigma exhibiting incompatible and aborted pollen-tube development from a poorly hydrated pollen grain (*Po*). (×900; Dzelzkalins et al. 1992)

the pectocellulosic bi-partite cell wall of each stigmatic papilla (Elleman et al. 1988; Elleman and Dickinson 1994). The outer layer will peel off from the inner layer to open the way to the compatible pollen tube at the proper moment (Sect. 3.2.2.2). Using the RFS (rapid freezing and substitution–fixation) method, Kishi-Nishizawa et al. (1990) studied the fine structure of the papillar cells, which were found to secrete abundantly on the day of anthesis. The distribution of S8 protein in the papillar cell walls of S8S8 homozygous plants of *Brassica campestris* could be visualized in sections prepared by the RFS method, with the help of an antibody against the S8 protein.

3.2.1.2 The Pollen Exine and the Pollen Coating

The first contact of a pollen grain with the stigmatic pellicle is established by a pollen membrane layer (CSL; coating superficial layer) with a thickness of approximately 10 nm; it has been described by Dickinson and Elleman (1985) and Elleman and Dickinson (1986). Throughout the entire pollen grain, it covers the exine and the substances carried by the exine cavities. Dickinson (1976), working with homomorphic, sporophytic, self-incompatible *Cosmos bipinnatus*, has shown how the morphology of the exine is first established within the cytoplasm of the young microspore and results from the organization by microtubules of dictyosome-like structures and their vesicles. The pattern is imprinted through an interaction between these vesicles and the plasma membrane. The young exine receives building material from the spore before liberation from the tetrad wall and thereafter begins to accumulate material (tryphine) of tapetal origin.

In *Brassica*, the mature pollen coat, which is lipidic, rich in enzymes and viscous, is ready to play a role in the attraction and adhesion of the grain to the stigma surface and to participate in the uptake of water from the stigma by the grain (Elleman and Dickinson 1990, 1994). Ruiter et al. (1997) developed a polyclonal antiserum that renders the detection of proteins located on the surface of *Brassica* pollen possible. The availability of an antiserum opens the opportunity to identify coat-protein-encoding sequences in a complementary DNA (cDNA) library of anthers.

3.2.1.2.1 Differences in the Pollen Exine Sculpturing between SSI and GSI.

From correlation analyses, Zavada (1984) concluded that pollen exine is perforate or reticulate in species with SSI and imperforate or macroperforate in species characterized by GSI. This thesis gave rise to a discussion between Gibbs and Ferguson (1987) and Zavada (1990) that essentially centered on the correct assignation of a number of taxa to the proper "exine" and "SI" categories and on complications arising from the multi-functional significance of exine perforations. If the controversy can be settled, more investigations on associations between exine sculpturing and reaction site could contribute to current research on the origin and possible filiations of different SI systems.

3.2.2 The Route of the Compatible Pollen Tube through the Stigma

3.2.2.1 Self-Incompatible *Brassica oleracea*

The water content of the pollen grain approximates 20% (Dumas and Gaude 1982) when it lands on the stigma. Water uptake from the stigma and from the atmosphere starts immediately, but with intensities that depend on the temperature and the level of atmospheric water in the growth chamber or the greenhouse. Two different modes of water uptake from the stigma by the pollen have been identified (Heslop-Harrison 1979): by means of a matric potential and, on formation of CSL, by a system of micro-channels based on turgor-pressure differentials (Sect. 3.2.1.1; Sarker et al. 1988).

Adhesion results from the first interactions between the CSL and the stigmatic pellicle, which appear to fuse (Elleman and Dickinson 1986; Gaude and Dumas 1987). The pollen coating undergoes several changes (including an increase in electron opacity and the apparition of membranous inclusions), probably associated with (1) water movement from the stigma to the grain and (2) structural modifications in the pollen proto-plast (Elleman and Dickinson 1990). It flows out to form a "foot" through which water and other materials, including enzymes, are absorbed by the grain.

Shortly after this activation phase, which requires the synthesis of numerous proteins and, in many instances, their phophorylation (Hiscock et al. 1995), the compatible pollen grain begins to germinate. As observed by Elleman and Dickinson (1996), the pollen tube emerges from the colpus closest to the stigma surface, penetrates the foot and grows through the cuticle after its degradation by a cutinase secreted by the pollen or the pollen-tube tip (Christ 1959; Linskens and Heinen 1962). Kroh (1966) studied the reaction of pollen grains in transfer experiments from compatible to incompatible stigmas and from incompatible to compatible stigmas. The results suggested a very rapid and irreversible activation of pollen cutinase by compatible stigmas. Hiscock et al. (1994), who reviewed the literature on the subject, identified and characterized an active cutinase in the pollen intine of *Brassica napus*.

As pollen germination proceeds, the outer layer of the stigma papilla loosens from the inner layer and separates from it to open a route for the pollen tube at the base of the papilla cell (Elleman and Dickinson 1994). The compatible tube enters the middle lamellae of cells that form the transmitting tissue of the pistil; the tube then continues its intercellular journey towards the ovary.

3.2.2.2 Self-Compatible *Arabidopsis thaliana*

A detailed analysis of pollination responses in *A. thaliana* has been performed by Kandasamy et al. (1994). The processes that precede or accompany adhesion of the pollen grain, foot formation, germination, investment of the papilla cell and growth of the tube between the outer and inner layers of the papilla cell wall have many similarities with those presented by the Dickinson group for *B. oleracea*. For the material they studied, the following timetable of events was established by Kandasamy and collaborators. The adhesion zone, consisting of tryphine-like material, becomes visible 5 min after pollination. Five to 10 min later, cytoplasmic changes that establish polarity and mark the point of tube emergence by the accumulation of vesicles occur in the grain. Twenty minutes after pollen capture, the pollen-tube tip reaches the papilla cell wall and grows down the length of the papillar cell in approximately 45–50 min. Kandasamy and co-workers (1994) also showed that pollen-tube growth was regulated throughout its course towards the ovary by stimulatory and targeting signals originating from stigmatic, stylar and ovarian tissues.

It is now suggested, from work in *Petunia* (Sect. 3.5.1), that the pollen coat in plants with dry stigmas, such as *Brassica* or *Arabidopsis*, contains lipids that direct pollen-tube growth by controlling the flow of water to the pollen and the penetration of pollen tubes (Wolters-Arts et al. 1998). The

Arabidopsis mutants *cer1*, *cer3* and *pop1* are defective in the production of long-chain lipids, and their pollen does not become hydrated on stigmas (Preuss et al. 1993 and Labarca et al. 1970, cited by Wolters-Arts et al. 1998).

3.2.3 Stigmatic Proteins Involved in the Recognition of Incompatible Pollen

It is not the aim of the present section to duplicate the expert (and usually very detailed) analyses and descriptions of the stigmatic S-proteins and S-related proteins of *Brassica* that were made during the 1990s, (Dickinson et al. 1992; Trick and Heizmann 1992; Hinata et al. 1993; Nasrallah and Nasrallah 1993; Franklin et al. 1995; Boyes et al. 1997; Dodds et al. 1997; Nasrallah 1997; Kusaba and Nishio 1999 b; Watanabe and Hinata 1999; Watanabe et al. 1999 c). The essential objective is to outline the importance of early research on the detection of S-gene products and to provide a summary of current knowledge regarding the two specific S glycoproteins (SLG, S-locus glycoprotein and SRK, S-receptor kinase) that appear to participate jointly in the recognition of incompatible pollen by the stigma.

3.2.3.1 Immunological Detection and Purification of SLG

An early and decisive contribution to the understanding of SSI was made by Nasrallah and Wallace (1967a, 1967b) and Nasrallah et al. (1970, 1972) when they injected rabbits with stigma homogenates of *B. oleracea* and collected sera with antibody reactions specific to the S alleles of the original material. With electrophoresis and double diffusion tests, Nasrallah and his associates demonstrated that a specific S allele could be correlated with a specific protein through successive generations. Evidence of genetic control of quantitative variations in the SI proteins detected by immunodiffusion was found (Nasrallah 1974). Similar results with other alleles were obtained by Sedgley (1974), Kucer and Polak (1975) and Sareen and Kakar (1977). Sedgley reported a direct relationship between allele dominance and high-titer S-protein antibodies.

However, Nishio and Hinata (1977, 1978) and Hinata et al. (1982) noted that some S genotypes produced more than one specific band. Analogous observations of differently expressed SLG sequences associated with the same S allele were made by Chen and Nasrallah (1990). It is possible that this "one genotype–two phenotypes" situation is remotely related to what appears to be the differential expression of the same S alleles in different flower morphs of a same plant in tristylic systems (Sect. 2.1.2.2). It is also possible that duplicated genetic structures are involved (Hinata et al. 1982). Concerning this last possibility, it has been found (Tantikanjana et al. 1993; Franklin et al. 1995; Giranton et al. 1995; Sect. 2.2.2.1) that, possibly as a result of differential splicing and a duplication–deletion process within the SLG–SRK tandem, the S2 haplotype of *B. oleracea* may encode three S-proteins corresponding to SLG, SRK and an SRK-related structure deprived of the kinase domain.

3.2.3.1.1 Purification of SLG. Purification of SLG was first achieved by Nishio and Hinata (1979), who used sepharose affinity chromatography and isoelectric focusing. The protein (57 kDa) had been obtained from *B. campestris.* Ferrari et al. (1981) confirmed the work of Nishio and Hinata when they purified a glycoprotein correlated to a specific S allele (S2) of *B. oleraceae* with features comparable to those of the SLG from *B. campestris.* This S2 glycoprotein, when applied to S2 pollen, prevented its germination on compatible stigma and had no action on the germination of pollen grains with different specificities on compatible stigmas. The effects of purified S2-SLG on S2 pollen suggest that they can occur, at least under certain circumstances, in the absence of S2-SRK (however, see Sect. 3.2.3.2).

3.2.3.2 Essential Features of SLG

SLG is abundant (5% of the total soluble protein) in the stigmas of flowers reaching anthesis and accumulates in the papilla cell walls (Kandasamy et al. 1989). Nevertheless, the contribution of SLG to the pollen–pistil recognition process has not been clearly established. Its central role in the expression of SI (Sect. 3.2.5.4) is supported by the identification (Nasrallah et al. 1992) of a mutation that drastically reduces the level of SLG and leads to the loss of the incompatibility response in the pistil in *B. campestris.* Its role in SI is challenged, however, by the finding in *B. oleracea* (Gaude et al. 1993, 1995) of self-incompatible plants with very low SLG levels and of self-compatible mutants with high SLG expression (Fig. 3.8) and by the observation (Kusaba et al. 1999) that SLG18 and SLG60 homozygotes produce defective proteins (a nonsense codon in SLG18 and a frame-shift deletion in SLG60) but remain self-incompatible (Sect. 3.2.5.4). Together with SLR1, SLG has been shown to have a function in pollen adhesion (Luu et al. 1999; Sects. 3.2.3.7, 3.2.5.1).

3.2.3.2.1 Cloning of the Gene Encoding SLG. Cloning of the gene encoding SLG was first achieved, in *B. oleracea*, by Nasrallah et al. (1985a). The clone (pBOS5) was obtained from cDNA libraries prepared from mRNA populations of S6S6 stigmas and submitted to differential screening (with radioactive stigma cDNA-hybridization probes) and counter-screening (with probes for non-stigma-specific sequences synthesized from mRNA provided by unrelated tissues). The identity of the clone was established with antiserum raised against the S6 glycoprotein. Descriptions of methods and reviews of knowledge for the screening, identification, linkage and co-segregation analyses of SLG have been prepared by Nishio and Hinata (1980), Nasrallah et al. (1987), Trick and Heizmann (1992), Dickinson et al. (1992), Hinata et al. (1993) and Franklin et al. (1995).

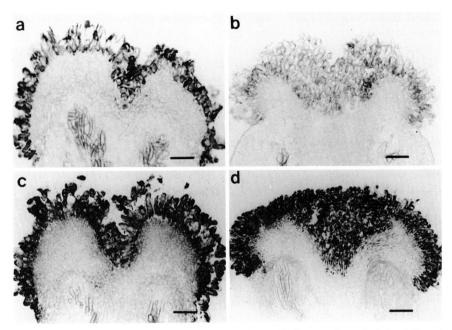

Fig. 3.8. Immunolocalization of SLG glycoproteins in longitudinal cryosections of pistils of Sc and class-II haplotypes. **a** Immunolabelling of stigma papillae in a cryosection of Sc pistil. **b** Weak immunolabelling of stigma papillae in a cryosection of S2 pistil. **c** Strong immunolabelling of stigma papillae in a cryosection of S5 pistil. **d** Strong immunolabelling of stigma papillae in a cryosection of S15 pistil. *Bars* = 0.1 mm. Cryosections were treated simultaneously with Mab 85-36-71. (Gaude et al. 1995, with kind permission from Kluwer Academic Publishers)

3.2.3.2.2 The SLG Sequence. The SLG gene encodes a primary protein of 431 AA, including a signal peptide (which allows the secretion of the protein) and a mature secreted protein of 405 AA. The cDNA sequence of S6 SLG was established by Nasrallah and co-workers (1985b), who derived from it the AA sequence of the protein. Shortly after, with the help of gas-phase high-pressure liquid chromatography of purified peptides, Takayama et al. (1986, 1987) determined the AA sequences of three S-specific glycoproteins and found that the cDNA sequence published by the Nasrallah group was incomplete. The corrected cDNA sequence of S6-SLG appeared thereafter, together with the sequences of two other alleles (Nasrallah et al. 1987). Sequence analyses of SLG alleles or of their products became more frequent, and sufficient data accumulated [in particular, the analyses by Kusaba et al. (1997), who reported the sequences of 31 alleles in *B. oleracea* and *B. campestris*] that not only rendered possible the attribution of different functions to specific regions of the gene but also permitted (as seen in Chap. 4) the comparison of alleles and the determination of the origin, extent and age of their variations.

3.2.3.2.3 Structure of SLG and Homologies between Alleles. All SLGs analyzed so far have 12 conserved cysteine residues in the C-terminal region of the protein; these are considered (Nasrallah et al. 1987) necessary to maintain the three-dimensional structure required for the folding of SLG. The protein also presents a number of potential N-glycosylation sites, of which four are possibly common to all alleles (Dickinson et al. 1992). There is some evidence, reviewed by Franklin et al. (1995), that glycosylation plays a functional role in the expression of SI but is not involved in the determination of allelic specificity (Sect. 3.2.5.3).

The mature secreted SLG protein has been divided (Nasrallah et al. 1987) into four regions, based on homology between the alleles. In region A and region C, different SLGs share the highest degree of homology, particularly if they belong to category-I haplotypes. According to Nasrallah et al. (1991), there is up to 80% homology at the DNA level and 90% at the peptide level. Regions B and D are highly polymorphic, with only 45–40% homology between alleles; therefore, they are more likely to be involved in S specificity (yet, see also Sect. 4.2.7.1).

3.2.3.2.4 Nature, Origin and Frequency of Sequence Variations between Different Alleles. The subject of the nature, origin and frequency of sequence variations between different alleles, taken up in Chapter 4, was intensively reviewed by Trick and Heizmann (1992) in the early 1990s who showed that the very important polymorphism of SLG largely results from nucleotide substitutions and insertion–deletion differences with the occurrence of small blocks of sequence changes. Their demonstration was recently contradicted by Kusaba et al. (1997), who provided evidence, after a large-scale examination of SLG hyper-variable (HV) regions, that intra-genic recombination (as measured by the accumulation of substitutions by different regions at relative rates that are not constant) also contributes to S polymorphism. Analogous views have also been expressed by Charlesworth and Awadalla (1998), who emphasized the significance of recombination (or its absence) on the production, circulation and accumulation of diversity. Nasrallah (1997) does not dismiss the possibility that recombination has contributed to the generation of new S specificities but considers that the frequent involvement of recombination in the evolution of the S locus would disrupt the link between stigma and pollen functions (and lead to the breakdown of the system).

The variability between alleles is reflected by a high proportion (~72%) of non-synonymous base changes (i.e., changes leading to the replacement of AAs). The total difference in base-pair composition between two alleles can be as high as 82% in the HV regions (Trick and Heizman 1992).

3.2.3.2.5 Co-Evolution of SLG and SRK. At the same time, through gene conversion, prevention of crossing-over, unequal crossing-over, RNA intermediates or other means (Trick and Heizmann 1992; Tantikanjana et al. 1993; Boyes et al. 1997; Nasrallah 1997; Chap. 4), within each haplotype, there appears to be a co-evolutive mechanism that prevents excessive divergence from occurring between SLG and the receptor S domain of SRK (Fig. 3.9; Sect. 3.2.3.3). Homologies between SLG and the S domain of SRK were observed to be very high (Watanabe et al. 1994; Yamakawa et al. 1995) in several haplotypes. However, Goring and Rothstein (1992) and Kusaba et al. (1997) have found that the maintenance of homology between the two genes in certain haplotypes is not as strict as initially expected. The finding, which needs to be complemented by the individual identification of variable and conserved domains in each of many SLG/SRK gene pairs (Nasrallah 1997), is important because, as shall be seen in Section 3.2.4 below, it suggests that SLG and SRK could bind different sites of the pollen determinant and contribute jointly to the determination of stigmatic specificity. Furthermore, Hatakeyama and co-workers (1998b) have shown that the strong sequence similarity between SLG and the receptor domain of SRK is not necessarily involved in higher dominance relationships in self-incompatible stigmas of *Brassica rapa*. The comparative analyses of SLG sequences between and within species of the Brassicaceae (Dwyer et al. 1991; Nasrallah and Nasrallah 1993; Hinata et al. 1995; Boyes et al. 1997; Kusaba et al. 1997; Charlesworth and Awadalla 1998) also demonstrate that some of the divergence between SLG alleles, which is often larger within species than between species, predated speciation (Sect. 4.3.1.5).

3.2.3.2.6 Structural and Functional Distinctness of SLG in Class-II Haplotypes.

Structural and functional distinctness of SLG in class-II haplotypes was first reported by Tantikanjana et al. (1993), who found that SLG in the S2 haplotype of *B. oleraceae* produced two transcripts that differ at their 3' ends. It has recently been reported by Hatakeyama et al. (1998c) for SLG29, SLG40 and SLG44 of *B. rapa* and by Cabrillac et al. (1999) in the S15 haplotype of *B. oleraceae*. An account of these observations is provided in Section 2.2.2.1.

3.2.3.3 The S-Receptor Kinase Gene, SRK

Walker and Zang (1990) found that a putative serine/threonine kinase (ZmPK1) from *Zea mays* shares approximately 27% of the AA sequence of S13 SLG of *B. oleracea*. ZmPK1 is the first trans-membrane receptor identified in higher plants. It is predominantly expressed in the shoots and roots of young maize seedlings and, to a smaller extent, in silks.

Walker and Zang's discovery of the relationship between ZmPK1 and SLG initiated a one-team effort by Stein and co-workers (1991), who identified, cloned and sequenced the cDNA of a gene, linked to the SLG gene (Boyes and Nasrallah 1993), that encodes a trans-membrane receptor-like protein kinase. The gene, named SRK for S-receptor kinase, was expressed for the first time in tobacco in 1991 (Stein et al. 1991). The kinase domain of SRK was shown to be potentially functional by expression in *Escherichia coli* (Goring and Rothstein 1992; Stein and Nasrallah 1993). The protein product of SRK was identified by Delorme et al. (1995). The genetic evidence (linkage to SLG, polyallelism, consequences of deletion mutations...) that supports the hypothesis of an SI mechanism involving the control of a signaling cascade by the SRK kinase and the arrest of incompatible pollen germination has been gathered by Nasrallah et al. (1994). It is now known that receptor kinases play an important role in plant physiology. In crucifers, particularly in self-compatible *Arabidopsis* (Sect. 3.2.3.7), many relatives of the *Brassica* S-gene family, among them several receptor-like kinase genes, have been identified in vegetative and reproductive tissues in the past (Walker 1993; Dwyer et al. 1994).

3.2.3.3.1 Essential Features of SRK. SRK, like its SLG relative, functions in the papilla cells of the stigma and is fully expressed in mature pistils. The level of expression is lower than that of SLG by a factor of 140- to 180-fold. The physical distance between the two genes is approximately 25 kb in *Brassica napus* (Yu et al. 1996) and 20 kb in the S8 haplotype of *B. campestris* (Boyes et al. 1997).

The SRK gene comprises seven exons (Fig. 3.9). Exon 1 encodes the potentially glycosylated extra-cellular S domain, which has been found to have a similarity as high as 98% with SLG. It is possible that, early during the evolution of SI, a duplication of exon 1 was the origin of SLG (Tanti-

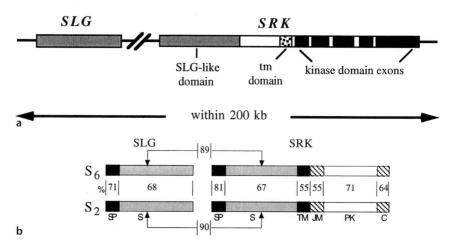

Fig. 3.9. The *Brassica* S-locus and the SLG/SRK gene pair. **a** Physical linkage and structure of the SLG and SRK genes. *Tm* trans-membrane. **b** Sequence relationship between the SLG and SRK genes in two haplotypes. The numbers refer to percent amino acid sequence identities between the predicted proteins and provide evidence for the concerted evolution of the gene pair within a haplotype and the divergence between haplotypes (Stein et al. 1991). *C* carboxyl-terminal domain, *JM* Juxta-membrane domain, *PK* protein kinase, *S* S domain, *SP* signal peptide, *TM* trans-membrane domain. SCR, the putative pollen determinant discovered by Schopfer et al. (1999), lies between SLG and SRK. (Nasrallah and Nasrallah 1993, with kind permission from the American Society of Plant Physiologists)

kanjama et al. 1993). Exon 2 encodes the trans-membrane domain that links the S domain to the cytoplasmic domain (encoded by exons 3 to 7).

The predicted transcript size for the fully spliced SRK product is 3.0 kb, with other transcript lengths of 2.3, 3.0, and 4.1 kb for the unspliced transcript (Stein et al. 1991). Using antibodies directed at the amino and carboxyl termini of SRK3, Delorme et al. (1995) verified this prediction and identified a membrane-associated, 120-kDa protein specific to stigmas. These results were abundantly confirmed by Stein et al. (1996), who demonstrated that SRK has the characteristics of an integral membrane protein. When expressed in transgenic tobacco, SRK is glycosylated and targeted to the plasma membrane.

Giranton et al. (1995) discovered truncated forms of SRK3, which corresponded to the extra-cellular domain of the putative receptor (Fig. 3.10). These truncated forms (eSRK3) were expressed specifically in stigmas and shared 82.6% similarity with SLG3 (from the same haplotype). Giranton and his co-workers consider the possibility that eSRK3 plays a role in the recruitment or canalization of ligand molecules for the SRK receptor (Sect. 3.2.5.4). They showed, very recently (Giranton et al. 1999a, 1999b),

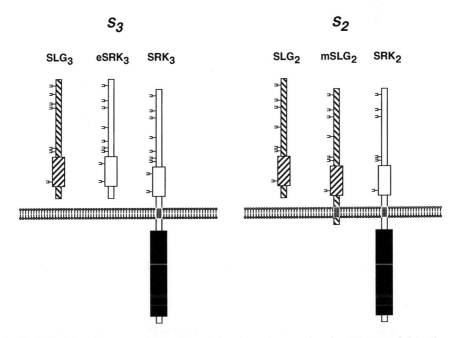

Fig. 3.10. Schematic representation of secreted and membrane-anchored proteins encoded by the S3 (class-I) and the S2 (class-II) haplotypes. The soluble, truncated SRK3 protein is represented as eSRK3, and the membrane-anchored form of SLG2 is represented by mSLG2. SLG proteins are *cross-hatched*. The extra-cellular domain of SRK is shown as a *white box*. Kinase and trans-membrane domains are shown as *black* and *stippled boxes*, respectively. Within the S domains, the cysteine-rich domain is represented by a *box*, and putative glycosylation sites are represented by branched side-chains. (Giranton et al. 1995)

through the study of the oligomerization and phosphorylation status of recombinant SRK expressed in baculovirus/insect cell systems and the co-expression of wild-type and mutated (non-functional) SRK proteins, that kinase activation occurred by trans-phosphorylation of SRK-kinase domains.

The cytoplasmic domain of SRK is characterized by sequence similarity to protein kinase and exhibits serine/threonine protein-kinase activity (Goring and Rothstein 1992). Stein and Nasrallah (1993) reported the auto-phosphorylation on serine and threonine residues of SRK expressed in *E. coli*. SRK, like SLG, is highly polymorphic (Stein et al. 1991; Delorme et al. 1995; Sect. 3.2.3.2 on allele diversity and the co-evolution of SRK and SLG). Nishio et al. (1997) found, through analysis by polymerase chain reaction/restriction-fragment length polymorphism (PCR-RFLP) and nucleotide sequencing of amplified fragments, that the third, fourth and fifth exons of SRK are strictly conserved, and the second and third introns are very variable.

3.2.3.4 A Direct Method for the Cloning of S Haplotypes

Neither chromosome walking by phage λ nor screening of genomic libraries constructed by partially digested DNA can be easily applied to the cloning of large genomic fragments containing both SLG and SRK genes (Susuki et al. 1997b). Chromosome walking (see Oliver and Ward 1985) is a technique used to identify a desired DNA sequence in a bank of overlapping DNA fragments from the chromosome under study. The fragments, produced by partial digestion, are first probed by an identified gene to pick up clones containing adjacent sequences that are used as probes for further detection and the constitution of successive generations of probes. The difficulties originate, in part, from the presence of repeated sequences (Song et al. 1991), which render chromosome walking difficult (Susuki et al. 1997b) or sometimes impossible (Boyes et al. 1997). The absence of reports on successful cloning from genomic libraries through the use of YAC, PAC, BAC and P1-phage vectors is also indicative of problems in the partial digestion of the region around the S locus, according to Susuki and his co-workers.

For these reasons, Susuki and his co-workers directly cloned a completely digested (80 kb) fragment containing the SLG9 and SRK9 genes of *B. campestris* into the P1-derived artificial chromosome (PAC). Over 990 independent clones had to be screened against the SLG9 probe, and one positive clone that contained both SLG9 and SRK9 was obtained. The method was found to be easier and more reliable than chromosome walking or the screening of DNA libraries constructed by partially digested DNA. It is well adapted to complex and large genomic regions spanning up to 300 kb.

3.2.3.5 SLG and SRK Are Present, Often as Traces, in Other Parts of the *Brassica* and Transgenic *Nicotiana* Flowers

3.2.3.5.1 In the Pollen. SLG- and SRK-like sequences have been found in *Brassica* microspores and pollen (Nasrallah and Nasrallah 1986; Guilluy et al. 1990, 1991; Nasrallah et al. 1991) in amounts so small (several-hundred-fold lower than in stigmas) that their participation as male determinants in the recognition reaction is unlikely (Sect. 3.2.4).

3.2.3.5.2 In Anther Walls. In anther walls, Watanabe et al. (1991) localized tapetum-borne proteins that were, in part, similar to pistil S glycoproteins but did not appear to be S specific. Thorsness et al. (1991) detected the activity of an S-locus gene promoter in both pistils (see below) and anthers of transgenic *Brassica*.

3.2.3.5.3 In the Transmitting Tissue of the Stigma, Style and Ovary. There is immunological evidence (Kleman-Mariac et al. 1995) that the SLG gene is expressed, at both the RNA and protein levels, in the stigma, the style and the septum of the ovary, exactly along the path of pollen tubes in the pistil. The observations, which apply to self-compatible and a self-incompatible lines, confirm and expand the work of Sato et al. (1991), who identified, through RNA blot analysis, low levels of endogenous RNA and protein SLG transcripts in the styles and ovaries of transgenic pistils. The possible significance of these results for the understanding of the evolutionary processes of SI systems is discussed in Chapter 4.

3.2.3.5.4 In Transgenic Tobacco. SLG is expressed in the stigma and, more abundantly, in the style (Kandasamy et al. 1990; Moore and Nasrallah 1990; Susuki et al. 1996). SRK is expressed in the stigma (Stein et al. 1991; Susuki et al. 1996) and in the style (Susuki et al. 1996).

3.2.3.6 A Putative Receptor Kinase Gene in *Ipomoea trifida*

A putative serine/threonine protein-kinase receptor (referred to as the IRK1 protein) has been detected by Kowyama et al. (1996) in *I. trifida*, a relative of the sweet potato; it is characterized by an SI system of the *Brassica* type. While several structural similarities were found between IRK1 and *Brassica* SRK, neither the distribution (predominantly in mature stigmas and anthers but also in leaves and roots) nor the pattern of expression suggested involvement in the control of SI. Furthermore, linkage tests through RFLP analyses revealed that the S locus and the IRK1 gene were not linked. Obviously, the search for an S-kinase receptor in *I. trifida* must continue, particularly because earlier findings of the Kowyama group (1995) indicated that several SLG/SRK-like genes were expressed in the reproductive tissues of *Ipomoea* (Table 3.2). However, further analysis (Kakeda and Kowyama 1996) suggests that these SLG-like genes could be,

Table 3.2. Groupings of polymerase chain reaction (PCR) clones based on the cross-hybridization of DNA dot blots. Messenger RNA from different reproductive tissues were amplified by reverse-transcriptase PCR using primers corresponding to conserved regions of *Brassica* SLGs. Four groups of PCR fragments were amplified. (Kowyama et al. 1995)

	Number of clones derived from Stigma			
	Group	Meiotic anthers	Pollen	Total
IPG1	15	5	2	22
IPG2	2	6	0	8
IPG3	0	2	0	2
IPG4	2	4	0	6
Total	19	17	2	38

in the same manner SLR3 is considered to originate from SRK (Cock et al. 1995; Sect. 2.2.2.3; Fig. 3.10), truncated derivatives of an IRK1-like receptor kinase gene that are, like SL3, essentially expressed in vegetative tissues.

S-locus-linked stigma proteins (SSPs) have been identified by two-dimensional gel electrophoresis of stigma extracts from S homozygotes (Kowyama et al. 1999). SSPs have a mass of approximately 7 kDa. From sequence analyses of cDNA clones encoding SSPs, a high homology was found with members of the short-chain alcohol dehydrogenase family. RNA blots and in situ hybridization experiments revealed that the SSP gene is expressed predominantly in mature papillar cells on the stigma surface and in the style. DNA gene blot analyses revealed that the SSP gene is present as a single copy in the haploid genome and is genetically linked to the S locus. The distance to the S locus, estimated from three recombinants detected in a population of 126 plants that had been subjected to RLFP analysis, is approximately 1.2 cM. Kowyama and co-workers are now using the SSP gene as a molecular marker for the identification of S genes.

3.2.3.7 SLG and SRK Have Many Relatives

The SLG/SRK haplotype belongs to a large family of *Brassica* genes. Its size was measured by Dwyer et al. (1991), who found that at least 15 related sequences can be detected in the *Brassica* genome when SLG probes are hybridized to genomic DNA. The S-related genes, including SR1, SR2, SR3 and an NS gene, which encode the non-S-specific proteins [first described by Isogai et al. (1988), analyzed for their variations and inheritance by Watanabe et al. (1992) and assimilated afterwards to SR1 (Heizmann 1992)], are presented above in Section 2.2.2.3. The list of S-linked genes (Sect. 2.2.2.4) was recently updated and was substantially enriched by Susuki et al. (1999). Among the new genes identified near the S locus, Susuki and co-workers found PCP-A1, the small, cysteine-rich pollen-coat protein that is a relative of SCR, the protein now strongly suspected to determine the pollen response to self-incompatible stigmas (Sect. 3.2.4.4) and, possibly, to incompatible stigmas from other species (Sect. 5.1.5.4). It seems that, among the SLG/SRK-related genes, only SLR1, through effects on (1) the interactions between the pollen and stigma (Lalonde et al. 1989) and (2) the control of pollen adhesion (Luu et al. 1997a, 1997b, 1999), could be (indirectly) involved in SI.

3.2.3.7.1 Members of the S Multi-Gene Family that are Linked to the S Locus.
Some of the genes mentioned in the linkage list of Susuki et al. (1999) and in Section 2.2.2.4 are very close neighbors of the S locus. This is, in particular, the case for three genes of *B. campestris*, BcRK1, BcRL1 and BcSL1, which are located less than 610 kb from SLG and SRK. BcRK1 is a putative functional receptor-kinase gene expressed in leaves, flower buds and stigmas; BcRL1 and BcSL1 carry frame-shift deletions, and BcRL1 appears to encode a truncated protein that contains only the S domain. The degree of nucleotide-sequence identity between the three genes and the other members of the S multi-gene family (Table 3.3) shows that they are most closely related to ARK1, an SRK-like gene in self-compatible *Arabidopsis* (see below) that may be involved in cell expansion and plant growth (Tobias et al. 1992; Tobias and Nasrallah 1996). Susuki and co-workers emphasize the similarity between the S locus of higher plants and the animal major histocompatibility complex, which also forms a large multi-gene (multi-allelic) family and occurs in tandem. Similar analogies have been reported for pathogen-resistance genes in angiosperms. The SCR gene (mentioned above), which presumably determines the S-phenotype of *Brassica* pollen (Sect. 3.2.4.4), is located between SLG and SRK.

Table 3.3. Degree of nucleotide-sequence identity[a] among the members of the S multigene family.[b] (Susuki et al. 1997a)

	BcR K1	BcR L1	BsS L1	SLG 9	SRK 9	SLR 1	SLR 2	SLR 3-1	SRK-like	AR K1	ZmP K1
BcR K1	–	88.6	77.1	66.2	66.6	57.4	64.3	59.0	61.6	78.0	47.8
BcR L1		–	77.4	66.7	67.3	52.2	62.1	59.4	55.8	77.5	46.8
BcS L1				66.8	69.2	52.7	65.4	60.2	61.2	79.4	46.2

[a] Nucleotide sequences corresponding to the S domains of BcRK1, BcRL1 and BcSL1 were compared with those of the multigene family. The degree of sequence identity (%) was calculated using DNA-SIS software (Hitachi Software Engineering)

[b] SLG9 and SRK9 are from the *B. campestris* S9 haplotype (Watanabe et al. 1994). SLR1 is from *B. campestris* (NS3; Yamakawa et al. 1995). SLR2 and SLR3-I are from *B. oleraceae* (Boyes et al. 1991; Cock et al. 1995). SRK-like is the S-linked SRK gene from the *B. oleraceae* S63 haplotype (Oldknow and Trick 1995). ARK1 is from *A. thaliana* (Tobias et al. 1992), and ZmPK1is from *Zea mays* (Walker and Zang 1990)

3.2.3.7.2 S-Locus-Related Sequences in *Arabidopsis*. Six sequences closely re-lated to the S proteins of *Brassica* were recognized in *Arabidopsis* by Dwyer et al. as early as 1994. Four functional genes could be identified, of which only one (AtS1) is specifically expressed in the papillar cells and could be involved in pollination. The other three genes, two of which encode puta-tive receptor-like serine/threonine kinases, are predominantly expressed in vegetative tissues. The conclusion of the study, made before sequence com-parisons between ARK1 in *Arabidopsis* and BcRK1 in *Brassica* (Susuki et al. 1999), is that there is no evidence for the evolution of the *Brassica* SI mechanism from a signaling system operative after compatible pollination. However, as emphasized by Dwyer and her co-workers, the sequences and expression data of the S-related proteins strongly suggest that the *Brassica* SLG/SRK tandem operative in SI was recruited from vegetatively expressed genes that have a function unrelated to pollination. It will be seen in Chap-ter 4 that the comparative mapping of the *Brassica* S-locus region and its homeolog in *Arabidopsis* is consistent with the hypothesis of the deletion of self-recognition genes as a mechanism for the evolution of autogamy in *Arabidopsis* (Conner et al. 1998).

3.2.3.7.3 Relationship of SRK/SLG to the Putative Kinase Receptor (ZmPK1) from Maize. Walker and Zhang (1990) found the extra-cellular domain of ZmPK1 to be partly homologous to the SLG 13 allele of *Brassica* (27% identity and 52% homology conservation) and, therefore, to the S domain of SRK. *Z. mays* is an autogamous species, and ZmPK1 is predominantly expressed in vegetative tissues. Nevertheless, the discovery of the relation-ship between SRK and ZmPK1 is very important, because it demonstrates the stability of these genes and emphasizes the very ancient origin of the S family and the S locus (Chap. 4).

3.2.3.7.4 ARC1, a Putative Downstream Effector for SRK. ARC1 is a stigma protein known to act in vitro as a substrate for SRK (Bower et al. 1996). When the expression of ARC1 is blocked through the use of ARC1 anti-sense oligonucleotides, the SI system of *Brassica* breaks down, suggesting that ARC1 is a downstream effector for SRK (Stone et al. 1999). Since the ARC anti-sense plants are most affected at the early stages of SI, it is possi-ble that ARC1 functions to inhibit pollen adhesion, hydration or germina-tion.

3.2.4 The S-Specific Pollen Determinant

3.2.4.1 Expected Features of the S Determinants

3.2.4.1.1 Allelism to the SRK or SLG Genes? At the time it was presented (Lewis 1960, 1965), the dimer model of the SI recognition reaction, such as that initially established for the gametophytic monofactorial system (Sect.

3.5), implied that pistil and pollen determinants were identical and were specified by the same genetic sequence. It is now known that such implications do not necessarily result. Separate sequences, such as the B and D regions of SLG and the S domain of SRK in a given S haplotype, may co-evolve during very long periods of time and may maintain a strong degree of homology (Sect. 3.2.3.2; Chapter 4). However, a single DNA sequence can generate non-identical S products by differential splicing of the same transcription unit or through different post-translational modifications (Clarke et al. 1985). Therefore, regardless of the correctness of the dimer hypothesis, there is no imperative argument for assuming that the pollen and pistil determinants are allelic. On the contrary, through the analysis (with GSI) of the effects of a deletion on the expression of SI in pollen and style (Sassa et al. 1997) and by the finding that S-RNases expressed in pollen are not involved in SI (Dodds et al. 1999), it has been shown that the S products in pollen tubes and styles are encoded by different parts of the S locus.

3.2.4.1.2 Likelihood of a Dimer Mechanism in SI Systems of the *Brassica* Type. Nevertheless, there are good reasons for considering that the dimer model is not representative of the situation in *Brassica*. The first is that only traces of SLG- and SRK-like sequences have been detected in *Brassica* pollen (Sect. 3.2.3.5). The amounts are so low that their involvement in SI would require very specific siting and targeting (Dickinson et al. 1992). Another argument against the dimer model is, of course, the recent discovery of the SCR gene linked (but not allelic) to SRK; in all probability, it encodes the pollen determinant of the SI reaction in *Brassica* (Sect. 3.2.4.4). However, in view of the strong homology of SCR-related PCP-A1 (Doughty et al. 1998) with SLG and its capacity to bind S products, it is possible that certain sequences of SCR are closely related to those of SRK. These sequences could perhaps be involved in some form of dimerization with SRK sequences after self-pollination when the level of homology is sufficiently high. The problem is apparently identical in the case of monofactorial GSI species belonging to the *Rosaceae* family, where, as stated above, Sassa and co-authors (1997) found that the unidentified pollen determinant was not allelic to the pistil determinant. For the time being, and in the absence of more information on the features and structure of SCR, the only (weak) argument in support of the dimer hypothesis comes from the observation, exposed in Section 3.2.4.1, that different sequences within a haplotype co-evolve with very little divergence for very long periods. However, there are also instances where significant variability between SLG and the S domain of SRK occurs within haplotypes (Sect. 3.2.3.2).

3.2.4.1.3 Linkage of Pollen and Stigma Determinants to the S Haplotype. Linkage of pollen and stigma determinants to the S haplotype remains a necessary condition for the preservation and transmission of any SI system based on the recognition of S-specific determinants in the male and female parts of the flower. The condition, fulfilled by SCR, which is sited between

SLG and SRK, had not been met by the defensin-like *Brassica* pollen-coat protein, PCP-A1 (Doughty et al. 1998).

3.2.4.1.4 Sporophytic Expression. It would be difficult to understand (1) the relationships of dominance that characterize pollen specificities in the *Brassica* system and (2) the recent demonstration (Stephenson et al. 1997) that S specificity resides in the pollen-coat protein if the determination of these specificities were not under sporophytic control. It has been suggested, however, that S substances of pollen (gametophytic) origin may migrate through the wall of the microspore into the thecal fluid of the loculus and combine with the other S determinant and tapetal products to form a mixture. In late stages of pollen maturity, this mixture is deposited on the pollen surface in a coating carrying the two S determinants and expressing a sporophytic-like S phenotype (Doughty et al. 1998). This hypothesis, still very speculative, was first formulated to take into account the typically gametophytic features of PCP-A1 transcripts, specifically localized on pollen grains late in development, with maximal expression at the trinucleate stage. It is now used to establish the origin of SCR as a sporophytic determinant of pollen grains (Schopfer et al. 1999). Furthermore, there may be a relationship between the apparent gametophytic features of cysteine-rich proteins like PCP and the G (gametophytic) gene concept introduced by Lewis and co-workers (Chap. 2) to explain anomalies in the segregation ratios of several plant families with an SI system of the *Brassica* type.

3.2.4.2 Contribution of the Tapetum to the Pollen Coat and to SI

3.2.4.2.1 Contribution to Pollen Coating. The tapetal origin of the materials deposited in the cavities of the pollen exine during pollen maturation was first demonstrated by Heslop-Harrison (1967, 1968), Dickinson and Lewis (1973a, 1973b) and Heslop-Harrison et al. (1973). They showed, in the Liliaceae, Cruciferae, Malvaceae and Compositae, that pollen kits (colored coatings of pollen grains containing carotenoids) and tryphines (heterogeneous pollen coatings composed of proteins and lipids) were transferred from the tapetum to the cavities of the pollen exine.

3.2.4.2.2 Evidence that the Pollen Coating Carries the Pollen S Determinant. During the 1970s, several demonstrations of the presence of proteins that play a significant role in the expression of SI were made in the pollen exine. In particular, Dickinson and Lewis (1973b) and Heslop-Harrison et al. (1974) found that, each time the pollen source was incompatible, extracts of the tryphine applied to the stigma could induce the symptoms of the SI reaction. The main evidence was the production, below the stigmatic papilla, of a callosic body considered at the time to prevent pollen-tube growth. The work of the Heslop-Harrison group was performed not only with pollen-wall materials (allowed to diffuse on agar or agarose gels) but also with the isolated fragments of the tapetum itself.

Stephenson et al. (1997) designed experimental strategies with biological goals other than callose deposition. In bioassays performed in vitro with *B. oleracea*, coating was removed from pollen grains, supplemented with protein fractions isolated from coatings of different S haplotypes, and interposed between individual pollen grains and the stigma surface. The criterion for measuring effects was the ability of pollen grains to germinate or to undertake initial tube growth. The results showed that the supplementation of coating by "cross" or "self" protein fractions could change the S phenotype of pollen grains initially compatible or incompatible with the stigma utilized in the bioassay. Pollen coating contains S-specific determinants that participate in the pollen–stigma recognition reaction. It would be interesting to know whether "induced" SC results from the presence of an active cross protein fraction in the supplemented pollen coating or is the mere consequence of the absence of a "self" protein fraction.

3.2.4.2.3 Tapetal Origin of Pollen S Determinants? The demonstrations, presented above, that the tryphine deposited in the cavities of the pollen exine has a tapetal origin and that the protein coating contains the determinants of pollen S specificities strongly suggest that these specific determinants are formed in the tapetum. The conclusion was particularly well substantiated by the group of Heslop-Harrison (1974), who, through the deposition of fragments of tapetum on the stigma surface, induced symptoms specific for the rejection process. However, since then, it has been shown (Singh and Paolillo 1990; Elleman and Dickinson 1994) that these symptoms (the formation of callose pads below the papillar cells) are not required for the incompatibility barrier and are induced by material leaking from already-inhibited tubes. Other arguments favoring the tapetal origin of S determinants were provided by Toryama et al. (1991a), who showed that a *Brassica* S-locus gene promoter directs sporophytic expression in the tapetum of transgenic *Arabidopsis*, and by Sasaki et al. (1998), who detected an SLG protein in the degenerating tapetum and pollen exine from immature *Brassica* anthers. This last discovery was made after the transfer of SLG9 from *B. oleracea* into self-compatible plants of *B. napus* that received a promoter (Osg6B) active in the tapetum of *Brassica* anther. It suggests the movement of the SLG protein from the tapetum to the pollen surface, but pollination tests indicated that the pollen of the transgenic *B. napus* did not express an SI phenotype, possibly because SLG, alone or associated with other products of the S locus, plays no role in the specification of the pollen S determinant. Now that the pollen determinant of the SI reaction seems to have been identified, further progress concerning the role of the tapetum in SSI should occur rapidly.

3.2.4.3 The Search for the Pollen S Determinant: Recent History

Several pollen proteins suspected to participate in the SI reaction were identified during recent years, but none of them satisfied all the requirements (outlined in Sect. 3.2.4.1) that are expected from an S-allele-specific pollen determinant.

3.2.4.3.1 The S-Glycoprotein-Like Anther Protein. A protein that is reactive with a polyclonal antiserum raised against the S-glycoprotein of S8S8 homozygous plants and does not react with polysaccharide residues similar to those of S-glycoproteins was identified by Watanabe et al. (1991) in anthers of *Brassica campestris*. However, no differences were found among S-glycoprotein-like anther (SA) proteins from plants with different S genotypes, and the stage at which SA was expressed (unicellular microspores) did not coincide with the periods during which SLG was present in the stigmatic papilla. The presence of SA in anther walls suggests that the tapetum (not the PMC or the tetrad) determines the SA phenotype of pollen grains.

3.2.4.3.2 The S-Locus Anther. The S-linked gene was detected (Boyes and Nasrallah 1995) in the S2 haplotype of *Brassica*. It is transcribed from two promoters to produce two complementary, anther-specific transcripts – one spliced and the other unspliced – that accumulate in microspores and anthers. Boyes and Nasrallah evoked the possibility that the spliced transcript of S-locus anther (SLA) was active in microspores and in the tapetum and had a dual pattern of gametophytic and sporophytic expression. The participation of SLA in SI was suggested by the fact that SLA could not be detected in an SC strain of *B. napus* carrying an S2-like haplotype.

There were, however, strong indications that SLA was not the standard pollen S-determinant. First, the SLA proteins or related substances had not been detected outside the S2 haplotype and either did not occur in other haplotypes or displayed very considerable polymorphism. Also, seven lines of *B. oleracea* were found to carry a SLA gene interrupted by an insert that prevents transcription (Pastuglia et al. 1997). These seven lines were self-incompatible.

3.2.4.3.3 Pollen-Coat Protein Class A. In their study on the effects of pollen-coating supplementation on the incompatibility phenotype of *B. oleracea* (Sect. 3.2.4.2), Stephenson et al. (1997) did not only demonstrate that S specificity resides in the pollen coating. In the water-soluble component of the pollen coat, which carries the male determinant, they also isolated a group of basic, cysteine-rich proteins, the pollen-coat protein class A (PCP-A), which are very similar to polymorphic anti-microbial proteins called defensins. One of the proteins was PCP-7 (now formally known as PCP-A1), a 7-kDa coat protein reported by Doughty et al. (1993) to bind in vitro with SLGs of *B. oleracea*. In a very similar form, it was also found to interact with SLGs from SC *B. napus* and SI *B. oleracea* and with SLR-1 from *B. napus*.

PCP-A1 was cloned and sequenced, and its putative structure established through homology modeling, by Doughty et al. (1998*)*. It can bind with monomeric and dimeric forms of SLG, and evidence was found for the existence of putative membrane-associated PCP-A1-binding proteins in stigmatic tissue. However, PCP-A1 was not eligible as a pollen S determinant, because its interactions with SLG were not S specific and because there is no linkage of the PCP-A1 gene with the S locus.

Conscious of these limitations, Doughty and co-workers (1998) considered the possibility that, due to both its ability to bind products of the S locus and its strong homology with SLG, PCP-A1 participates (as a co-factor with the still-unknown S-specific determinant) in the activation of the S-receptor complex to elicit the rejection of self-pollen. The group of Dickinson (Stephenson et al. 1997) also suggested that PCP-A1 plays a central role in the expression of pollen–stigma compatibility among SI species (Chap. 5). Research is underway (Wingett et al. 1999) to find out if PCP-A1 is required for normal adhesion and hydration and/or the correct expression of SI.

At an earlier date, the group of Dickinson (Stanchev et al. 1996) cloned another PCP: PCP1, a member of the large families of PCP genes. PCP1 is closely related to PCP-A1 but lacks the ability to bind stigmatic S proteins. The gene contains a single intron and encodes a small basic protein of 83 AA.

3.2.4.3.4 Pollen-Coat Protein A2. PCP-A2 displays all the main features of PCP-A1 but varies considerably over surface loop regions potentially involved in protein–protein interactions (Doughty et al. 1999). PCP-A2 binds specifically to SLR1, a protein that plays a role in pollen adhesion (Sect. 3.2.5.1).

3.2.4.3.5 SLL2-S9 and S-Locus Anther-Expressed S9 Gene. These two anther-expressed genes of *B. campestris*, were discovered by Watanabe and co-workers (1999b) in the 3′ flanking region of SLG9. The expression pattern and the high homology of SLL2-S9 with other haplotypes did not suggest its participation in SI recognition as a pollen determinant. SAE1-S9, however, displayed characteristics (specific expression in anthers, close proximity to SLG9 and, probably, high degree of sequence diversity) that suggested a possible involvement in the recognition reaction.

3.2.4.3.6 The Systematic Analysis of S and S-Related Regions. The search in Japan, mentioned above, for anther genes likely to play a role in SI forms part of a large effort (Susuki et al. 1999; Watanabe et al. 1999 a) to identify expressed sequences in the 6 kb SLG-SRK region of the S9 haplotype of *B. campestris*. Ten genes were detected in this region, which displays a gene density of one gene per 5.4 kb. An En/Spm transposon-like structure was found downstream of SLG 9. Genes encoding a J-domain protein and an anti-silencing protein homolog were also among the genes identified by Susuki et al. (1999). Several efforts (in addition to those outlined above in the present section) are presently underway in other laboratories. At Cornell University, where the group of Professor Nasrallah has been successful in its attempts to identify the pollen determinant, the systematic analysis of genes implied in the SI reaction has not been restricted to the search for recognition sequences; it extends to the detection and examination of genes that may modify signal perception and transduction (Nasrallah 1999). At Guelph, the S-locus region of the S910 and A14 S haplotypes of *B. napus* were sub-cloned in cosmid vectors and sequenced by Brugière (1999). New transcription units could be identified, either by hybridization of a subtraction cDNA pool to the genomic fragments or by computer analysis and PCR amplification. Large-scale research efforts for finding the pollen S determinant or identifying new members of the S gene family were also performed in species with GSI (Kao et al. 1999; Ride et al. 1999; Sects. 3.3.3.4, 3.5.8.5).

3.2.4.4 Finding the Pollen Determinant

This spectacular contribution of Cornell University (Schopfer et al. 1999) to the understanding of stigmatic SSI resulted from the systematic sequencing of the 13-kb region that separates SLG8 from SRK8 in *B. campestris*. SCR, the putative pollen determinant, is encoded by a gene consisting of two exons (110 bp and 300 bp) separated by a large intron of 4.1 kb.

3.2.4.4.1 The Gene Fulfils the Requirements for the Hypothesized Pollen Determinant. SCR is specifically expressed in anthers only after the generation of haploid microspores, with transcripts accumulating in the microspores. It is highly polymorphic and, as demonstrated by Schopfer et al. (1999), can be transformed (together with pollen specificity) by genetic engineering. Furthermore, loss of function experiments showed that the absence of the SCR gene product leads to SC.

3.2.4.4.2 SCR Is a Relative of PCPs. In the past, PCPs basic 6- to 8-kDa proteins with a cysteine pattern related to that of plant defensins have been proposed (Doughty et al. 1998; Sect. 3.2.4.3) as co-factors (non-specific and unlinked to S) participating in the recognition reaction. It is remarkable that SCR, despite differences in polymorphism, linkage and cysteine patterns, resembles PCPs in charge, molecular size and, to some extent, cysteine content.

3.2.4.4.3 Origin (Sporophytic and Tapetal) of SCR. At first, the time of expression (and possibly the site of origin) of the SCR protein appears to be that of a gametophytic product. However, the general view (Doughty et al. 1998; Schopfer et al. 1999) is that the SCR molecules produced by each S allele combine in the pollen coating to form mixtures of the two parental proteins. These then translocate into the cell walls of the stigma epidermal cells and express a phenotype typical of a sporophytic system. Whether the SCR proteins join the pollen coat from the tapetum or are produced by the microspore is not known.

3.2.5 What Happens after an Incompatible Pollination?

3.2.5.1 Pollen Capture by the Stigma

Capture after stigma landing is the fate of most pollen grains, compatible or incompatible, provided that the grain is placed on the stigma papilla so that one of its operculate apertures is in contact with the pollen protein pellicle (Heslop-Harrison et al. 1975). If the aperture of the grain is not in contact with the pellicle of the papilla, it may undertake further migration or roll until an aperture finally touches the pellicle. After contact between the pollen grain and the pellicle, hydration usually starts immediately, but at a flow rate and under conditions that vary considerably with the nature ("self" or "cross") of the pollination, the environment and the S genotype of the pistil (Sect. 3.2.5.2; Dickinson and Lewis 1975; Roberts et al. 1980; Sarker et al. 1988).

Adhesion results from the first interaction between the CSL (Sect. 3.2.1.2) and the stigmatic pellicle, which appear to fuse (Elleman and Dickinson 1986; Gaude and Dumas 1987). Through a statistical analysis of the stigmatic expression of SLR1 and SLR2 and of different pollination variables in 11 different species or genera of the Brassicaceae, Luu et al. (1997a, 1997b) showed that expression of SLR1 may be a factor in pollen adhesion. Through the use of a biomechanical assay that measures pollen adhesion forces (Fig. 3.11), this hypothesis was supported by the finding that transgenic suppression of SLR1 and/or the pre-treatment of wild-type stigmas with anti-SLR1 antibodies, anti-SLG antibodies or pollen-coat protein extracts reduce pollen adhesion (Luu et al. 1999). The results indicate a common adhesive function for the SLR1 and SLG proteins. However, the contribution of SLR1 to rejection and acceptance processes is not essential because, through the specific ablation of SLR1 by *Agrobacterium*-mediated transformation with anti-sense construct, Franklin et al. (1996) demonstrated that the gene was not required for the expression of SI or for successful cross-pollination between compatible lines.

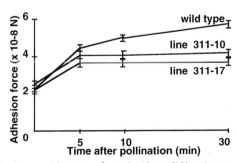

Fig. 3.11. Kinetics of pollen–stigma adhesion in *B. napus* plants transformed with an SLRI anti-sense construct. The first points (time of initial capture) were measured a few seconds after the pollination of pistils with non-transformed pollen. Measurements were repeated with 30 pistils to determine pistil variability. *Error bars* indicate standard deviation. (Luu et al. 1999, with kind permission from the American Society of Plant Physiologists)

3.2.5.2 Relationships between Pollen Hydration and SI

Such relationships are obvious but still poorly understood. A clear stand on the matter was taken during the early 1980s by Roberts et al. (1980), who analyzed each of the different steps leading to pollen acceptance or rejection in compatible and incompatible situations. Their conclusion was that the recognition reaction between pollen and stigma determinants results in the formation of a biologically active complex that turns off water supply to the incompatible grain. All other manifestations of SI were, according to them, a consequence of this initial response.

Current thinking maintains an essential role for the availability of water in the metabolism of compatible pollen and emphasizes the necessity of protein synthesis and of at least some water supply to the incompatible pollen grain for the launching of the recognition reaction. This is supported by two sets of experimental results that provide evidence that the control of pollen hydration and the manifestation of SI require the continued synthesis of unidentified proteins (Sarker et al. 1988 working with the protein inhibitor cycloheximide) and the normal activity of MOD, an aquaporin-like gene. Concerning this second requirement, Ikeda et al. (1997) showed that a recessive mutation at the MOD locus leads to the disappearance of water channels from stigma to pollen and the elimination of SI. It is also possible (Sect. 3.2.3.7) that the ARC1 gene, a putative downstream effector for SRK, also participates in the control of water supply after an incompatible pollination.

3.2.5.3 Stigmatic S Glycoproteins Are Glycosylated

The potential N-glycolysation sites of SLG were first identified by Nasrallah et al. (1987), who suggested that their number and position could be specific for each allele. This particularity has not been demonstrated (Franklin et al. 1995), but Sarker et al. (1988) found that the treatment of excised stigmas with tunicamycin, reported by Lord (1985) to inhibit the glycosylation of glycoproteins, prevented the occurrence of the SI reaction.

3.2.5.4 The Recognition of Incompatible Pollen

Only assumptions on the nature and features of the recognition reaction can be made for the time being. However, in view of the presence of a S-receptor kinase in stigmatic papillae, it is likely that, when the pollen and stigma are incompatible, the pollen gene SCR encodes sequences or presents folding configurations that are similar or complementary to portions of SRK and are able to act as ligands to this receptor (Fig. 3.12).

In stigmas, the role of SRK or its truncated form (Sect. 3.2.3.3) appears to be essential, but little is known about the function of SLG. Stein et al. (1996) considered that the expression of both SRK and SLG is required for pollen recognition. The task of SLG, a water-soluble molecule distributed

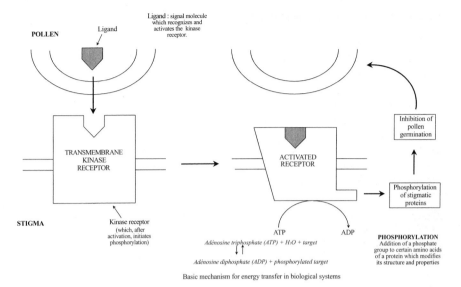

Fig. 3.12. The probable basis of pollen–pistil recognition and rejection in SI species of the *Brassica* type (prepared in cooperation with Professor M. Boutry, Catholic University of Louvain). It is possible that SCR, the putative pollen determinant discovered in *Brassica* by Schopfer et al. (1999), functions as ligand for the S-kinase receptor

throughout the volume of the papilla cell wall, is to shuttle the pollen determinant from the surface of the outer cell and to present it to the plasma-membrane-localized SRK. In this manner, SRK is able to bypass the obstacle of the thick papilla cell wall which, according to Stein and his co-workers, may prevent membrane-based signaling. This hypothesis is weakened by the discovery that SLG is expressed at low levels in certain self-incompatible plants (Sect. 3.2.3.2; Gaude et al. 1995) and by reports from Kusaba et al. (1999), Nishio and Kusaba (1999) and Kusaba and Nishio (1999b) that normally self-incompatible plants carry SLG alleles that at best encode only truncated proteins unlikely to function properly. However, different conclusions have been reached in a number of transgenic studies designed to observe the effects of SLG ablation or the transfer of a new SLG allele (Takasaki et al. 1999; Sect. 4.2.3). In an attempt to conduct gain-of-function approaches in the absence of interference from interspecific incompatibility, Takasaki and co-workers (1999) undertook transfer experiments involving donor and recipient plants from a same species (*B. rapa*). Their results show that the introduction of a new SLG allele alters the phenotype of the endogenous S haplotype but confers no new S-haplotype specificity to the stigma.

3.2.5.5 Rejection of Incompatible Pollen

On activation by the pollen ligand (presumably SCR), the kinase receptor switches on a signaling pathway in the papillar cell which leads to the rejection of incompatible pollen (Fig. 3.12). The process is postulated to precipitate a phosphorylation/de-phosphorylation cascade within the papillar cells (Nasrallah 1997) and results in the loss of one or several properties of the papillar cell surface that are essential for the proper hydration, adhesion and germination of pollen grains (Sects. 2.5.1, 3.2.3.7). In the incompatible pollen, probably as a direct consequence of changes on the papilla surface or possibly as a result of decentralized reactions within the pollen, the switching process is associated with a very important reduction of pollen protein synthesis and the appearance of phosphorylated or de-phosphorylated proteins not always found in compatible pollen (Hiscock et al. 1995). The focussing of the stigma response to the incompatible grain is so extraordinarily precise that a single papilla is simultaneously able to accept cross pollen and reject self pollen (Dickinson 1995).

Very little information is available regarding the molecular nature of the SRK transduction pathway. The results of pistil treatment with okadaic acid show that type 1 or type 2a phosphatases are involved (Rundle et al. 1993; Scutt et al. 1993). In *Brassica*, Bower et al. (1996) identified two members of the thioredoxin family that interact with the kinase domain of SRK in a yeast two-hybrid library. The discovery may be important, because the standard role for thioredoxins is modulation of enzyme activity (by reducing disulfide bridges; Holmgreen 1989) and because a protein with thioredoxin activity is linked to the S locus in the gametophytic stigmatic bifactorial system of *Phalaris* (Sect. 3.4.2). However, the nature of the interaction between the products of the thioredoxin gene and the kinase domain of SRK does not seem to be phosphorylation dependent in *Brassica* (Bower et al. 1996).

It is also probable that further work with aquaporins or related genes (Sect. 3.2.5.2) will, as predicted by Nasrallah (1997), contribute to the elucidation of the relationships between the SRK signaling pathway and the regulation (and deregulation) of water supply. In particular, the search for genes that, like ARC1, show a phosphorylation-dependent interaction with SRK will probably reveal the identity of other downstream effectors of the SI rejection reaction.

3.2.5.6 The Role of Callose

In the Crucifereae, the formation of callosic pads immediately below the tube tip of incompatible pollen probably represents the consequence rather than the cause of the arrest of pollen-tube growth by the stigma (Sect. 3.5.2.3). In *Brassica*, Singh and Paolillo (1990) and Elleman and Dickinson (1994) observed that the deposition of callose in the papillar cells of the stigma, which is associated with the SI response (Nasrallah and Nasrallah

1993), is not required for the incompatibility barrier (Bell 1995) and is induced by material leaking from already-inhibited tubes. Transgenic plants of *B. napus* that expressed an enzyme (β-1,3-glucanase) known to degrade callose contained little or no callose in the papillar cells of their stigmas but remained self-incompatible (Sulaman et al. 1997).

▪ 3.3 Stigmatic Monofactorial Multiallelic GSI in *Papaver rhoeas*

3.3.1 Compatible and Incompatible Pollinations

3.3.1.1 Morphology and Growth of Compatible Pollen Tubes

A complete description, which is summarized below, has been provided by Elleman et al. (1992). The pistil of *P. rhoeas* is composed of a stigmatic cap featuring from 10 to 14 rays of papilla. There is no style. After pollen landing, the pollen coating (electron lucent, layer delimited) flows onto the relatively dry stigma surface to establish a clearly visible "foot". The pollen hydrates and germinates. As seen by Elleman et al. (1992), the cuticle below the foot detaches itself from the papilla cell wall and forms a floating structure (which the growing pollen tube penetrates) at a point close or within the foot. This separation between cuticle and cell wall is probably stimulated or provoked by a pollen secretion. The tube then grows in what appears to be a much-enlarged sub-cuticular space and, without establishing contact with the stigmatic cell wall, travels towards the base of the cell. After passing between the papillae and, horizontally, between the stigmatic rays, it reaches the central transmitting tract.

3.3.1.2 Incompatible Pollen Grains and Pollen Tubes

The inhibition of incompatible pollen has been described by Franklin-Tong and Franklin (1992). It occurs during or just after germination and before tube emergence. Callose (for research results on the role of callose in stigmatic and stylar SI, see Sects. 3.2.5.6 and 3.5.2.3) appears inside the grain (at the colpal aperture from which the pollen tube, if germination had taken place, would have probably emerged). The tubes that appear may have different lengths; typically, however, they are short and distorted, with a heavy deposition of callose at the tube tip. From the fact that the stigmatic cuticle also detaches itself after an incompatible pollination, it would appear that the signals that stimulate pollen secretion and the secretory response are independent of the SI system (Elleman et al. 1992).

3.3.2 An In Vitro Bioassay for the Study of Stigmatic S Proteins, Pollen Metabolism and Pollen–Stigma Interactions after Self-Pollination

Franklin-Tong and co-workers (1988) developed an in vitro bioassay as the starting point of a program concerning the molecular biology of SI in the field poppy (Franklin et al. 1991). The initial work was focussed on the determination of a suitable medium for pollen germination and growth in vitro and on different methods (application or incorporation to the medium of stigma prints and of eluates or macerates of purified stigmatic extracts) for reproducing the phenotype of specific stigmatic S alleles in the medium. The final results were remarkable, as pollen reactions in vitro were quantitatively indistinguishable from those obtained with in vivo pollinations. With the bioassay, it is possible to distinguish between incompatible (S1S2×S1S2), semi-compatible (S1S2×S1S3) and compatible (S1S2×S3S4) pollinations.

From there on, the test could be used for the study of compounds, reactions or conditions likely to disturb the germination and growth of compatible pollen or to prevent the recognition or inhibition of incompatible pollen. As shall be seen below, it was with the help of this in vitro system that a stigmatic component that expressed specific S-gene activity (Franklin-Tong et al. 1989) could be isolated. A study of pollen metabolism was undertaken by testing the effects of transcription, translation and glycosylation inhibitors on pollen-tube growth and the SI reaction. The system has also been adapted to allow the analysis of changes in gene expression (Franklin-Tong et al. 1990), protein phosphorylation (Franklin-Tong and Franklin 1992) and cytosolic calcium levels (Franklin-Tong et al. 1993) that occur in the pollen as a result of the SI reaction. The test meets most, if not all, of the prerequisites and conditions listed by Jackson and Linskens (1990) in their review of "bioassays for incompatibility".

3.3.3 Characterization of Stigmatic S Proteins and Cloning of the Stigmatic S Gene

3.3.3.1 Isolation and Characterization of the Stigmatic S Proteins

3.3.3.1.1 Isolation and Testing of Function. Through the use of the in vitro test presented above (Sect. 3.3.2), the stigmatic S products were isolated and analyzed by Franklin-Tong et al. (1989), Foote et al. (1994) and Walker (1994) and were submitted to improved purification procedures by Foote et al. (1994). They were found, in the bioassay, to inhibit self pollen in vitro and did not interfere with the germination and growth of pollen carrying S alleles other than those present in the stigmatic prints or extracts.

3.3.3.1.2 Co-Segregation with S Alleles. On an isoelectric focusing gel (IEF), examinations of stigmatic proteins from more than 250 plants segregating for the S1 allele revealed the presence of glycoproteins co-segregating with this allele.

3.3.3.1.3 Characteristics of the Protein. The S products in question were represented by two isoforms (S1a and S1b) with pIs of 7.5 and 6.9. It was found, after sodium dodecyl sulfate polyacrylamide-gel electrophoresis (SDS-PAGE) analysis, that each isoform separated into two proteins with molecular masses of 16.7 kDa and 14.7 kDa for S1a and 16.8 kDa and 14.8 kDa for S1b. The occurrence of these different forms probably results from glycosylation, which increases the molecular mass, and from other post-translational processing. N-terminal sequences (12–17 AA) were identical for the a and b glycosylated forms and for the a and b non-glycosylated forms (Foote et al. 1994).

3.3.3.1.4 S Activity, S Specificity and the Role of Glycosylation. Foote and his co-workers obtained high expression of the active S1 protein in *E. coli* and tested the biological activity of purified S1e against S1 pollen grown in vitro. S1e exhibited the same S-specific activity as that found in partially purified preparations of the native protein from stigma tissue. The fact that S1e displays S-specific activity suggests that glycosylation and the processing of S-products are not essential prerequisites for S activity and S specificity [however, see Franklin-Tong et al. (1990), who showed that pollen components need to be glycosylated de novo for the SI reaction to take place, and Section 4.1.4.2 on the effects of transcription and translation inhibitors].

3.3.3.1.5 Polymorphism of S Sequences. The analysis of other S alleles indicated a molecular weight of approximately 17.5 kDa for the (glycosylated) protein, and allelic variations ascribed to differences in isoelectric points. Evidence from the Southern-blot data and from other unpublished results showing that many S-alleles do not cross-hybridize led Foote and his co-workers to suggest important sequence polymorphism at the S locus of *Papaver*. A comparison of S alleles of *P. rhoeas* and *P. nudicaule* also revealed a high degree of AA-sequence polymorphism (51.3–63.7%) between the alleles of the two species [which, nevertheless, share very similar secondary structures (Kurup et al. 1998)]. Significantly (Sect. 4.2.7.1), inter-allelic variation is not restricted to the HV regions but is scattered throughout the S-protein, among numerous short, strictly conserved regions.

3.3.3.1.6 The S Proteins Are not Major Proteins of the Stigma. The S proteins represent less than 1% of the total soluble fraction and are first expressed in stigmas 2 days before anthesis. Their appearance is followed by a rapid burst of activity from 1 day before anthesis until several days after flower opening. The S1 proteins were not detected in pollen grains.

3.3.3.1.7 The S Protein is not a Ribonuclease. Unlike the stylar proteins involved in monofactorial stylar GSI (Sect. 3.4), the *Papaver* stigmatic S protein is not a ribonuclease. Franklin-Tong and Franklin (1993) and Franklin et al. (1995) have presented arguments to the effect that its ribonuclease activity is 100-fold lower than that of the S proteins in *Nicotiana alata* and that there is no detectable RNase activity that correlates with the functional stigmatic gene product in *P. rhoeas*. In addition, *Papaver* pollen is insensitive to ribonuclease activity, and the *Papaver* stigmatic S gene, as sequenced by Foote et al. (Sect. 3.3.3.2), shows no homology with the S products of other SI systems in general or with S ribonucleases in particular. The SI system of the field poppy is basically different from that of any other known system, but its mechanism is the inverse of that of *Brassica*, where the putative receptor is thought to operate in the pollen after activation by a ligand provided by the stigma. One cannot make conclusions regarding the completely unique nature of the *Papaver* SI mechanism before the discovery of the stigmatic and pollen determinants operating in the grasses (Sect. 3.4) and before more information regarding the function and features of SCR in *Brassica* becomes available. In addition, one must consider the fact that the stigmatic product of the S gene of *Papaver* belongs to a large family of proteins probably present in numerous families of angiosperms (Ride et al. 1999; Sect. 3.3.3.4).

3.3.3.2 Cloning and Nucleotide Sequencing of the Stigmatic S Gene

The Sl allele was cloned (Foote et al. 1994) by screening a cDNA library using an oligonucleotide based on the N-terminal nucleotide sequence of the S1 protein. Twelve independent clones with DNA inserts up to 900 bp in size, which hybridized to the S oligonucleotide and had identical nucleotides, were identified. Nucleotide-sequence analysis revealed an open reading frame (ORF) of 417 bp, corresponding to a mature 14.5-kDa polypeptide of 120 AA. Foote and co-workers found that the ORF is preceded by a putative signal-peptide sequence with a relatively hydrophilic C-terminus, with glycine at the cleavage site and a central hydrophobic region. The sequence also predicts the position of a single N-glycosylation site. The protein is relatively rich in charged AAs. Southern analysis of genomic DNA confirmed that the S1 sequence co-segregates with the S1 phenotype and is probably a single-copy gene. The molecular analysis of two functional homologs of the S3 allele isolated from different populations was performed by Walker et al. (1996).

3.3.3.3 Biological Activity of Mutant Derivatives of the S Protein

Through the use of site-directed mutagenesis, Kakeda et al. (1998) constructed variations in AA sequences at different sites of the S1 protein and tested the mutant S proteins for their ability to specifically inhibit S1 pollen. The majority of the mutations (AA substitutions that were conservative or led to exchanges corresponding to known positions in other alleles) were targeted to predicted hydrophilic loops, particularly loop 6, which may be readily available to pollen and where mutations could affect the folding of the protein. The results are discussed, together with their inferences, in Section 4.2.7.1. They indicate that certain AA residues are essential for the S-specific inhibition of incompatible pollen. This is particularly the case for loop 6, where mutation of the only HV AA situated in this loop resulted in the complete loss of the ability of the S protein to reject S1 pollen. The residue is obviously involved in pollen recognition and, most probably, in allelic specificity. Both variable and conserved AAs in the loop-6 region participate in the recognition and rejection of self pollen.

3.3.3.4 Large Numbers of ORFs with Homology to the Stigmatic S Gene of *Papaver* Are Present in the *Arabidopsis* Genome

The products of these potential genes, discovered by Ride et al. (1999) in the course of an in-depth analysis of the currently available *Arabidopsis* genomic sequence, are predicted to be relatively small, basic, secreted proteins with similar predicted secondary structures. They were named SPH (S-protein homolog) genes by Ride and co-workers, who believe they belong to a family that may contain more than 100 genes. The genes do not appear to be present in human, microbial, *Drosophila* or *Caenorhabditis elegans* databases and could have a function specific to the plant kingdom. Preliminary reverse transcriptase PCR (RT-PCR) analysis reveals that at least two members of the SPH family (1 and 8) are expressed, with expression greatest in floral tissues. These results show that the S alleles of *Papaver* have a (large) family of their own and perhaps have near relatives, overlooked until now, in other SI systems. They also demonstrate the utility of genomic sequencing and the systematic search for S-related proteins in the pollen and pistil (Sects. 3.2.4.3, 3.5.8.5).

3.3.4 Pollen Genes that Participate in the SI Response

3.3.4.1 Inhibition of Incompatible Pollen Tubes Depends on Pollen-Gene Expression

The effects of the transcription inhibitor actinomycin D on the in vitro growth of pollen tubes in the presence of self S proteins show that de novo RNA synthesis in the mature pollen is required for the full inhibition of pollen-tube growth during an incompatible response (Franklin-Tong et al. 1990). While comparative analyses between pollen grains before and after germination indicate that this synthesis does not occur during early pollen growth, study of the translation products from the RNA participating in an incompatible reaction clearly reveals the presence of novel proteins not found in other samples. The function of these proteins of molecular weight

approximating 21–23 kDa, possibly a cluster of isoforms from a single species, could not be ascertained by Franklin-Tong and her co-workers. Because the transcription of these genes is initiated after specific pollen–pistil recognition, their effects are evidently restricted to the rejection phase
(Sect. 3.5.3.2; Van der Donk 1974a). In pollinated styles of *Petunia*, Van
der Donk (1974a) reported proteins that are not present in virgin styles
and have the ability to inhibit the growth of self pollen. Franklin-Tong et
al. (1990) showed that the pollen component of the SI reaction in *Papaver*
had to be glycosylated de novo for proper functioning of the SI mechanism
(Sect. 4.1.4.3).

3.3.4.2 Involvement of a Signal-Transduction Mechanism in the SI Response

Several lines of evidence (Franklin-Tong et al. 1997) that indicate that a
Ca^{2+}-dependent signal-transduction pathway mediates the SI response
emerge from the analyses of the effects in vitro of stigmatic S proteins on
pollen tubes:

▦ The growth of pollen tubes in *P. rhoeas* is regulated by a slow-moving
  calcium wave propagated by inositol 1,4,5-triphosphate (Franklin-Tong
  et al. 1996).
▦ A challenge of pollen with incompatible S proteins leads to transient increases in cytosolic free Ca^{2+} $[Ca^{2+}]_i$, followed by cessation of pollen-
  tube growth (Franklin-Tong et al. 1993, 1995, 1997).
▦ Artificial increase of $[Ca^{2+}]_i$ also causes the inhibition of pollen-tube
  growth (Franklin-Tong et al. 1993, 1996; Malhó et al. 1994).
▦ The SI response induces the increased phosphorylation of the 26-kDa
  (Rudd et al. 1996) and 68-kDa (Rudd et al. 1997) pollen proteins.

Although such data does not necessarily implicate the participation of a kinase receptor, it is tempting to speculate that such a receptor (operating in
the pollen and not, as in *Brassica*, on the stigma) governs the incompatibility reaction in *Papaver* (Fig. 3.13). As far as the 26-kDa protein is concerned, the task of Rudd et al. (1996) was to study phosphorylation events
induced in the pollen as a consequence of SI and to analyze the response
to different inhibitors (Figs. 3.14, 3.15) of the two phosphoproteins that
comprise the protein, p26.1 and p26.2. Their observations clearly suggest
that increased phosphorylation of p26.1 is implicated in the manifestation
of SI and appears to be Ca^{2+} and calmodulin (CaM) dependent. Involvement of a Ca^{2+}- and CaM-dependent protein kinase in the *Papaver* SI system is very plausible (Franklin-Tong et al. 1997). However, the situation is
quite complex and may involve the intervention of another transduction
pathway, because the kinases responsible for the phosphorylation of p68
are not Ca^{2+} dependent (Rudd et al. 1997). As the p68 response occurs
after that of p26 it is likely that a second wave (Ca^{2+} independent this
time) of signaling also participates in the expression of SI. This "second-
messenger" signaling system, present in the transduction processes of sev-

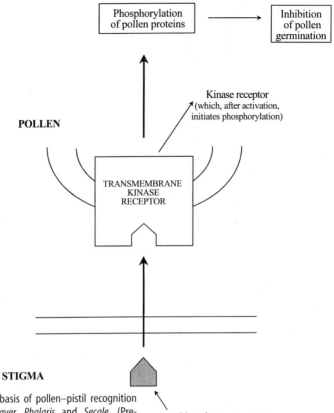

Fig. 3.13. A possible basis of pollen–pistil recognition and rejection in *Papaver, Phalaris* and *Secale.* (Prepared in cooperation with Professor M. Boutry, Catholic University of Louvain)

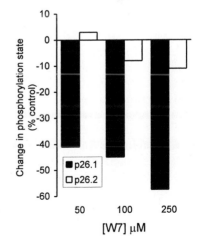

Fig. 3.14. Quantification of the relative changes in phosphorylation of p26.1 and p.26.2 in the presence of the calmodulin inhibitor W7. (Rudd et al. 1996, with kind permission from the American Society of Plant Physiologists)

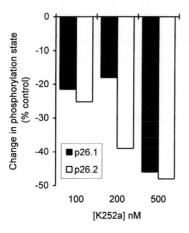

Fig. 3.15. Quantification of the relative changes in phosphorylation of p26.1 and p26.2 in the presence of the serine/threonine protein kinase inhibitor K252a. (Rudd et al. 1996, with kind permission from the American Society of Plant Physiologists)

eral plants (Trewavas and Gilroy 1991; Gilroy et al. 1993, cited by Rudd and co-workers) is suggested to participate in the phosphorylation of some of the pollen-gene products responsible for the final, irreversible effects leading to pollen-tube growth inhibition.

As concluded by Rudd et al. (1997), we must now "(characterize) further p 68 and p26 and their respective kinases and interacting proteins". A clearer comprehension of SI in *P. rhoeas* and of signal transduction in plants should result from such an effort. Although SI in the field poppy is probably basically different from that of other known systems, the *Papaver* mechanism, like that of grasses, seems to be the inverse of that of *Brassica*. One cannot make definitive conclusions regarding the unique nature of the S-specific recognition sequences in *Papaver* stigma before the discovery of the pollen ligands of *Brassica* or of the stigmatic ligand(s) possibly operating in grasses.

3.3.4.3 A Membrane Glycoprotein that Binds Stigmatic S Proteins in Pollen

At first, the story of this 70- to 120-kDa S-protein-binding protein (SBP) does not appear to end happily, because it seems that the binding occurs in a non-S-allele-specific manner (Hearn et al. 1996). Nevertheless, SBP deserves to be mentioned, because its characteristics (timing of expression that mirrors that of the stigmatic S proteins; integration in the plasma membrane and availability to interact with secreted S proteins) suggest that SBP may act as an accessory receptor or co-receptor for the interaction of stigmatic S proteins with an S-specific pollen receptor (Hearn et al. 1996). Its functions could be analogous to those of the much smaller molecule PCP-A1, which was suggested by Doughty et al. (1998) to act as a non-specific co-factor of the male determinant in *Brassica* (Sect. 3.2.4.3).

Furthermore, data that unveil the functional significance of the interaction between S proteins and SBP have recently accumulated at the University of Birmingham (Franklin and Franklin-Tong, personal communication; Jordan et al. 1999). These data indicate a functional role for SPB in the SI reaction of *Papaver rhoeas*. The work (Kakeda et al. 1998) involved the construction of mutant derivatives of the S1 protein and the testing of their SBP-binding activity and their biological activity. The results provide evidence that AA residues in predicted loops 2 and 6 of the SI protein (Sect. 3.3.3.3) cause significant reductions in their SBP-binding activities. These same mutants show a concomitant reduction in their ability to inhibit incompatible pollen. Therefore, there is a direct link between the SBP binding and inhibition of incompatible pollen, and the implication of SBP as a pollen component that plays a key role in the SI reaction. SBP could be the pollen receptor with high- and low-affinity binding sites, and yet display S-specific binding activity not detected by the current SBP-binding assay.

3.3.4.4 Programmed Cell Death Is the End Point of the SI Response in *Papaver rhoeas*

Programmed cell death (PCD), often referred to as apoptosis or necrosis, can result from the triggering of a genetically programmed suicide process by an external biochemical stimulus (Geitmann 1999). An example of this type of PCD in plants is the degeneration of the stylar transmitting tissue by compatible pollen tubes (Greenberg 1996, cited by Geitmann). A Birmingham group (Jordan et al. 1999) has recently discovered another example of post-pollination PCD. The team found, through the use of the FraGel assay, that S proteins elicit DNA nicking in incompatible pollen tubes but not in compatible ones (Franklin-Tong and Franklin, personal communication). The concomitant ability of the S proteins to inhibit pollen in an in vitro bioassay and the detection of DNA nicking constitute the first evidence that PCD is involved in the rejection phase of SI.

■ 3.4 Stigmatic Bi-Factorial GSI in the Grasses

SI is widely distributed in the grasses, which contribute so much to agriculture around the world. However, perhaps because the rejection system (bi-factorial, multi-allelic, complementary, apparently restricted to monocots) lacks the simplicity that has wrongly been attributed to the mechanisms operating in the Solanaceae and Brassicaceae, relatively few laboratories are contributing to the understanding of SI in grasses. This has been clearly exemplified in Chapter 2 through reviews of the work of Lundqvist and co-workers in Sweden and Denmark. In the current section on the physiology and molecular biology of SI in the grasses, it will be confirmed by an outline of the research results, which often originated from three universities (Wales, Hannover and Adelaide) and Kew Garden.

3.4.1 Flowers and Pollination

3.4.1.1 Stigma and Pollen

The stigma and pollen have been studied in detail by Heslop-Harrison (1982). A brief and incomplete abstract of his descriptions is outlined below. The dry stigma is composed of two stylodia covered by secondary stigma branches made of up to five ranks of cells, the papillate tips of which support the pollen-receiving surfaces. The surface proteinaceous and pectigel layers carry a thin secretion and clothe a discontinuous cuticle, below which lies a stratum of thinly dispersed, microfibrillar material with a considerable pectic component corresponding morphologically to the middle lamellae. The inner-wall layer of the papillate cells is cellulosic.

3.4.1.2 Compatible Pollination

The fate (very rapid) of the compatible pollen grain after landing on the receptive surface of a stigma papilla, such as reported in depth by Heslop-Harrison (1982), may be summarized as follows. The grain attaches to the "sticky" receptive zone, hydrates and lifts the operculum at the germination aperture. The tube tip emerges and penetrates the stigma cuticle. Guided by tract geometry, the tube travels through the intercellular spaces of the stigma branch and grows through the transmitting tracts in the stylodium and in the upper ovary wall. It then enters in the ovarian cavity and proceeds towards the micropyle, the nucleus and the embryo sac.

3.4.1.3 Self-Pollination

In certain species like *Gaudinia fragilis* or *Phalaris minor*, rejection occurs on the stigma surface as rapidly as 30 s after the emergence of the tube tip, i.e., less than 2 min after pollen landing. In other cases, such as that of *Hordeum bulbosum*, rejection takes place within the intercellular spaces of the stigma branch. However, the time and site of the reaction response are very variable and change with the genotype of individual plant (Heslop-Harrison 1982). It seems that the SI response depends on contact of the emerging tube tip with the surface of the stigma papilla; exine contact alone does not seem to be sufficient to initiate a reaction (Heslop-Harrison 1976, personal communication). In *G. fragilis*, the contact of the tube tip is accompanied by a rapid outflow of the antigens held in the intine, near the germination aperture. At the same time, the ordered deposition of wall microfibrils in the tube-tip region ceases, and the inordinate callosic wall begins (for information regarding the role of callose in stigmatic and stylar SI, see Sects. 3.2.5.6 and 3.5.2.3). In *H. bulbosum*, the reaction is essentially similar, but further down, in the stigma branch (Heslop-Harrison 1982). It will be seen in Section 3.4.2 that the S proteins are probably not transferred across the plasma membrane during pollen germination and that, as in *Papaver*, the incompatibility reaction occurs within the germinating pollen grain (Li et al. 1994). Lundqvist (1961) elaborated a routine Petri-dish test that enables an accurate scoring of emptied (compatible) and still-filled (incompatible) pollen grains on excised stigmas after staining with Cotton blue.

3.4.2 SI in *Phalaris coerulescens*

SI in *P. coerulescens* was discovered and its genetic system understood by Hayman in 1956. Through this work and others, Hayman established a collection of S–Z genotypes and of "pollen" and "complete" (pollen–pistil) SF mutants in the species (Hayman and Richter 1992; Chap. 4). At the University of Adelaide, the collection provided the starting point for the analysis of the molecular biology of SI in *P. coerulescens* and, in particular, for the identification of the S–Z pollen–pistil gene foursome that participates in the recognition reaction.

3.4.2.1 Identification of Restriction Fragments Linked to the Pollen S Gene

The strategy followed in Adelaide (Li et al. 1994) for isolating the S gene was to look for pollen-specific cDNA clones that could identify RFLPs co-segregating with the S or Z genotype. A clone library was prepared from mature pollen RNA of a plant with the genotype S1S2-Z1Z1 (i.e., segregating for S1Z1 and S2Z1 pollen) and with a differential screen established with ^{32}P-labeled cDNA from the pollen of plants with a S1S2-Z2Z2, SFSF-Z1Z1 or S1S2-Z1Z1 genotype. Restriction fragments co-segregating with S genotypes were identified by one cDNA clone, Bm2. Differences between the S1 and S2 alleles are evenly distributed among the five regions of Bm2, with a majority of variations in the coding regions in the second exon. Tests for the evaluation of linkage indicated that Bm2 represents the S gene or is very close to it. Bm2 is only expressed in mature pollen and displays strong thioredoxin activity.

3.4.2.2 Bm2 is not the S Gene

Early results (Li et al. 1994, 1995) had indicated that Bm2 was the product of the S2 allele with a variable N-terminus (signaling S specificity) and a conserved terminus with catalytic activity. The thioredoxin activity of Bm2 was demonstrated by Li and co-workers (1995), who expressed the C-terminal domain of the gene in *E. coli* and discovered that it was recognized as a thioredoxin by the bacteria. It is a substrate for *E. coli* thioredoxin and can act as an effective disulfate reductase.

However, further work with Bm2 homologs from three self-incompatible grasses (*Lolium perenne, Secale cereale* and *Hordeum bulbosum*) and two other alleles from *Phalaris* showed that their ORFs differed from that predicted for Bm2 by Li et al. in 1994 (P. Langridge, personal communication). Moreover, as reported by Langridge et al. (1999), most of the newly isolated sequences translate into thioredoxin-like protein only and do not contain an allelic domain. Similar conclusions were reached for *Secale cereale* at the Institute of Agricultural Crops in Germany (P. Wehling, personal communication), where the Bm2 probe from *Phalaris* and rye sequences homologous to Bm2 were used to test Bm2 as a candidate for the S gene in rye. In addition, recombination tests indicated that the thioredoxin gene sequences of rye homologous to Bm2 are linked (but not allelic) to the S gene of rye. Bm2 and S may be as much as 2 cM away from each other (Langridge et al. 1999).

3.4.2.3 Involvement of Thioredoxins in the SI Mechanism?

Thioredoxins are small, heat-stable molecules with catalytic activities that depend on several factors, such as the sequence beyond the protein core, the assembly of the protein and post-translational modifications (Li et al. 1995; Holmgreen 1989, cited in Li et al. 1995). Their functions are not well known. In the species (animals, plants and bacteria) where they have been identified, thioredoxins act in many important biological reactions as hydrogen donors, substrates for reduction enzymes or receptors, and as subunits of viral DNA polymerase. Bower et al. (1996) have suggested that a thioredoxin-like region of the S gene in *Brassica* regulates the activity of a protein-kinase signaling cascade (Sect. 3.2.5.5).

Despite the finding that the Bm2 gene is not the S gene in either *Phalaris* or rye, its involvement in the rejection reaction appears to be plausible from its close linkage to the S locus and in view of the fact that a "complete" mutant (which, when crossed with wild type, expresses SC in both the pollen and stigma) displays lesions at the C-terminus of Bm2. These lesions result from three AA substitutions (leucine to valine, glutamine to glutamic acid and serine to arginine). Current information regarding the predicted and observed effects of such replacements suggests that, possibly through structural modifications of the protein, the serine/arginine change leads to the significant reduction of thioredoxin activity associated with the SC character (Li et al. 1994).

A third argument favoring the hypothesis that the SI mechanism interferes with thioredoxin activity is supported by the occurrence of thioredoxin genes (some of which are perhaps present as relics) related to Bm2 in all grasses tested so far. DNA gel-blot examinations (Li et al. 1997) in 15 different SC or SI grass species for sequences homologous or partly homologous to those of the *Phalaris* S-pollen gene revealed that all grasses examined contained several genes related to the Bm2 gene of *Phalaris*. Their number and degree of divergence from the *Phalaris* sequence varies considerably among species. Ten of these species were submitted to RNA gel-blot analysis. In rye, an abundant Bm2-like transcript becomes visible after an exposure of 3 days, compared with 1 day in the case of *Phalaris*. All the other species tested had to be submitted to RT-PCR for successful amplification of the Bm2-like sequences. Sequencing of amplified regions from wheat, barley, rye and *Dactylis* indicated that these regions are highly conserved and share 94–97% sequence similarity with *Phalaris*.

3.4.3 SI in Rye

SI in *Secale cereale* was described by Lundqvist in 1956. The S and Z loci were localized, with the help of isozyme markers, on chromosomes 1R and 2R (Wricke and Wehling 1985; Gertz and Wricke 1989). Heslop-Harrison (1982) showed that the SI reaction was accomplished within the 2 min following pollen landing on the stigma, i.e., before the occurrence of any transcription or translation resulting from contact between the pollen and stigma.

The tubes, usually short, may vary in length from one genotype to the next; they are distorted and occluded with callose deposition (Wehling et al. 1995). The association of callose with SI in rye was first reported in incompatible grains and tubes by Vithanage et al. (1980), who observed a substance staining with aniline blue, resorcin blue and calcofluor, which they identified as callose.

3.4.3.1 Evidence that the SI Mechanism Involves Phophorylation and Is Ca^{2+} Dependent

Wehling et al. (1994a, 1995) undertook experiments to assess the extent of pollen phosphorylation in "cross" and "self" stigma eluates; they identified differentially phosphorylated proteins and studied the effects of kinase inhibitors and Ca^{2+} on the occurrence of SI.

3.4.3.1.1 In Situ Pollen Phosphorylation. A striking increase in the incorporation of ^{33}P was observed after the pre-incubation of pollen grains with self-pollen eluates (53% and 36% higher after the 1930s and 1990s, respectively, than for cross-treated pollen). These differences are due to a stimulation of protein phosphorylation within the pollen grains, not within the stigma eluates. Basic phosphorylation was significantly reduced in self-compatible pollen.

3.4.3.1.2 Gel Electrophoresis of Pollen Phosphoproteins. Separation of ^{33}P-phosphorylated pollen proteins by SDS-PAGE revealed four major proteins in the range of 43–82 kDa; they were differentially phosphorylated in SI and SC genotypes and in self-treated and cross-treated pollen.

3.4.3.1.3 Effects of Inhibitors. Application of different tyrosine-specific protein-kinase inhibitors and of Ca^{2+} antagonists to isolated intact stigmas led to an inhibition of the SI response after incompatible pollination. These experimental results do not provide specific information regarding the site of the recognition reaction (presumably the pollen grain) or on the nature of the S and Z products. They clearly suggest, however, that protein kinases and Ca^{2+}-dependent protein phosphorylation are involved in a signal-transduction pathway that renders possible the recognition and rejection of incompatible pollen in rye. Analogies with the situation in *Papaver* (where the SI-signaling pathway is also thought to be located in the pollen) and, to some extent, *Brassica* (for which the stigma plays the active part) are obvious. This could mean that, at least for stigmatic systems, the recognition mechanisms are less heterogeneous than usual.

3.4.3.2 A Model for the SI Mechanism in Rye

Wehling and co-workers (1995) propose a mechanism for the SI reaction in rye; if it turns out to be correct, it will probably be applicable to all grasses and, perhaps, to other multi-genic stigmatic SI systems. The model foresees the presence in pollen grains of an SI-specific pathway managed by S- and Z-specific molecules (possibly receptor protein kinases) on the plasma membrane and presenting "extra-cellularly exposed" receptor domains. On the landing of an incompatible grain, the S and Z gene products formed in the stigma would associate with their "self" S and Z homologs of the pollen grain to trigger the phosphorylation of pollen proteins and an influx of Ca^{2+} (acting as second messenger). The Hannover group suggests that a simple, independent, additive interaction, similar to the one predicted by Larsen (1977) for four-loci systems, is established between structurally similar S and Z receptor molecules to trigger the SI response of the pollen. According to this model, the growth of any given pollen tube is inhibited only if each of the two types of receptors (S and Z) borne by the grain is activated individually but "in concert" with the other receptor (Wehling et al. 1994a, 1995). For reasons perhaps related to dilution effects, activation of only half the receptor molecules (for instance, S1Z6 pollen on an S1S2-Z4Z5 stigma) does not lead to the inhibition of growth. The signal-transduction mechanism leading to this inhibition induces a cascade of events mediated by a Ca^{2+} influx and subsequent increases in cytosolic calcium. It involves the participation of many pollen genes, including those Wehling et al. (1995) were able to ascribe to chromosomes 3R, 5R and 6R (Fig. 2.10).

3.4.4 Applicability of the Model to All SI Species of Grasses

The main postulates (independence of S×S and Z×Z interactions, additivity of their effects, similarity of the S and Z structures) satisfy most of the known requirements. The relative simplicity of the mechanism is well adapted to the criteria of speed and accuracy typical of the SI reaction in grasses and compensates for the genetic complexity of systems with multifactorial SI. The model also seems to account for the possibility of a duplicative origin of the S and Z genes, such as proposed by Lundqvist (1954; but see below). The available evidence of an involvement of phosphorylation and protein kinases in the manifestation of SI in the grasses establishes a link with the systems operating in *Brassica* and in *Papaver*.

3.4.4.1 The S and Z Loci Are Not Interchangeable

Wehling et al. (1994a, 1995) note that the simple relationships between S and Z molecules their model foresees explains why SI in the grasses is also expressed at the tetraploid level. Matching of only one allele of both the S and Z locus (i.e., activation of the full haploid complement of S and Z receptors) is sufficient to trigger the SI response. However, and as noted by the Hannover group, the model does not explain why the interaction in diploid pollen must involve S and Z alleles and cannot occur within one allelic series. The possibility that S and Z supplement each other suggests structural differences, which could invalidate the hypothesis of their duplicative origin.

3.4.4.2 Conserved S Sequences of *Brassica* Amplify S-linked Fragments in Rye

Wehling et al. (1994b) developed a method (denaturing gradient gel electrophoresis; DGGE) of forming PCR amplification products; they applied it to rye in combination with primers derived from conserved S sequences of *Brassica*. In this manner, a 280-bp fragment was identified from genomic rye DNA; the fragment produces a polymorphic DGGE banding pattern entirely correlated to the S genotype in two inbred families (Wehling et al. 1994b). Recombination between the fragment and the S genotypes did not occur in a population of 46 individuals, and all four alleles present in the two inbred lines could be identified by their DGGE banding pattern. This finding could be very important, because it suggests that the 280-bp fragment represents part of the S gene of rye (Wehling et al. 1995). Furthermore, in view of the partial homology found between the *Brassica* allele SLG13 and the S locus of rye (or a gene closely linked to it), it demonstrates a relationship between gametophytic and sporophytic stigmatic SI. Finally, as noted by Franklin et al. (1995), the method has great potential value for the determination of S genotypes in the absence of flowers and without tedious pollinations, and for the identification of new genes in other grasses. Comparable approaches to the identification of S sequences have been tried successfully (with other materials) through chromosome walking, the screening of DNA libraries and other methods (Sect. 3.2.3.4).

■ 3.5 Monofactorial Stylar GSI with Multiple Alleles: the *Nicotiana* Type

As stated above (Sect. 2.3.4), monofactorial stylar GSI with multiple alleles operates in the Solanaceae, the Leguminosae (however, see the case of *Lotus tenuis* in Sect. 2.3.6), the Rosaceae and the Scrophulariaceae. It was discovered by East and Mangelsdorf (1925) in *N. sanderae* and is frequently referred to as the *Nicotiana* type of SI. The system is typical of species with bi-nucleate pollen and wet stigmas that receive (from the stylar transmitting tract) exudates rich in the arabinogalactans necessary for pollen-tube growth through the style.

3.5.1 Pollen-tube Morphology and Growth in Compatible Styles

3.5.1.1 Observation under the Light Microscope

Depending on the species involved and various internal and external factors, the distance that incompatible pollen tubes may reach via the style varies from a few pollen diameters to the entire length of the flower. Growth of the tube is pulsed and governed by Ca^{2+} channels (Geitmann and Cresti 1998). The level of growth can be assessed with the light microscope through the use of various dyes applied (after pollination and fixation) to removed strands of stylar conducting tissue in species with large flowers (*Lillium*, *Nicotiana*) or to entire styles before crushing or after sectioning. The choice of the proper stain will vary with the species and the objectives of the examination, but one can name acid fuchsin with light green, Cotton blue, Lacmoid-Martius yellow and various fluorochromes (such as water blue and aniline blue) as classic chemicals for differentiating the outlines of pollen tubes in their stylar backgrounds. The technique most commonly used is probably the aniline-blue fluorescence method, which was specifically developed by Linskens and Esser (1957) and Martin (1958) to detect the callosic lining of pollen tubes.

One of the recent studies in the fluorescence of compatible and incompatible pollen-tube growth in *N. alata* was that of Lush and Clarke (1997). They observed that the walls of most compatible tubes fluoresce faintly and evenly with aniline blue, except at the tip. Callose plugs, twice as long (8–12 µm) above the entrance to the ovary than in other parts of the style, were spaced at regular intervals (on average, 800 µm apart). Growth rates averaged 20 mm/day and 25 mm/day for S6 and S2 alleles, respectively, between day 1 and day 3. Lush et al. (1997) presented the details of an in vitro assay they developed to assess the effects of growth factors on *N. alata* pollen tubes.

3.5.1.1.1 Role and Specificity of the Stigmatic Exudate.

In a recent paper, Wolters-Arts et al. (1998) showed that the lipids contained in the stigmatic exudate of *Petunia* play an essential role in the penetration and directional growth of pollen tubes. The exudate of *Petunia* stigma also has the ability to restore stigma function and fertility to transgenic, female-sterile tobacco plants in which the secretory zone of the stigma had been ablated by the expression of a cytotoxic gene. The finding by the Nijmegen-Melbourne team that the applications of lily exudates (which are rich in carbohydrates) stimulated pollen hydration and germination but failed to restore stigma function to the tobacco flowers suggests that differences in the basic requirements of individual plant families in exudate composition probably play a significant role in the manifestation of incongruity between distantly related species (Sect. 5.2.1).

Lush et al. (1998) reproduced the environment on the stigma of *N. alata* by immersing pollen in stigma exudate or oil close to an interface with an aqueous medium. The behavior of pollen in this system was similar to that of pollen on a stigma: the grains hydrated, germinated and grew towards the water source. Lush and co-workers observed that the rate-limiting step was the movement of water towards the exudate and proposed that a gradient of water in the exudate guides the pollen tubes on the stigma. The composition of the exudate needs to be sufficiently permeable for pollen hydration and to prevent the supply of water from becoming directional.

3.5.1.1.2 Mitosis in the Generative Nucleus of *Petunia hybrida*.

Ünal (1986) found that pollen-tube chromosomes are regularly arranged on the metaphase plate and separate normally at anaphase. The sperms usually are elliptical and display no nucleoli. Bonig, in a personal communication to Read et al. (1992), notes that at least some of the generative nuclei in incompatible tubes divide after 48 h.

3.5.1.2 Electron Microscopy

The studies in *P. hybrida* by Van der Pluijm and Linskens (1966) and in *Lycopersicum peruvianum* (de Nettancourt et al. 1973a) and *N. alata* (Cresti et al. 1985) by the Siena group headed by Professor Cresti revealed that the cross-sections of compatible tubes were round in the apical area and infolded in the remaining portion. The tube wall clearly displays a bi-partite structure with an outer layer consisting of loose fibrils and an inner layer that is homogeneous, callose-rich and less electron dense (Fig. 3.16).

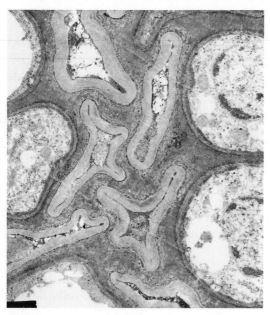

Fig. 3.16. Cross sections of upper portions of the style of *Lycopersicum peruvianum* (monofactorial stylar GSI) 24 h after compatible pollination. The cross-sections of the pollen tubes are flattened and have numerous in-foldings. The thick tube wall is clearly bi-partite, with an outer portion consisting of loose fibrils and an inner one that is homogeneous and less electron dense. (de Nettancourt et al. 1973 a)

3.5.2 Morphology and Growth of Incompatible Tubes

3.5.2.1 Incompatible Tubes of *N. alata* under Epifluorescence Illumination

Lush and Clarke (1997) report that there is no difference between the growth rates of compatible and incompatible tubes as they travel through the stigma. This apparently unaffected ability of incompatible tubes to grow normally in stigmatic tissue suggests that the S proteins present in the stigmatic region are inactive or are slowly incorporated by the tubes (Sect. 3.5.7.4). Lush et al. (1998) grew S2 and S6 pollen in exudate from S2S2 plants and confirmed that the exudate itself has no role in the rejection of incompatible pollen.

In the style, incompatible S2 and S6 tubes grow 1.00 mm and 1.50 mm per day, respectively, during the first 3 days after pollination, i.e., approximately 5% of the rate of compatible S2 and S6 tubes (Lush and Clarke 1997). Lush and Clarke also found that many incompatible tubes continue to grow until flower senescence; only a small proportion die as a consequence of tube bursting. This information, which was confirmed and complemented by grafting experiments, is important because, as shall be seen in Section 3.5.7.2, it suggests that the growth of many incompatible pollen tubes in *N. alata* and related SI species is neither completely nor irreversibly stopped by the SI reaction. The discovery implies that the effect of S-RNases on the physiology of incompatible tubes may be more discrete than initially thought (Sect. 3.5.7.2).

3.5.2.2 Electron Microscopy

The work of Van der Pluijm and Linskens (1966) has shown that the SI reaction in *Petunia* is accompanied by a thickening of the pollen tube wall and the degeneration of the cytoplasm. In tube tips at the first third of the style, in *L. peruvianum*, the Siena group (de Nettancourt et al. 1973a) found that the rough endoplasmic reticulum appears as a whorl of concentric parallel membranes (Fig. 3.17) similar to those described in inactive cells of resting potato tubers and *Betula* buds (Dereuddre 1971; Shih and Rappaport 1971). The presence of this concentric endoplasmic reticulum in incompatible tubes may correspond to an inhibition of protein synthesis (de Nettancourt et al. 1973a; Parry et al. 1997a) resulting from the degradation of ribosomal RNA (rRNA) by S-RNases in incompatible tubes. Thereafter, the inner wall of the tube becomes thinner, and numerous particles (approximately 0.2 μm in diameter and often polyhedral in shape) begin to accumulate in the cytoplasm. To some extent, these particles, formed by an outer shell and a dense granular core, resemble the spheres liberated by the compatible pollen tubes at the time they discharge the spermatid nuclei in the degenerated synergid within the ovule (de Nettan-

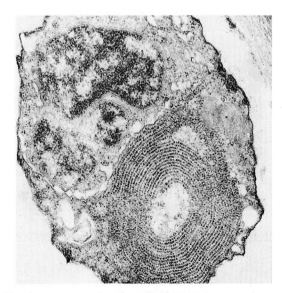

Fig. 3.17. Cross-section of incompatible tube in upper region of a style of *Lycopersicum peruvianum* 24 hours after pollination. The endoplasmic reticulum shows a concentric parallel configuration (Professor D. Cresti, Univ. of Siena; for a slightly, different cross-section, see de Nettancourt et al. 1973a)

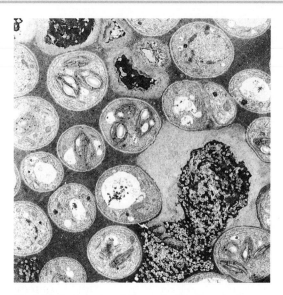

Fig. 3.18. Cross section of the conducting tissue in the upper region of the style after an incompatible pollination. An incompatible pollen tube has burst open, and a great mass of particles have been released into the intercellular space of the stylar tissue. (de Nettancourt et al. 1973 a)

court et al. 1973 a). A number of tubes where the cytoplasm at the apex was completely loaded with particles could be observed. At this stage, some of the tubes had burst and released particles in the intercellular spaces of the stylar conducting tissue, which is obviously filled with a substance of sufficient fluidity to allow free dispersion of the tube content (Fig. 3.18). Under ultraviolet illumination and after staining in aniline blue, the open extremity of the tube appears as a large vesicle practically devoid of callose (de Nettancourt et al. 1973 a).

3.5.2.3 The Role of Callose

The electron microscopy (EM) studies of de Nettancourt et al. (1973 a) reported above showed that tube tips of incompatible pollen tubes are particularly rich in callose. Through alterations of the culture medium of pollen grains grown in vitro, Cresti and co-workers (1985, 1986) found that, as in the case of stigmatic SI (Sect. 3.2.5.6), the changes in callose deposition observed in incompatible pollen tubes are probably an effect rather than an origin of growth inhibition.

3.5.2.4 Mitosis in the Generative Nucleus of *P. hybrida*

Ünal (1986) found that mitosis occurred later than in compatible pollen tubes. At prophase and metaphase, the chromosomes stuck to one another and failed to form a regular metaphase plate. The chromosomes did not separate evenly at anaphase, and the sperm produced were of irregular size and shape.

3.5.3 Early Research on the Nature of the SI Reaction

3.5.3.1 SI as a Process of Growth Inhibition

The first explanation of the positive inhibition of pollen-tube growth in incompatible styles was that of East (1926, 1929), who compared SI to the immunity reaction in animals. His hypothesis implied that the product of the specific S allele in the pollen acts as an antigen that is recognized during growth by an antibody produced by the same specific allele in the style. He also implied that pollen-tube inhibition follows such an immunological reaction. The theory of East received confirmation from the work of Lewis (1952), who showed that, after injection in rabbits, the pollen extracts from known S genotypes of *Oenothera organensis* (a species with stigmatic GSI) gave rise to antisera that produced precipitin reactions with pollen extracts of the same genotype. Lewis (1952) and his associates (Makinen and Lewis 1962; Lewis et al. 1967) reported the detection of this reaction at the level of individual grains. The search for information on the specific products of S alleles in pollen and pistil, the stage at which these substances were formed, and the mechanisms leading to their mutual recognition had started. In its early phase, the search was often complicated by uncertainty.

3.5.3.2 Is the S Phenotype of Mature Styles Determined before Pollination?

3.5.3.2.1 Evidence from In Vitro Tests. Improving a technique elaborated by Picard and Demarly (1952), Brewbaker and Majumder (1961) showed that, in *Petunia inflata*, artificial styles containing extracts of unpollinated pistils are able to recognize and inhibit, in vitro, the growth of pollen tubes carrying one of the two S alleles present in the styles from which the extract had been made. The results were reproducible, and clear-cut differences between several different compatible and incompatible pollinations could be obtained. Similar results were reported for *Lotus corniculatus* (Miri and Bubar 1966) and *N. alata* (Tomkova 1959). Because it is doubtful that the substance able to identify the S genotype of pollen grains cultured in vitro and to inhibit their growth in incompatible combinations is synthesized in the culture medium itself, it seems that S-gene action occurs in the pistil before pollination, probably at the stage where the style becomes able to prevent the growth of incompatible pollen.

3.5.3.2.2 Diverging Results. Diverging results originated from the findings (Van der Donk 1975a, 1975b) that RNA and proteins from virgin styles or proteins produced in the egg cells of *Xenopus laevis* from the RNA of virgin styles had no effects on pollen-tube growth. In contrast, proteins originating directly (or indirectly, through *Xenopus*) from pollinated styles inhibited pollen-tube growth when the S genotype of the pollen matched that of the style from which the proteins originated. Unexpectedly, the proteins "stimulated" growth when the S allele in the pollen was different from those present in the style used for obtaining the proteins.

From the work with artificial styles described above and from research performed during the last 20 years (Sects. 3.3.4, 4.1.4), it appears that the observations of Van der Donk (1974 a, 1974 b) did not concern the proteins involved in the recognition phase but were related to sequences of post-recognition events that participate in the control of pollen-tube growth. Another possibility, which has not been confirmed by contemporary research, could be that the stylar S proteins (now known to be S-RNases) are not activated in virgin styles or in *Xenopus* eggs.

Due to the observations by Franklin-Tong et al. (1990) that new pollen proteins in *Papaver* (Sect. 3.3.4.1) are produced after self-pollination, it seems that the different RNAs detected in "selfed" and "crossed" styles originated essentially from the pollen and not from the style. Van der Donk had no way of knowing this.

3.5.3.3 First Models of the Gametophytic Stylar SI Mechanism

Monofactorial GSI systems are particularly well adapted for the detection and selection of function-loss mutations (induced in PMCs or tetrads) that lead to the ability of compatible pollen able to cross the stylar barrier on selfing and to transmit the SC character to the subsequent generation. Such mutations provide valuable information regarding structure–function relationships at the S locus. Thus, it is not surprising that research regarding the organization of the incompatibility locus and the recognition mechanism essentially concentrated on species with GSI, particularly before the advent of site-directed mutagenesis. The spectrum of function-loss mutations obtained at the S locus (Sect. 4.2) covers three types of changes towards SC: changes in the pollen (pollen mutants produce pollen accepted by all pistils), changes in the style (stylar mutants accept all pollen) and changes in both the pollen and style ("complete" mutants). The occurrence of such mutations was taken into account by the majority of models proposed in the period covering the 1960s and the 1970s. Depending on the nature of the male and female determinants, these models usually fell into one of two categories.

In the models of the first category, SI is assumed to result from the recognition of unlike substances, supposedly encoded by different genes or sub-units of the S locus. This concept, specified in 1965 by Linskens (Fig. 2.8) from the early suggestions of East (Sect. 3.5.3.1) and Lewis (1960), predicts the formation of an inhibiting complex between a stylar antibody and a pollen protein. It is analogous to a model, proposed by Van der Donk (1975 b), that also predicts the matching (on self-pollination) of different substances in the pollen and style and the subsequent inactivation of stylar polypeptides necessary for pollen growth. The concept does not necessarily imply the involvement of different genetic structures, because non-identical gene products can be generated by the same transcription unit (Sect. 3.2.4.1). However, after an analysis of the effects of a deletions of the stylar S gene (S-RNase) of *Pyrus serotina*, Sassa et al. (1997) suggested that separate genes specify the S phenotypes of the pollen and style.

The concept basic to the models of the second category postulates that the male and female determinants participating in the recognition process are identical. It was first proposed by Lewis (1965), who reconsidered his initial proposal for a bi-partite structure of the S locus with a segment governing pollen specificity and a segment governing stylar specificity. The reasons that led Lewis to modify his views derived essentially from the consideration that the independent occurrence of mutations into new specificities in each segment would unavoidably lead to the breakdown of the system. In his new model, based on the "dimer–tetramer hypothesis", Lewis (1965) then suggested that identical polypeptides, coded by the same S allele, first dimerize in the pollen and style. On self-pollination, they tetramerize to form a repressor complex that switches off one or several genes responsible for the proper metabolism of the pollen tube (Fig. 2.7). To account for function-loss mutations affecting only the pollen or only the style (and which could no longer be attributed to separate mutations in different pollen and style specificity segments), Lewis predicted a tri-partite structure for the S locus. This structure consisted of a single specificity segment expressed in the pollen and style and two distinct activity parts regulating the expression of specificity in the pollen and style, respectively. This model, simple and logical, was adopted by most of the scientific community in its initial version (the dimer hypothesis), its revised form (which introduced the formation of "tetramers") or a modified scheme proposed by Ascher (1966). It does not absolutely require that a common genetic structure encode the S product in the pollen and pistil, because tandem repeats of S-sequences (such as in the SLG and SRK of *Brassica* haplotypes) often appear to co-evolve without any substantial loss of their initial homology (Sects. 3.2.3.2, 4.3.1.1, 4.3.1.2).

3.5.3.4 Towards the Detection of Stylar S Proteins

The first step was taken 40 years ago by Linskens (1961), who detected, through serological methods, S-specific antigens in the transmitting tissue of unpollinated *Petunia* styles. Decisive evidence that S-specific proteins occurred in both the stigma and style parts of virgin flowers was provided by Bredemeyer and Blaas (1981). The S-allele-protein relationship was established for three different S alleles after comparison by isoelectric focusing of the stigma protein patterns in several inbred and cross progenies in *N. alata*. At approximately the same time, by immunoelectrophoresis, Raff et al. (1981) detected the presence of one antigen correlated with a specific

S3S4 genotype in the style extracts of *Prunus avium* (GSI). The same team (Mau et al. 1982) from the laboratory of A.E. Clarke at the University of Melbourne described the major soluble components of the style of *P. avium* and succeeded in isolating and partially characterizing an antigenic glycoprotein ("Antigen S"). This glycoprotein contained at least two main components and was a powerful inhibitor of pollen-tube growth in vitro (Williams et al. 1982). This enabled a new era of research on the molecular genetics of monofactorial gametophytic stylar SI to begin (Sect. 3.5.4).

3.5.4 Isolation, Cloning and Sequencing of a Stylar Protein Segregating with the S2 Allele of *N. alata*

A few years after the work of Bredemeyer and Blaas (Sect. 3.5.3.4), the group of Adrienne Clarke reported the isolation of a major glycoprotein segregating with the S2 allele and present in extracts of mature styles of *N. alata* (Clarke et al. 1985). The protein had a pI higher than 9.5, an apparent molecular weight of 32 kDa and was found to be, at the concentration used (25 µg/ml), a strong inhibitor of S2 pollen grown in vitro (and, to a lesser extent, of S3 pollen and of different pollen genotypes from *L. peruvianum* and *P. avium*). Style proteins of 53 plants resulting from reciprocal crosses involving the genotypes S2S2, S2S3, S3S3 and S1S3 were analyzed using two-dimensional gel electrophoresis. The 32-kDa component always only co-segregated with the S2 allele (Anderson et al. 1986).

The cloning (Anderson et al. 1986) of the cDNA encoding the putative S2 glycoprotein was based on N-terminal AA-sequence data obtained from the protein (see also Mau et al. 1986 for sequence data on allele S6). The origin of the cDNA was confirmed by the correspondence found between the predicted protein encoded by the cDNA and the sequence of a number of isolated peptides. The *N. alata* S2 clone contains an ORF of 642 nucleotides, which is sufficient to encode a protein of 24.8 kDa. Using S2 cDNA as a probe for in situ hybridization analyses of the pistil, Cornish et al. (1987) found that the gene was expressed in the transmitting tissue of the stigma and style and in the epidermis of the placenta. Anderson et al. (1986) had earlier reported high concentrations of the protein in the part of the style in which the inhibition of incompatible pollen occurs.

The cDNA encoding two other S alleles of *N. alata* (S3 and S6) was obtained subsequently by M.A. Anderson et al. (1989). Comparisons of the predicted AA sequences of the three alleles revealed approximately 65% AA identity among the different parts of the alleles, with 56% of the AA identical in all three alleles. The distribution of variation covers the sequence but concentrates on variable regions interspersed with five conserved regions. Through cation-exchange fast protein liquid chromatography (FLPC), Jahnen et al. (1989) purified five S glycoproteins of *N alata* and confirmed the high degree of homology among them and with sequences of *L. peruvianum* (Mau et al. 1986; N.O. Anderson et al. 1989).

The S protein is glycosylated (four potential N-glycosylation sites were found) and possesses a signal peptide, suggesting (as will be demonstrated later) that the protein is secreted. Cysteine residues (ten in S3 and S6, nine in S2), which play an important role (through disulfide-bond formation) in the maintenance of the tertiary and quaternary structures of proteins (McClure et al. 1990; Oxley and Bacic 1996; Parry et al. 1997b), are conserved between the alleles, even in the variable regions. Any interaction between the protein and pollen or pollen tubes would presumably involve features of the surface of the protein (Parry et al. 1997b).

The primary structural features of these S-specific stigmatic and stylar proteins, such as have been established for a large number of *N. alata* alleles and for many species with an SI system of the *Nicotiana* type, are presented in Section 3.5.7 together with recent information regarding the distribution of the protein throughout the flower. Aspects dealing with the polymorphism and evolution of the S alleles are reviewed in Chapter 4.

3.5.5 The S-Associated Glycoproteins Are Ribonucleases, and SI Involves the Degradation of Pollen RNA

Sequence analysis by McClure et al. (1989) reveals significant homology between the S-associated glycoproteins of *N. alata* and the extra-cellular ribonucleases that are typical of *Aspergillus oryzae* (ribonuclease T2) and *Rhizopus niveus* (ribonuclease Rh). Among other similarities, the two histidine residues participating in the catalytic activity of T2 are also present in the S proteins of *N. alata*. Of the 122 AA conserved among the three S proteins available for analysis, 30 are aligned with identical AAs in the fungal ribonucleases, and 22 others are aligned with closely related AAs. Between the S-associated proteins and the T2 ribonuclease, half of the cysteine residues are conserved. It was the conclusion of McClure and co-workers that the similarities between the active domain of the fungal ribonucleases and the homologous region of the glycoproteins indicate a close structural relationship. Indeed, assays (by the perchloric-acid precipitation method) for ribonuclease activity with the glycoproteins encoded by five different alleles of *N. alata* showed that the S-associated proteins account for most of the ribonuclease activity recovered from style extracts. The ribonuclease activity of these extracts is 100- to 1000-fold that of the related SC species, *N. tabacum*. In two experiments, extracts from styles homozygous for two different alleles were mixed and analyzed. Ribonuclease activity always co-eluted with the individual S-associated glycoprotein and was S-allele specific. The RNase activity associated with the S glycoproteins in *N. alata* accounts for 40–80% of the total RNase activity of stylar extracts (Haring et al. 1990).

This work was followed by the demonstration that S-allele specific degradation of pollen RNA occurs in vivo (McClure et al. 1990). The Melbourne team grew *N. alata* plants in the presence of ^{32}P. Agarose-gel fractionation of the [^{32}P]RNA recovered from styles after compatible and incompatible pollinations showed that less radioactive pollen-derived RNA is recovered from incompatible pollinations and that the rRNAs are intact in the styles of compatible crosses and are degraded in incompatible crosses. This specificity was not observed in vitro (using isolated S-ribonucleases and pollen RNAs), because digestion of pollen RNA was performed by each of the S ribonucleases tested. Thus, it became evident that the pollen RNA is not specifically "tailored" to the corresponding stylar ribonuclease. It would appear that the S specificity of individual ribonucleases is recognized by the pollen at the time of entry in the pollen tube and that a ribonucleases is allowed (or not de-activated) only when its specificity matches that of the pollen grain.

Similar conclusions were reached by Gray et al. (1991) when they failed to find evidence (with in vitro-grown pollen tubes) that specificity involves a specific mRNA substrate. They found that pollen tubes of three different genotypes (S2, S3 and S6) take up S2 ribonuclease and that the inhibition of protein synthesis in S2 tubes results from both S2 and S6 ribonucleases. They attributed these findings to the partial expression of allelic specificities reported by Jahnen et al. (1989) under in vitro conditions.

However, recent observations cast doubt on the exact role of S ribonucleases in the rejection phase of SI. The evidence, which deals essentially with the rate of rRNA degradation in incompatible tubes, is reviewed in Sections 3.5.7.2 and 3.5.7.3.

3.5.6 Evidence that the S Proteins of *Petunia* and *Nicotiana* Are Responsible for the S-Allele-Specific Recognition and Rejection of Self Pollen

As emphasized by Kao and Huang (1994) and Kao and McCubbin (1996), the fact that S proteins co-segregate with S alleles does not prove that they are products of S alleles, because they could also be produced by genes very closely linked to the S locus. In transgenic experiments, Huang et al. (1994), Lee et al. (1994) and Murfett et al. (1994) demonstrated that the S proteins are necessary and sufficient for the rejection of self pollen by the pistil. The results of their work are outlined below and in Section 4.2.4.2.

3.5.6.1 Induction of Loss and Gain of Functions at the S Locus of *P. inflata*

Lee and co-workers (1994), from the group of Professor Kao in Pennsylvania, tried loss-of-function and gain-of-function approaches to find out whether *P. inflata* S proteins, previously identified by Ai et al. (1990), control the SI behavior of the pistil. In a set of transformation experiments involving the use of an anti-sense S3 gene also active against S2 alleles, they prevented S3S2 plants from producing S2 and S3 proteins in their pistils, and introduced a cDNA encoding the S3 protein into an S1S2 plant. With such material, they showed that eliminating the S2 and S3 proteins from *P.*

inflata styles led to the failure of the transgenic plants to reject S2 and S3 pollen. They also observed that stylar expression of the encoding S3 protein in S1S2 genotypes conferred the ability to reject S3 pollen on the transgenic plants.

3.5.6.2 S-Allele-Specific Pollen-Tube Rejection in Transgenic *Nicotiana*

At the same time, work towards similar objectives was performed by Murfett et al. (1994) who, through the construction of transgenic hybrids and the use of a powerful promoter, found a way to overcome difficulties with the engineering of *N. alata* and the expression of ribonucleases. They analyzed the direct relationship between pollen recognition, pollen rejection and the role of S ribonucleases in F1 and F2 populations of transgenic interspecific hybrids between SC *N. alata* and SC *N. langsdorfii* transformed with S-RNases and the promoter of a style-expressed tomato gene ensuring high-level S-ribonuclease expression. Recognition and rejection of *N. alata* pollen occurred faithfully in transgenic plants that carried the same S allele and expressed S ribonucleases at relatively high levels. In fact, this work goes a step further than that of Lee and associates (1994; Sect. 3.5.6.1), because it demonstrates that S ribonucleases contain all the information necessary for pollen recognition.

3.5.6.3 Proof that Ribonuclease Is Involved in the Rejection of Self Pollen

However, the rigorous proof that ribonucleases participate in the rejection of incompatible pollen tubes in species with monofactorial GSI was provided by Huang et al. (1994) with *Petunia*. They showed that S1S2 styles acquired the ability to produce S3 ribonuclease and to reject S3 pollen after transformation with a wild-type S3 allele but failed to express such properties if transformed with a mutant S3 that produces a mutant S3 protein with no detectable ribonuclease activity. Complementary evidence was found in *L. peruvianum*, where a self-compatible mutant displays the substitution of the conserved histidine codon in the C1 region of the S-ribonuclease gene by an asparagine (Royo et al. 1994) and produces an S protein with no detectable ribonuclease activity (Kowyama et al. 1994).

3.5.7 Main Features of the S-Ribonuclease Gene and of Ribonucleases

3.5.7.1 Distribution and Structural Features of the Gene

The S-ribonuclease gene operates in three plant families: the Solanaceae, the Rosaceae and the Scrophulariaceae, where it determines, as seen above, the incompatibility phenotype of the pistil. Attempts to detect an S ribonuclease in *Trifolium pratense* (Poaceae) using a variety of heterologous probes and PCR primers were unsuccessful (Sims et al. 1999).

3.5.7.1.1 Solanaceae. Knowledge from the cDNA sequences of *N. alata* produced by M.A. Anderson et al. (1986, 1989), particularly the putative signal sequence and a number of conserved regions, allowed the cloning of the S alleles in several species of the Solanaceae during the early 1990s. A list of these species, prepared by Nass et al. (1997), refers to *N. alata* (Kheyr-Pour et al. 1990), *P. hybrida* (Clark et al. 1990), *P. inflata* (Ai et al. 1990), *Solanum chacoense* (Xu et al. 1990a, 1990b), *S. tuberosum* (Kaufmann et al. 1991) and *L. peruvianum* (Tsai et al. 1992; Rivers et al. 1993; Royo et al. 1994). Partial S-allele cDNA sequences were obtained by RT-PCR for *S. carolinense* and *Physalis crassifolia* (Richman et al. 1995, 1996a).

The ribonucleases include five conserved regions (Fig. 3.19), two of which show (Sects. 3.5.5, 3.5.6) a high degree of sequence similarity with the corresponding regions of two fungal ribonucleases and with a number of other ribonucleases (McClure et al. 1989; Singh and Kao 1992; Green 1994). The three other conserved regions contain mostly hydrophobic AAs and are believed to be involved in the core structure of the protein. Two HV regions, designated as Hva and Hvb by Ioerger et al. (1991), are also part of the protein and display the highest degree of sequence diversity. They may constitute the recognition (specificity) region of the S gene in mono-factorial GSI (however, see Chap. 4). In *Petunia* (Coleman and Kao 1992) and in *Nicotiana* (Matton et al. 1995), the flanking region around the S-ribonuclease gene contains numerous repetitive sequences and completely lacks homology. In *L. peruvianum*, four of the most variable regions of the S-RNase are clustered on one surface of the molecule and may constitute an important part of the allele-specific binding site for the pollen product of the S locus (Parry et al. 1998). Digestion of the cDNAs of S alleles from *N. alata* with restriction enzymes produces mostly single fragments not large enough to contain one or a few copies of the coding sequence (750 bp). This suggests that the S gene is probably a single-copy genetic system (Bernatzky et al. 1988).

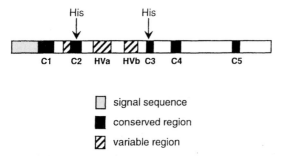

Fig. 3.19. Generalized features of S ribonucleases. A comparison of S alleles from different solanaceous species reveals a common structure. All contain a secretion signal sequence (*gray*) and share five conserved domains (C1–C5; *black*) that include two histidine residues (His) that are required for ribonuclease activity in related fungal RNases. Two hyper-variable regions (Hva and Hvb; *hatched*) are also shown. (Dodds et al. 1997, with kind permission from John Wiley & Sons, Inc.)

3.5.7.1.2 Rosaceae. S ribonucleases have been found to operate in the SI systems of several species of Rosaceae (a family that is not related to the Solanaceae), including apples, Japanese pears, cherries (Hiratsuka 1992a, 1992b; Sassa et al. 1992, 1993, 1994, 1996; Broothaerts et al. 1995) and, more recently, almonds (Tao et al. 1997) and apricots (Burgos et al. 1998).

Sassa et al. (1992) were the first to establish that polymorphism in stylar ribonucleases of Japanese pears is associated with S genotypes and to describe a self-compatible mutant where stylar ribonuclease is barely detectable but is normally produced in the original variety from which the mutant developed as a sport. In 1993, they showed that the N-terminal sequences of the ribonuclease from Japanese pears are similar to those of the Solanaceae. Broothaerts et al. (1995; with apples) and Sassa et al. (1996; with apples and Japanese pears) isolated cDNA clones and deduced the AA sequences. The position of C4 in the Solanaceae is occupied by region RC4, which shares no homology with C4 (Ushijima et al. 1998a). Other regions present striking similarities with the Solanaceae. These similarities are meaningful. In particular, throughout the two families, the conservation of the eight cysteine residues that form the four disulfide bridges of the S ribonucleases demonstrates the importance of these cross-links for the stabilization of the tertiary structure of the protein (Ishimizu et al. 1996). Using cryosections from apple pistils and specific antibodies against S proteins, Certal et al. (1999) showed that the S ribonucleases of apples are localized in the intercellular space of the transmitting tissue along the pollen-tube pathway in both the stigma and style. Non-S-specific intracellular labeling confined to one layer of the nucleus was observed in all ovary sections.

3.5.7.1.3 Scrophulariaceae. In the Scrophulariaceae, close relatives of the Solanaceae, research has concentrated on *Anthirrhinum*. Xue and co-workers (1996) characterized the cDNAs encoding polypeptides homologous to S ribonucleases. Through co-segregation studies with DNA gel blots probed with cDNAs, determination of the period of gene expression at anthesis and the observation of sequence polymorphism typical for S allele, they demonstrated that these cDNAs are derived from functional S alleles. Region C4 is missing, and considerable inter-allelic diversity can be observed for regions Hva and Hvb.

3.5.7.2 What Are the Effects of S Ribonucleases on rRNA and mRNA?

It has been shown, through EM observations of ribosomes and polysomes in pollen tubes formed after self-pollination in *Datura suaveolens* and *N. alata*, that S-RNases do not act as intracellular cytotoxins that degrade rRNA and perhaps mRNA. Walles and Han (1998), who undertook that study, found that there is no decrease over time in the number of bound ribosomes per unit of rough endoplasmic-reticulum membrane after incompatible pollination. They concluded that the substrate for S-RNases is probably far more specific than expected. The discovery implies that the effects of S-RNases on the physiology of incompatible pollen tubes may be more discrete than initially considered (Walles and Han 1998; Sect. 3.5.7.2). In her review of the different causes for the death of incompatible pollen tubes, Geitmann (1999) lists some of the reasons, essentially based on the presence of large quantities of ribosomes in incompatible tubes, that led her to reconsider the exact function of S ribonucleases. However, she agrees with Clarke and Newbigin (1993) that "it seems unlikely that the RNase function would have been so tightly conserved if it were not functional in SI." Indeed, as suggested by the observations of Lush and Clarke (Sects. 3.5.2.1, 3.5.7.3), the possibility also exists that the rRNAs degraded by specific S ribonucleases are, at least to some extent, replaced by the pollen tube through de novo synthesis. The final fate (death through bursting of its apex, no growth or slow growth) of the incompatible tube could be determined by the rate of replacement.

3.5.7.3 Are the Effects of Ribonucleases Irreversible?

Stylar grafts, which associate the top of a style (the scion) bearing known S alleles with the base of another style (the stock) having a different S genotype, have been used in the past for different purposes (Sect. 4.1.5). The technique was utilized recently by Lush and Clarke (1997) in a study of the behavior of incompatible pollen-tube growth in *N. alata* (Fig. 3.20). Through the grafting of incompatible and compatible scions and stocks, they found that the tubes (which continue to grow very slowly in an incompatible scion) could revert to normal growth if they managed to cross the graft junction and reach compatible territory. Not all incompatible tubes are able to perform in this manner, because some burst open and others stop growing; however, a substantial proportion of them finally reach the ovary. The observations require very precise measurements of growth rate. They suggest that either S-ribonuclease action is not instrumental in the expression of SI (Sect. 3.5.7.2) or – and this is more likely – that pollen tubes have the ability to synthesize fresh rRNA when degraded RNA needs to be replaced, at least under SOS conditions. This second ex-

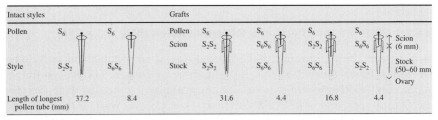

Fig. 3.20. The growth of pollen tubes (genotype S6) in intact styles and in style grafts. Grafts were made 5 h after the pollination of styles. Scions were 6 mm long. Pollen-tube lengths were measured 24 h after grafting. Figures are the means of five pollinations or grafts. (Lush and Clarke 1997)

planation does not agree with conclusions from McClure et al. (1990) and Mascarenhas (1993) that rRNA is not expressed in angiosperm pollen tubes. However, the explanation is supported by results from earlier work by Campbell and Ascher (1975), Tupý et al. (1977) and Bagni et al. (1981). A third possibility, resulting from the work of Walles and Han (1998), is that there is no such thing as a general degradation of rRNA in incompatible pollen tubes (Sect. 3.5.7.2).

3.5.7.4 Why Pollen Tubes Are Not Inhibited in the Stigma

The responsibility probably lies on the side of the pollen tube, because McClure et al. (1993) clearly established that S ribonuclease purified from the stigma is indistinguishable (by SDS-PAGE, chromatographic behavior and ribonuclease activity) from S ribonuclease purified from the style. Furthermore, although the stigma transcripts are shorter and more heterogeneous that those of the style, sequence analyses of cloned cDNA showed that stigmatic and stylar S transcripts derive from the same gene. It is possible that ribonucleases cannot have access to the pollen or to pollen RNA before the tube has reached a particular length or developed a certain receptivity. Another explanation could be that the concentration of rRNA in the tube at the onset of pollen germination is sufficient to sustain protein synthesis during the first period of growth through the stigma.

3.5.7.5 What Determines the S Specificity of Stylar Ribonucleases?

S ribonucleases are glycoproteins, and their S specificity must lie in the carbohydrate moiety, the AA sequence or a combination of the two.

3.5.7.5.1 The Role of the Carbohydrate Moiety. Five potential N-glycosylation sites occur in the S ribonucleases of *N. alata*; these sites are present in variable numbers in different S alleles (five in S3; four in S2, S6, S7; one in S1...) and may play a role in S specificity (Oxley et al. 1996 for information and detailed descriptions of the structure of N-glycans on S alleles).

Taking into account the important role played by the glycan chains of many animal cell proteins in cell–cell recognition, Karunanandaa et al. (1994) studied the possibility that the glycan chain of the S3 ribonuclease of *P. inflata* is involved in self/non-self recognition. They replaced the codon for Asn-29 by a codon for Asp at the only potential N-glycosylation site of the S3 protein and introduced this mutated S3 gene into S1S2 plants via *Agrobacterium*-mediated transformation. Six transgenic plants produced a normal level of the non-glycosylated S3 protein and expressed the ability to reject S3 pollen completely (Sect. 4.2.4.2). In other words, the carbohydrate moiety of the S3 ribonuclease does not participate in the recognition or rejection of self pollen. The possibility, suggested by Oxley et al. (1996), that the lack of N-glycosylation had abolished the specificity of the SI reaction in the transgenic plant and allowed it to reject any pollen is unlikely because, in this case, one would not expect the non-glycosylated S3 allele to reject all pollen. Instead, it would act as an SC allele that accepts all pollen.

In a discussion of the role of the N-glycans on the S-ribonuclease of *N. alata*, Oxley et al. (1998) noted that all cDNA sequences of the S-RNase products of functional S alleles contain at least one potential N-glycosylation site, with site I conserved in all cases. While not dismissing the conclusions of the Karunanandaa group, they observe that site I is not present on the S-like ribonucleases that do not participate in SI (Green 1994). As a consequence, they maintain their opinion that site I is possibly involved in SI. Ishimizu and co-workers (1999) identified the structures of the N-glycosylation sites of seven S-RNases in *Pyrus pyrifolia* (Rosaceae). The presence of N-acetylglucosamine and chitobiose in the putative recognition sites of the S-RNases led Ishimizu and co-workers to suggest that these sugar chains may interact with pollen S products.

The debate on the possible role of the glycan chains in pollen–pistil recognition is important, because the current efforts to identify the recognition determinants of S-ribonucleases through domain swapping and site-directed mutagenesis (Sect. 4.2.7.1) are based on the hypothesis that S specificity results from differences in the AA sequences of the S domains.

3.5.7.5.2 The Role of HV Regions. One naturally assumes that the two regions Hva and Hvb, which display a high degree of sequence diversity and are hydrophilic, are directly involved in the recognition interaction of S ribonucleases with pollen determinants. In order to assess this role, three sets of mosaic constructs among S ribonucleases have been realized. These constructs involved the swapping of HV regions in *Petunia* (Kao and McCubbin 1996), of the different domains of the protein in *Nicotiana* (Zurek et al. 1997) and of nucleotides in the HV regions of *Solanum* (Matton et al. 1997). These experiments (Sect. 4.2.7.1) led to diverging conclusions.

The work in *Petunia* and *Nicotiana* showed that the transfer of HV regions and domains between alleles resulted in the production of hybrid ribonucleases unable to cause the rejection of any of the two parental alleles. The sequences necessary for pollen recognition are apparently not confined only to the HV regions but are scattered throughout the entire ribonuclease. In contrast, the elimination (through an appropriate replacement of nucleotides) of differences between the HV regions of two closely related alleles of *S. chacoense* conferred the recognition capacity of the second allele to the modified alleles (Matton et al. 1997). In other words, it seems that the HV regions (and only they) determine the specificities of stylar ribonucleases.

While the controversy is not settled (Verica et al. 1998; Matton et al. 1998), it appears that at least two reasons can be advanced to account for the differences of results by the Montreal group and the two American universities. First, the two *Solanum* alleles compared by Matton and co-workers are nearly identical (with a variation of four AAs between the HV regions and six among the other regions). It is therefore likely that other parts of the ribonuclease sequence, possibly involved in the determination of S specificity, are identical in the two alleles. It is also possible, if only the HV regions control S specificity, that the swapping of HV regions in *Petunia* and *Nicotiana* interfere with their activities and functions.

3.5.8 S-Gene Products in Pollen Grains and Pollen–Pistil Recognition

3.5.8.1 The S-Ribonuclease Gene Is Expressed in Developing Pollen Grains ...

It has been found, in conformity with the dimer hypothesis of Lewis (1965), that S-ribonuclease transcripts accumulate in young anthers and developing pollen grains.

3.5.8.1.1 *N. alata*. The S ribonuclease is expressed at low levels in developing microspores and could be detected by immunochemistry in the intines of mature, hydrated pollen grains Dodds et al. (1993). A transcript in young microspores hybridized to a cDNA encoding the S2 ribonuclease of the parent plant and did not hybridize to cDNAs encoding other S-ribonuclease alleles. Furthermore, two cDNAs for the S2 ribonuclease were cloned from a library derived from the anthers of a plant homozygous for the S2 allele; both corresponded to the coding sequence of the S2 ribonuclease.

3.5.8.1.2 *P. hybrida*. In *P. hybrida* also, the S-ribonuclease gene was found (Clark and Sims 1994) to be expressed at low levels in non-stylar tissue, including immature anthers. Amplification products of anther RNA were used to establish that S-ribonuclease sequences identical to the stylar RNA accumulated in developing anther tissue. No amplification was observed from mature anthers; pollen germinated in vitro or when primers specific to one of the two S alleles studied (S1) was used.

3.5.8.1.3 *L. peruvianum*. In *L. peruvianum*, an S-ribonuclease gene was found to be expressed (at low levels) during microsporogenesis. The product of this gene was detected by immunochemistry in the intines of mature, hydrated pollen (Dodds et al. 1993; Sect. 3.5.8.2).

3.5.8.2 ... but Pollen S Ribonucleases Do Not Determine the Pollen S Phenotype

To test the possibility that the ribonuclease (S3) detected during pollen development in *L. peruvianum* determines its SI phenotype, Dodds et al. (1999) transformed the cotyledon explants of defined S genotypes through *Agrobacterium* mediation with S3. S3 sense constructs were introduced into S1S2 backgrounds, and S3 anti-sense constructs were introduced into S1S3 and S2S3 backgrounds. The promoter chosen to direct transgene expression was LAT52, a promoter previously shown by Twell et al. (1990) to direct expression of the β-glucuronidase reporter gene in developing pollen of transgenic tomato, tobacco and *Arabidopsis*. The presence of the transgene in regenerated lines was confirmed by DNA gel-blot analysis. Pollen from the transgenic plants accumulated S3-ribonuclease transcripts, and the S3-ribonuclease protein was detected immunologically in the sense transgenic plants. However, neither the sense nor the anti-sense S3-ribonuclease constructs altered the SI phenotype of pollen from the transgenic plants (Sect. 4.2.4.4).

The earlier results of Dodds and co-workers (Sect. 3.5.8.1) had not entirely disqualified the tri-partite model of the S locus (Lewis 1965), because pollen S ribonucleases or parts of their sequences, if they determine pollen specificity, may be masked in some way in the pollen intine, possibly by the biosynthesis and deposition of other pollen components. The new results of the Melbourne group indicate that S-ribonuclease expression in pollen is not sufficient to modify its S phenotype. They confirm the demonstration made by Sassa et al. (1997), through the analysis of the effects of gene deletions, that the pollen and pistil determinants are encoded by non-allelic gene sequences (Sects. 1.3.5.1, 3.5.3.3).

Before these Australian and Japanese experiments, other laboratories had reported results that suggested that the S-RNase gene do not govern the S specificity of the pollen determinant. This was particularly the case, if expression problems did not interfere, of the work by Lee and co-workers (1994). They showed that the sense and anti-sense constructs they used so spectacularly to regulate the production of stylar S ribonucleases (Sect. 4.8.4.2) had no effect on the SI pollen phenotype displayed by the transgenic plant. Similarly, the successful attempts of Kirch et al. (1995) to express a *Solanum* S ribonuclease in tobacco pollen did not affect its fertility and ability to grow through stigmas and styles.

3.5.8.3 Involvement of Protein Kinase?

3.5.8.3.1 A Pollen Receptor-Like Kinase 1 in *P. inflata*. Pollen receptor-like kinase 1 (PRK1) was characterized and the activity of its encoded kinase studied by Mu et al. (1994). It was first detected in bi-nucleate microspores and was found to reach its highest level in mature pollen and to remain at a high level in pollen tubes germinated in vitro. The absence of PRK1 from the pistil, the lack of S-allele-associated polymorphism and the absence of any similarity in the extra-cellular domains of *Brassica* SRK and PRK led Mu and co-workers to conclude that the function of PRK1 is different from that of SRK. The gene appears to belong to the class of "late-pollen" genes (Mascarhenas 1990, 1993) required for pollen maturation or pollen-tube growth; it appears to participate in signal-transduction processes not necessarily related to SI. Further analyses have been made by Lee et al. (1997); these analyses show that embryo sac development is affected in plants of *P. inflata* transformed with an anti-sense gene encoding the extra-cellular domain of PRK1.

3.5.8.3.2 In Vitro Phosphorylation of the S Ribonucleases from *N. alata*. Kunz and co-workers (1996) have found, in a follow up of earlier work by Polya et al. (1986), that protein kinases of *N. alata* use S ribonucleases as substrates in vitro. These water-soluble Ca^{2+}-dependent kinases (Nak-1) were partially purified and were observed to phosphorylate the S ribonucleases of *N. alata* but not *L. peruvianum*. It is not impossible, if the discovery can be confirmed in vivo and extended to other species and families with mono-factorial GSI, that a kinase receptor and the phosphorylation of S ribonucleases participate, at some stage of the SI reaction, in the processing and/or screening of S ribonucleases. However, the site(s) of phosphorylation have not been determined, and there is no evidence for allele-specific phosphorylation. The authors rightly conclude that their work "does not prove that protein kinases are involved in mediating the SI interaction but does indicate directions for further exploration".

3.5.8.4 The Role of Pollen Determinants

Similar sets of models of the SI interactions that reconcile current knowledge on S-ribonucleases have been presented simultaneously by Kao and McCubbin (1996, 1997) and by Dodds et al. (1996, 1997). A recent description of these models was proposed by Kao and McCubbin (1997). In the first hypothesis (Fig. 3.21 model 1 and Fig. 3.22 A), it is predicted that the only stylar ribonucleases that can pass the pollen plasma membrane, pene-

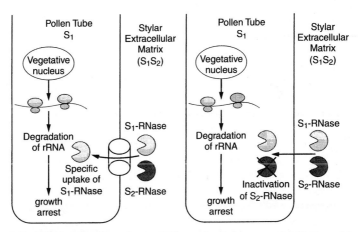

Fig. 3.21. Allele-specific inhibition of pollen tubes by S-RNases: the Melbourne models. Both models describe events in the inhibition of S1 pollen tubes in an SI S2 style. Model 1 (*left*) proposes that rejection results from the specific uptake of the S1 RNase. S-RNases encoded by other alleles of the S locus (S2 RNase, in this case) do not enter the pollen tube. Model 2 (*right*) proposes that S-RNases enter pollen tubes non-specifically but are inactivated in the cytoplasm. The product of the S1 allele in pollen prevents inactivation of the S1 RNase. In both cases, the presence of active S1 RNase within the pollen tube causes growth inhibition by degrading rRNA. (Dodds et al. 1997, with kind permission from John Wiley & Sons, Inc.)

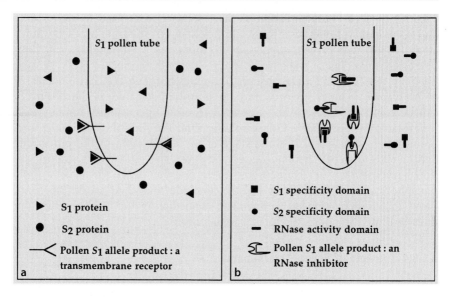

Fig. 3.22. Allele-specific inhibition of pollen tubes by S-RNases: two models from University Park. **A** Receptor or gatekeeper model. **B** RNase-inhibitor model. Each model depicts an S1 pollen tube growing in the extra-cellular space of the transmitting tissue of a style that produces S1 and S2 proteins. In **A**, only self S protein (S1 protein) is able to enter the S1 pollen tube whereas, in **B**, both self (S1) and non-self (S2) S proteins are able to enter the S1 pollen tube, but only S1 protein is able to function as an S ribonuclease. In both models, the RNase activity of S1 protein is responsible for the growth inhibition of the S1 pollen tube. (Kao and McCubbin 1996)

trate in the cytoplasm and inhibit pollen-tube growth have an S specificity identical to that of a membrane receptor in the pollen grain. The second model (Fig. 3.21, model 2 and Fig. 3.22 B) predicts that all S proteins can enter the pollen-tube cytoplasm and bind either *indiscriminately* (S1 protein in S2 tube) with "cross" pollen products or *discriminately* (S2 protein in S2 tube) with "self" pollen products. *Indiscriminate binding*, which occurs after a compatible pollination, brings together the catalytic domain of the stylar ribonuclease and the catalytic "end" of the pollen product: *it leads to the inactivation of the stylar ribonuclease. Discriminate binding* takes place after self-pollination and associates the specificity domain of the ribonuclease to the specificity "end" of the pollen product: *the stylar ribonuclease is not inactivated.*

In the course of a cytogenetic and biochemical analysis of pollen-part mutants in *N. alata*, Golz et al. (1999a, 1999b) showed that the two models are consistent with what is known regarding the mutability of the S locus. However, the occurrence of clear-cut cases of complementation (Pandey 1967) suggests that the "receptor" model is the more likely (Sect. 4.2.2.1).

3.5.8.5 Current Research Regarding the Identification of Pollen S Determinants

Many laboratories, as can be seen throughout this chapter, concentrate their efforts on the identification of the pollen determinant that participates in the recognition phase. Examples, among several others, of the large-scale work in progress, can be briefly outlined as follows.

3.5.8.5.1 A Functional Genome Approach to Search for the Pollen S Gene of *Petunia inflata.*

At Pennsylvania State University (Kao et al. 1999), a systematic attempt was made to identify genes that satisfy the necessary requirements (sequence diversity, expression in pollen and close linkage to the S locus). To date, 14 cDNAs corresponding to genes located within 1 cM of the S-ribonuclease gene have been identified through the approaches of RNA differential display and substractive hybridization. Although none of these show the high sequence diversity expected from S genes, the materials studied so far should be useful as molecular markers of the pollen determinant. The construction of BAC libraries is underway, and BAC clones containing large-sized genomic DNA fragments of the S locus will be isolated using the 14 pollen cDNAs and cDNAs for S ribonucleases as probes. The BAC clones will then be examined for the presence of the pollen S gene.

3.5.8.5.2 Towards the Fine-Scale Mapping of the S Locus in *Petunia hybrida.*

A population of 83 independent transformants of self-compatible *P. hybrida* was used to test individual T-DNAs for linkage to the S locus (Robbins et al. 1999). Seven T-DNA insertions linked to the S locus were physically mapped on the long arm of chromosome III, some of them in a sub-centromeric position consistent with the centric fragments (typical of certain pollen-part mutations) obtained by irradiation mutagenesis (Sects. 4.2.2.1, 4.2.2.2). The selectable markers carried by the insertion can be combined with a semi-compatible pollination to force recombination around the S locus. The approach could facilitate fine-scale mapping of the S locus and the identification of the pollen S determinant.

3.5.8.5.3 Use of a Two-Hybrid System to Identify the Pollen S-Component in *S. chacoense.*

A complex procedure has been established by Luu and co-workers (1999). It involves the cloning of a modified (His→Leu at position His 114 in the C3 conserved region) S11-ribonuclease cDNA in the yeast expression vector pBD-GAL4 and the formation of a S11-ribonuclease/GAL4 binding-domain fusion protein (Luu et al. 1999). At the same time, the mRNA of pollen grains from an S11S12 plant was used to construct an expression library of pollen-tube proteins fused with GAL4-activating domain. The two constructions were thereafter co-expressed in yeast. Luu and co-workers are presently attempting to identify, by growth on a selective medium and through β-galactosidase assays, potential positive clones in which the S11 ribonuclease has interacted with a pollen S component.

CHAPTER 4

Breakdown of the Self-Incompatibility Character, S Mutations and the Evolution of Self-Incompatible Systems

One of the purposes of this chapter is to describe the different types of modifications which can lead – spontaneously or after manipulation, from either physiological or genetic changes – to a breakdown, mutation or transfer of the self-incompatibility (SI) character. The physiological changes are always temporary and cannot be transmitted from one generation to the next. However, their actions, when they contribute to the promotion of inbreeding or to variations in mating relationships, may have important implications for the genetic structure and fitness of the population in which they occur. Genetic changes, however, may or may not be permanent and lead to a variety of different effects, including:

- The breakdown of the incompatibility barrier through inactivation, silencing or deletions (partial or complete) of the S locus or the modification of other major genes, polygenes or ploidy levels.
- The increase in S-gene expression.
- The modification of S specificity or the substitution of autogamy by allogamy.

Physiological and genetic changes may eventually become inducible and their effects studied in depth, because molecular biology provides, as shown in the present monograph, an impressive assortment of techniques for:

- Preventing, promoting or amplifying the expression of individual S genes
- Operating site-directed mutagenesis at the level of specific codons.
- Swapping genetic domains between different S alleles.
- Transforming the breeding behavior of self-fertilizers.

The second purpose of the chapter is to review current knowledge regarding the origin of the main SI systems and their ages, distributions, relationships and adaptive values. SI is one of the recognition processes that significantly contributed to the expansion of angiosperms throughout the world, to multi-allelism and to the distinction between self and non-self. It is important for our understanding of plant life (and our ability to protect it and exploit it) that we find out how primitive mechanisms of cell–cell communication and signal transduction evolved (once or several times) to

give rise to the present diversity of pollen–pistil identification and rejection procedures
.

▦ 4.1 The Physiological Breakdown of SI

There are a variety of factors and environmental circumstances that, in most species, can prevent the incompatibility reaction from occurring or can enable the incompatible pollen tube to escape the pistil barrier and accomplish illegitimate fertilization. Such effects may result from separate treatments of the microspores or pistil before pollination or during the recognition or rejection phases after pollination. Depending on the type of treatment performed, the incompatibility system involved and the stage treated, the effects obtained may correspond to:

▦ An absence or inhibition of translation or transcription.

▦ An inactivation or destruction of S-gene products.

▦ The transmission of a growth stimulus enabling the incompatible pollen tube to continue its journey towards the ovary.

▦ An increase of the time interval granted to slow-growing pollen tubes to reach the ovary before floral abscission.

Research on the various conditions and treatments known to lead to the manifestation of such effects was particularly active during the period that preceded the discovery of the specific inhibitors of gene translation, transcription and expression. These conditions and treatments are briefly presented below.

4.1.1 Age Factors

Because the incompatibility phenotype of pollen and pistil is fully determined in the mature flower, several researchers have attempted to overcome incompatibility barriers by using immature material (flower buds) in which the S phenotype is not yet expressed or by using old flowers and aged pollen, where a weakening of the incompatibility components may be expected to occur. Similarly, experiments have been conducted to test the possibility that the SI character loses its strength in plants at the end of the flowering period, i.e., at a time characterized by sudden modifications in environmental conditions and metabolic activity.

4.1.1.1 Bud Pollination

Bud pollination has been tried successfully on gametophytic and sporophytic systems (Yasuda 1934; Attia 1950; Lewis 1951; Pandey 1959; Shivanna and Rangaswamy 1969) and is still in current use all over the world for the production of S homozygotes needed for research purposes or for hybrid-seed production. Its success results from the fact that the immature pistil, at least in SI systems of the *Brassica* or *Nicotiana* type, where bud pollinations (with mature pollen) are usually very effective for the production of large quantities of seed on self-pollination, has not yet formed active S proteins. Shivanna and Rangaswamy (1969) found that bud pollination can be even more productive in *Petunia axillaris* if the stigma is first smeared with exudates from the stigma of an open flower. This exudate does not appear to contribute to pollen nutrition (Konar and Linskens 1966) but simply reinforces, particularly under dry conditions, adhesivity on the immature stigma, which lacks the ability to produce its own exudate. The optimal stage for pollination corresponds to that of buds in *P. axillaris* within 2–4 days before anthesis, and there is, from this stage onwards, a positive correlation between the age of the bud and the degree of pollen-tube inhibition (Linskens 1964). However, experimenters must be aware of the possibility that the offspring issued from self or outcross bud pollinations differ from those obtained after mature-flower outcross pollinations. This was demonstrated by Cabin et al. (1996) when they compared seed yield per fruit and dry mass in the progenies derived from flower and bud pollination in different species of mustard.

4.1.1.2 Delayed Pollination, Use of Stored Pollen and End-of-Season Effects

Aged pistils pollinated with self pollen from freshly open flowers have been found to set a certain proportion of seed in *Brassica* (Kakizaki 1930) and *Lilium* (Ascher and Peloquin 1966a) but not in other genera (Stout and Chandler 1933; Yasuda 1934; Shivanna and Rangaswamy 1969).

The effects of storage on the incompatibility phenotype of pollen grains were analyzed by Shivanna and Rangaswamy (1969) in *P. axillaris*. They studied storage periods ranging from 7 days to 28 days and failed to observe any incidence of the treatment on the incompatibility of the pollen or its capacity to germinate and grow in compatible pistils.

End-of-season effects, which many experimenters failed to observe, have been reported for *Nicotiana alata* and *N. sanderae* (East 1934), *Petunia violacea* (Yasuda 1934) and *Abutilon hybridum* (Pandey 1960). In a study of the effects of pollination dates, Linskens (1977) showed that the pollen-tube/style-length ratio of *Petunia* after cross- and self-pollination was not affected by the flowering period.

4.1.2 Irradiation

In addition to their interest for radiation biology during the 1960s (study of dose effect and dose-rate effect relationships on a specific biological end point), the physiological effects of irradiation on the SI character have, to a certain extent, contributed to our understanding of SI as an active process (which can be inhibited by ionizing rays), not as the simple absence of growth stimulation. As shall be seen in Chapter 5, they have also been found to be very convenient for killing the compatible pollen used as an inter-specific mentor to promote the growth of incompatible pollen.

4.1.2.1 Chronic Exposure to Low Dose Rates of Radiation

Low dose rates of gamma rays applied chronically during the entire flowering season significantly increase fruit setting and seed set on selfing in plants of *Lycopersicum peruvianum*, which are normally allogamous and only rarely produce a few seeds after self-pollination (de Nettancourt and Ecochard 1968). The effect is two-fold and involves:

■ An inhibition of the floral abscission process, which normally occurs a few days after self-pollination.
■ An increase in the capacity of the incompatible pollen tube to bypass the incompatibility barrier and fertilize.

These two categories of effects are cumulative but, as shown by reciprocal pollinations between irradiated and non-irradiated members of a same clone, they are distinct and independent in origin. The inhibition of floral abscission does not result from the ability of incompatible pollen tubes to fertilize the egg and occurs even in instances where the irradiated flowers are left unpollinated. It simply enables slow growing incompatible tubes to reach the ovary. The compatibility character obtained in these experiments was never observed in the subsequent generation and clearly belongs to the category of temporary physiological changes.

4.1.2.2 Acute Irradiation of Styles

Linskens et al. (1960) demonstrated that acute irradiation of *Petunia* styles could attenuate the ability of the style to reject incompatible pollen. They showed that a dose of 2000 rad of X-rays applied to the style immediately before selfing could break down the incompatibility reaction and induced approximately 50% of the treated flowers to yield seeds. Only a weak effect was observed when pollination was done 24 h after style irradiation, whereas no effect could be recorded when pollination was performed 20 h before irradiation. The finding that a relatively low dose (far lower than the dosages necessary for the direct destruction of a protein) was sufficient to produce the desired effect, coupled with the fact that maximum sensitivity occurred when pollination immediately followed irradiation, is highly suggestive of temporary inactivation or immobilization of the S ribonucleases in the style.

Results similar to those of Linskens and co-workers (1960) have been reported by Hopper and Peloquin (1968), who exposed styles of *Lilium longiflorum* to different dosages of X-rays. However, the effective dose (24 000 rad) was much higher than the 2000 rad estimated by Linskens et al.

(1960). Hopper and Peloquin (1968) noted that their results appeared to be identical to those obtained after heat treatment and suggested enzyme inactivation as a possible cause of the breakdown of SI.

4.1.2.3 High Temperatures

Several workers have shown that heat treatment at temperatures ranging from 32 °C to 60 °C leads to a breakdown of SI. The technique has been reported to be applicable to many plant genera with gametophytic SI (GSI), including *Malus* and *Pyrus* (Lewis 1942; Modlibowska 1945), *Oenothera* (Lewis 1942; Hecht 1964; Linskens and Kroh 1967) *Prunus* (Lewis 1942), *Trifolium* (Leffel 1963; Townsend 1965, 1966, 1968, 1971; Kendall and Taylor 1969), *Lilium* (Ascher and Peloquin 1966b; Hopper et al. 1967), *Lycopersicum* (Hoffmann 1966; de Nettancourt et al. 1971; Hogenboom 1972b), *Nemesia* (Campbell and Ascher 1972) and *Secale* (Wricke 1974). SI breakdown after high-temperature treatment was also observed in genera with homomorphic or heteromorphic sporophytic SI (SSI), such as *Brassica* (Gonai and Hinata 1971; Johnson 1972; Okazaki and Hinata 1987), *Raphanus* (Matsubara 1980), *Ipomoea* (Prabha et al. 1982, cited in Hinata et al. 1994) and *Primula* (Lewis 1942). In most cases, and particularly from the work with *Lilium*, *Trifolium*, *Lycopersicum*, *Raphanus* and *Brassica*, it appears that the sensitive site is the pistil, and the sensitive period is the first 2 days following pollination.

In contrast, with *N. alata*, Gray et al. (1991) found that hot-water treatment of S2 ribonuclease (15 min in boiling water and cooled on ice for 5 min before use) prevented its activity but increased its inhibitory effect on in vitro pollen-tube growth. This effect does not result from an increased uptake by the pollen but occurs in association with an enhanced accumulation of S2 ribonuclease on the outer surface of the pollen grains. Earlier work by the same group (Clarke et al. 1989) had already shown that S specificity in the interaction between the heated proteins and pollen-tube growth was lost.

Very little is known regarding the mechanism leading to the temperature effect. Ascher and Peloquin (1966b) and Pandey (1972, 1973) suggested that temperature-induced self-compatibility (SC) results from the inactivation or denanuration of proteins participating in the SI reaction. Okazaki and Hinata (1987) recorded a regular correspondence between the duration of the temperature effect and the reversibility of ultrastructural damage to the plasma membrane of the stigma papilla of *Brassica oleracea*, *B. campestris* and *Raphanus sativus*. They consider the possibility that high temperature does not affect the recognition phase but facilitates, through these damages, the penetration of pollen tubes in incompatible stigmas. Surprisingly, the S glycoproteins isolated from *N. alata* styles become more potent (though non-S-specific) inhibitors of pollen-tube growth in vitro after heat treatment. Jahnen and his co-workers (1989), who observed this temperature effect, suggested that it results from the induction of conformational changes to the S glycoprotein (now known to be a ribonuclease).

Plants expressing temperature-sensitive SI and genes responsible for SI temperature sensitivity have been detected in *Oenothera*, *Trifolium* and *Lycopersicum*. With *O. organensis*, Lewis (1942) provided evidence for linkage between the S locus and a temperature-sensitive gene. In tetraploid alsike clover, Townsend (1968) isolated a clone that is strictly self-incompatible under normal greenhouse conditions but becomes self-compatible at temperatures above 32 °C. The inheritance of this response to temperature is due to a single locus (T) with one dominant allele and one recessive allele. In *Nemesia strumosa* (GSI), plants that share a certain S allele fail to react to heat treament and remain self-incompatible (Campbell and Ascher 1972). Richards and Thurling (1973b) found that the groups of polygenes that influence the action of the S locus in *B. campestris* express different temperature requirements for optimal efficiency.

4.1.3 Application of CO_2

Nakanishii and co-workers (Nakanishii et al. 1969; Nakanishii and Hinata 1973; Nakanishii and Sawano 1989) promoted SC in *Brassica* by applying CO_2. The sensitive stage is the so-called second phase, during which approximately 40% of the pollen germinates and a "foot" (Sect. 3.2.2.1) is established between the pollen and stigma. The most active concentrations of CO_2 ranged between 3% and 5%, which is close to the level at which the formation of callose in the stigmatic papilla is blocked (O'Neil et al. 1984, cited in Hinata et al. 1994). A thickening of the walls of pollen tubes germinated in the presence of CO_2 has been reported by Nakanishii and Sawano (1989) together with an increase in phosphoenol pyruvate carboxylase activity (Dhaliwal et al. 1981). Uncertainties regarding the nature of the CO_2 effect have not prevented its success in the production of inbred seeds (Hinata et al. 1994).

4.1.4 Hormones and Inhibitors

4.1.4.1 α-Naphtalene Acetic Acid and Indole Acetic Acid

Inhibitors of floral abscission, such as α-naphtalene acetic acid (NAA) and indole acetic acid (IAA), when applied to the calyx of the flower enable slow-growing incompatible tubes to reach the ovary before flower dropping. The technique is comparable, in its effects, to the one based on chronic irradiation (Sect. 4.1.2.1) but avoids any associated damage to the genetic structures of the male and female gametes. It has been used successfully on a diversity of plant genera (Eyster 1941; Lewis 1942; Emsweller and Stuart 1948; Martin 1961). Application of NAA to a strictly self-incompatible clone of *L. peruvianum* resulted in a yield of 0.65 seeds per treated flower (de Nettancourt et al. 1971). Henny and Ascher (1973) injected IAA into the styles of *Lilium* and found that the only observable effects (inhibition of pollen-tube growth) were restricted to compatible tubes. This observation is perhaps indicative of basic differences in the metabolic sensitivities of compatible and incompatible tubes and demonstrates that the production of selfed seeds after NAA or IAA treatment results from the inhibition of floral abscission and does not involve any stimulation of incompatible pollen-tube growth. This finding is consistent with the recent conclusion by Lush and Clarke (1997) that the growth of many incompatible pollen tubes is not totally inhibited by S ribonucleases. It is likely that all substances that prevent floral abscission, such as kinetin and morphocitin (Prakash 1975), will also allow slow-growing incompatible pollen tubes to reach the ovary.

4.1.4.2 Effects of Transcription and Translation Inhibitors

Research on the effects of metabolic inhibitors on SI is not only important for the production of inbred seeds by self-incompatible species but also provides essential information regarding the nature of the SI response and the involvement of gene activity in the recognition and rejection phases. Examples of such fundamental work for the understanding of SI are particularly numerous (below and Chap. 3).

For instance, by means of pre-pollination injection of the RNA-synthesis inhibitor 6-methylpurine into detached *Lilium* styles, Ascher (1971) discovered that RNA synthesis is a prerequisite for the establishment of the stylar incompatibility phenotype. Working with *L. peruvianum*, Sarfatti et al. (1974) found that actinomycin D (a transcription inhibitor that acts on RNA-polymerase II) partially inhibits the expression of incompatibility in self-pollinated styles. Similarly, injection of RNA-synthesis inhibitors into *Petunia* buds rendered possible the production of seeds on selfing (Kovaleva et al. 1978). A somewhat comparable situation occurs in *Papaver rhoeas* (stigmatic GSI). Using an adaptation of the Birmingham bioassay (Chap. 3) specially prepared for the identification of events taking place in the pollen, Franklin-Tong et al. (1990) applied Actinomycin D to stigma extracts of known S specificity. These extracts were then "tested" with compatible and incompatible pollen. A majority (60–70%) of the incompatible tubes (which, in the absence of actinomycin would not exceed a mean length of 22.5 μm) were able to reach control length (82.08 μm in H_2O). These in vitro observations confirm and complement the in vivo results of Van der Donk (1974a), who showed that RNA synthesis took place after pollination but were not able to distinguish "new" pollen RNA from "new" stylar RNA in pollinated styles (Sects. 3.3.4.1, 3.5.3.2). The work of Franklin-Tong and co-workers fully demonstrates that pollen-gene expression is induced during the SI response.

The effects of inhibitors of protein synthesis have also been studied in several self-incompatible plants. Ascher (1974) presented evidence that puromycin applied to lily styles enables the growth of incompatible pollen. At an earlier date (Ascher 1971), he had considered the hypothesis that the stylar incompatibility (recognition) substance in *L. longiflorum* was a short-lived RNA and that puromycin treatment, when applied before anthesis, blocked the synthesis of the RNA polymerases. Application of the protein inhibitor cycloheximide to the stigma of *B. oleracea* 2 h before pollination prevented the occurrence of SI and the regulation of pollen hydration, two processes that may involve the same molecular species (Sarker et al. 1988). More recently, Hiscock and Dickinson (1993) reported that application of cycloheximide to the pistils of SI species of the Brassicaceae 2 h before pollination leads to a substantial hydration of incompatible pollen, which overcomes not only SI but also, to a slightly lower extent, unilateral SC×SI inter-specific incompatibility (Chap. 5). Elleman and Dickinson (1996) provided evidence that the inhibition of SI by cycloheximide in *B. oleracea* does not affect callose synthesis (which, as seen in Chap. 3, is probably independent of SI). In *P. rhoeas*, cycloheximide is such a powerful inhibitor of tube growth (but not of germination, a process apparently independent of growth) that its possible effects on the inhibition of SI could not be studied (Franklin-Tong et al. 1990).

4.1.4.3 Effects of Proteinase and Tunicamycin

Pronase is a proteinase known to disrupt the proteinaceous pellicle of the stigma and to alter the integrity of non-specific esterases and glycoproteins contained by the pellicle, these substances possibly participate in recognition events or activate the pollen-held cutinase needed for the penetration of the stigmatic cuticle (Hiscock and Dickinson 1993). Pronase treatments on the stigmas of self-incompatible species from the Brassicaceae prevent pollen tubes from penetrating the stigma after normally compatible intra- or inter-specific pollination. They do not promote the production of selfed seeds or of SC×SI inter-specific hybrid seeds (Chap. 5).

Application of tunicamycin (which prevents the glycosylation of glycoproteins) on *Brassica* stigmas indicated that the glycosyl groups of the S glycoproteins (SLGs) are required for the operation of SI (Sect. 3.2.5.3) but not for the regulation of hydration (Sarker et al. 1988). In *P. rhoeas*, prepollination applications of tunicamycin to stigma extracts considerably alleviated the SI response in vitro and suggested the importance of de novo glycosylation of pollen components for the SI reaction (Sect. 3.3.4.1).

4.1.4.4 Effects of Inhibitors of Protein Phosphatase

Application of okadaic acid to *Brassica* pistils leads to the inhibition of SI, probably through the inhibition of type 1 or type 2A phosphatase (Rundle et al. 1993; Scutt et al. 1993). These observations confirm the importance of the phosphorylation of stigma proteins for the proper functioning of SI in *Brassica* (Sect. 3.2.5.5).

4.1.5 Pistil Grafting

In the course of attempts to identify the site (stigmatic) of the SI reaction in *Oenothera*, Hecht (1960, 1964) developed a technique for assembling various portions of stigmas and styles collected from different plants. Holding such assemblies on a slide in a solution of gelatin and lactose covered by lens paper, Hecht found that pollen tubes compatible with the scion (stigma and stigmatic region of the style) could cross the junction zone and continue to grow in an incompatible stock (basal end of the style) for distances ranging from 22 mm to 55 mm. Hecht also proceeded to graft together different upper parts of pistils (stigmas to stigmatic regions of the style) or to place a stigmatic lobe below the upper region of a style. He was able to demonstrate that growth inhibition is essentially governed by the incompatible stigma and can be partly overcome by removing the incompatible stigma or by replacing it by a compatible stigma. This approach to the modification of incompatibility relationships by pistil grafts does not appear to have been used for producing seeds on self-pollination.

Manipulations somewhat similar to those of Hecht had been tried many years before by Straub (1946, 1947) in an attempt to characterize stylar incompatibility in *P. violaceae*. Straub made grafts of compatible and incompatible stylar portions and found that the amount of growth in incompatible tissue depended on the length of the compatible portion that the tubes had previously traversed. Straub did not consider his method as a possible means of overcoming incompatibility; instead, he saw it as an experimental approach to test a theory regarding the incompatibility reaction, based on the inactivation of growth substances in incompatible tubes (Lewis 1943). Recently, Lush and Clarke (1997) also used style grafting to find out whether the effects of S ribonucleases on pollen-tube growth in *N. alata* are reversible (Fig. 3.20; Sect. 3.5.7.2).

4.1.6 Mutilations, Injections and the Effects of Castration on Pollen-Tube Growth

The work of Hecht suggested that the removal of the stigma in species with stigmatic SI should result in the production of seeds on selfing. The method was successfully tested by Roggen and Van Dijk (1972), who showed that SI could be bypassed by mutilating the stigma with a steel-wire brush before pollination. Whether this result implies that the elimination of the protein pellicle that covers the cuticle constitutes a method for overcoming stigmatic SI is doubtful, because the finding (Heslop-Harrison and Heslop-Harrison 1975) that the removal of the pellicle in the Caryophyllaceae (under conditions which do not kill the papillae) prevents or greatly delays the entry of compatible pollen tubes. More radical approaches to overcoming the GSI barrier in *P. axillaris* were used, with some success, through placenta and "two-site" pollinations in vitro (Rangaswamy and Shivanna 1971, 1972).

With regard to stylar incompatibility, interesting experiments have been performed by Ascher and Drewlow (1971), who filled the stylar canal of excised pistils of *L. longiflorum* with stigmatic exudates from other species and observed that, when injected 24 h before pollination, this promoted the growth of compatible and incompatible tubes but did not modify the relationships of incompatibility and did not prevent growth inhibition from ultimately taking place. Similar conclusions were reached with the same species by Rosen (1971), who failed to detect any concordance between the origin (compatible or incompatible flower) of the stigmatic exudate and pollen-tube growth. Lush et al. (1998) compared the growth of S2 and S6 pollen in exudates from S2S2 plants of *N. alata* and confirmed that stigma exudates have no role in the recognition and rejection of incompatible pollen.

4.1.7 Mentor Effects

Mentor effects, first suggested by Michurin (cited in Stettler 1968) and reported by Glendinning (1960), Grant et al. (1962), Miri and Bubar (1966), Stettler (1968), Opeke and Jacob (1969) and Knox et al. (1972), result from pollination with mixtures of compatible (mentor) and incompatible pollens. Mentor effects have been exploited to overcome inter-specific pre-zygotic barriers (Sect. 5.3.3) and SI. In this last case, the technique was applied, with variable success, to a wide range of different genera, such as *Citrus* (Yamashita et al. 1990), *Cola* (Glendinning (1960), *Crocus* (Chichiricco 1990), *Cucumis* (Den Nijs and Oost 1980), *Helianthus* (Rieseberg et al. 1995; Desrochers and Rieseberg 1998), *Lilium* (Kunishige and Hirata 1978; Vantuyl et al. 1982), *Lotus* (Miri and Bubar 1966), *Malus* and *Pyrus* (Dayton 1974; Visser 1981; Visser and Oost 1982; Visser and Marcucci 1986; Marcucci and Visser 1987), *Nicotiana* (Pandey 1978), *Paspalum* (Burton and Hanna 1992) and *Theobroma* (Falque 1994).

4.1.7.1 Mentor Pollen Is More Efficient when Inactivated or Killed

In order to avoid competition effects and ovule mobilization by the compatible mentor pollen tubes, Stettler (1968) and Knox et al. (1972) used mentor pollen that had been inactivated or killed by high radiation doses or methanol; this considerably increased the efficiency of the technique. Stettler and Guries (1976) improved the application of the method to black cottonwood by delaying the administration of irradiated but viable mentor pollen until incompatible pollen, applied some time before, had a chance to initiate its start. Visser and Verhaegh (1980) modified this "two-step" procedure to take into account the features of incompatibility and incongruity in the Rosaceae (where pollen-tube inhibition occurs in the upper part of the style) and launched what is known as the "pioneer-pollen" method. This method foresees a first pollination with mentor pollen, followed by the application of a mixture of incompatible and mentor pollen 1–2 days later.

4.1.7.2 Nature of the Mentor Effects

Very little is known about such effects, which are variable and not always reproducible. In the sporophytic systems of the Compositae, Knox and co-workers (1972) were able to induce a mentor effect with wall proteins separated from the mentor pollen and added to the incompatible pollen at the time the self-pollinations were made. They concluded that the mentor effect is essentially manifested in sporophytic systems of self- and inter-specific cross-incompatibility and is less likely to occur in the case of stylar GSI, where intine-held proteins are not suspected to participate in massive interactions among different pollen grains (however, see Doughty et al. 1998; Sect. 3.2.4.1). It is hoped that the incompatibility-recognition substances of pollen grains will be identified in the near future and that this identification will improve our understanding of the effects of mentor pollen.

▦ 4.2 Genetic Breakdown of SI and S-Gene Mutations

Genetic breakdown of SI and S-gene mutations occur in nature and can be induced by man (through an arsenal of traditional and modern methods, including site-directed mutagenesis, the use of anti-sense DNA, exploitation of the silencing effects of certain duplications or the swapping of DNA sequences) to mutate, duplicate, transform or silence the S gene(s). It is the purpose of the present section to describe the spectrum of effects that can result from such changes and to emphasize their importance for applied and basic research.

The changes most frequently observed are those that lead to SC. They result from a loss of S function in pollen grains (Sects. 4.21, 4.2.2) and pistils (Sects. 4.2.3, 4.2.4). Several cases of the occurrence of SC through gene mutations outside the S locus (Sect. 4.2.5) or resulting from polyploidy and the production of di-genic hetero-allelic pollen (Sect. 4.2.6) are known. "Constructive" mutations, that do not lead to SC but cause the generation of new S alleles also take place at the S locus (Sect. 4.2.7).

4.2.1 Loss of S Function in Pollen Grains of Species with SSI

SSI is not well suited for the detection of genetic changes induced at the S locus by conventional physical or chemical mutagens, because the pistil sieve, so useful for the mass screening of pollen grains bearing SC mutations (or new S specificities), does not necessarily lead to the recovery of SC pollen-part mutations (PPMs) in the generation immediately following treatment. Recovery problems and/or uncertainties concerning the presence of a mutation in a pollen grain that expresses it depend on the origin [tapetum, tetrads, pollen mother cells (PMCs) or their primordia] of the sporophytic tissue that determines the pollen S phenotype and on the interactions and dominance relationships of the two S alleles present in each cell of the sporophyte. In certain cases of traditional mutation breeding, particularly if S specificity is determined by the tapetum, the search for self-compatible mutants may have to be performed at the level of entire plants one generation after mutagenic treatment, with only little chance of success if the rate of mutation to pollen SC is particularly low.

In cabbage (*B. oleracea* var. *capitata*), Kakizaki (1930) found a self-compatible mutant that produced an offspring segregating in an approximately 1:1 ratio (SC:SI). In *B. campestris*, Zuberi et al. (1981) detected self-compatible plants that displayed SC alleles recessive to wild-type SI alleles in the stigma and dominant (in the pollen) only to the wild-type S alleles classified at the the bottom of the dominance series.

Self-compatible plants were also reported by Bateman (1954) in a wild, self-incompatible population of *Iberis amara* and in *B. oleracea* by Thompson and Taylor (1971) and by Thompson (1972). However, these two authors, basing their views on the hypothesis of S-allele interactions (Sampson 1960), conclude that the SC plants they found in *B. oleracea* do not bear a mutation at the S locus. Instead, they carry different S alleles that produce such a small quantity of incompatibility substances that their association in a same genotype leads to the production of self-compatible (or partially self-compatible) pollen grains. The relationship between two such weak alleles, presumably lacking powerful promoters, can be considered one of competitive interaction or of complementary weakening, because all plants homozygous for any of the two S alleles involved display a normal SI phenotype. Van Mal (cited by Thompson 1972) found similar instances of mutual weakening for several S-allele combinations in Brussels sprouts. However, from surveys made by Howard (1942) and Thompson (1967) in diploid and colchicine-induced tetraploids, it appears that competitive interaction occurs only rarely in the sporophytic system of SI. As shall be seen in Section 4.2.2, the situation is very different in plants with GSI. In the course of their analysis of the properties of SCR, the hypothesized pollen determinant in *B. campestris* (Sect. 3.2.4.4), Schopfer et al. (1999) produced (through loss-of-function approaches with SCR anti-sense DNA) mutant plants that could not express their pollen specificity and were compatible (as males) with untreated material bearing the same initial S alleles.

4.2.2 Loss of S Function in the Pollen of Species with GSI

S-function loss in pollen grains has been extensively studied in the past in mono-factorial GSI systems that are particularly well adapted for mutation analyses and large-scale screening tests. These studies have used the pistil sieve for the selection and transmission of all viable mutations leading to the loss of S function in individual pollen grains to the next generation. Experimental procedures that offer the ability to handle very large quantities of haploid cells and to submit them to a mechanism of self-selection after mutagenic treatment (usually applied to PMCs) are very close to those offered in microbiology. However, under in vivo conditions, the changing and diversified environment of the pistil and the genetic heterogeneity usually typical of the experimental material do not provide the reproducibility and reliability of testing procedures with micro-organisms.

The segregation of S genotypes and S phenotypes among the progenies of plants heterozygous for a deletion or base-pair (BP) mutation leading to the loss or inactivation of one of the two pollen determinants is illustrated in Table 4.1. The far more complex segregation tables established by Van Gastel (1976) for pollen function losses that result from sequence duplications within or outside the S region in *N. alata* can be found in Table 19 and Table 22 of Van Gastel (1976) or Tables 12 and 13 of de Nettancourt (1977). These pollen function losses are described in Sections 4.2.2.1 and 4.2.2.2. Information on spontaneous and induced rates of PPMs and the utilization of pollen-part mutants in plant breeding appears in Sections 4.2.2.3, 4.2.2.4 and 4.2.2.5. As can be seen below, depending on the species considered, four main categories of SC mutations may occur in mono-factorial GSI.

4.2.2.1 Function Loss of the Pollen Determinant Associated with the Presence of a Free Centric Fragment

Function loss of the pollen determinant associated with the presence of a free centric fragment, specific to the Solanaceae, is poorly understood. The different theories (competitive interaction, complementation, restitution) formulated approximately 30 years ago to explain the role of the centric fragment have been presented by Van Gastel (1976) in a detailed analysis of the segregation ratios of S chromosomes and centric fragments in the progenies of different pollen-part mutants submitted to self-pollination. The karyotype of a pollen-part SC mutant of *N. alata* with a centric fragment is presented in Fig. 4.1c.

4.2.2.1.1 Competitive Interaction. Pollen mutations with a centric fragment were first observed in *Petunia inflata* by Brewbaker and Natajaran (1960), who considered that the centric fragment that characterizes self-compatible pollen has an additional S allele different from the S allele present in the haploid complement of the pollen grain. As in hetero-allelic diploid pollen,

Table 4.1. Segregation for S genotypes and S phenotypes among the progenies of stylar and pollen-part mutants submitted to self-pollination and to test-crosses with different accession lines. (de Nettancourt 1977)

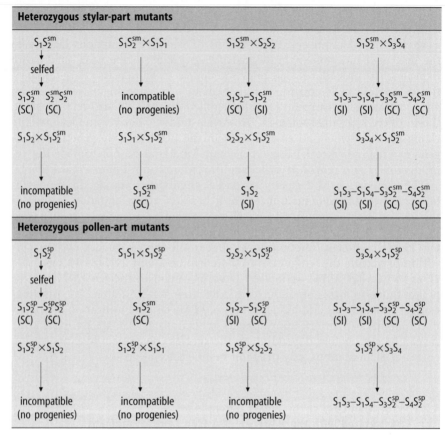

Heterozygous stylar-part mutants

$S_1S_2^{sm}$	$S_1S_2^{sm} \times S_1S_1$	$S_1S_2^{sm} \times S_2S_2$	$S_1S_2^{sm} \times S_3S_4$
↓ selfed ↓	↓	↓	↓
$S_1S_2^{sm}$ $S_2^{sm}S_2^{sm}$ (SC) (SC)	incompatible (no progenies)	$S_1S_2-S_1S_2^{sm}$ (SC) (SI)	$S_1S_3-S_1S_4-S_3S_2^{sm}-S_4S_2^{sm}$ (SI) (SI) (SC) (SC)
$S_1S_2 \times S_1S_2^{sm}$	$S_1S_1 \times S_1S_2^{sm}$	$S_2S_2 \times S_1S_2^{sm}$	$S_3S_4 \times S_1S_2^{sm}$
↓	↓	↓	↓
incompatible (no progenies)	$S_1S_2^{sm}$ (SC)	S_1S_2 (SI)	$S_1S_3-S_1S_4-S_3S_2^{sm}-S_4S_2^{sm}$ (SI) (SI) (SC) (SC)

Heterozygous pollen-art mutants

$S_1S_2^{sp}$	$S_1S_1 \times S_1S_2^{sp}$	$S_2S_2 \times S_1S_2^{sp}$	$S_3S_4 \times S_1S_2^{sp}$
↓ selfed ↓	↓	↓	↓
$S_1S_2^{sp}-S_2^{sp}S_2^{sp}$ (SC) (SC)	$S_1S_2^{sm}$ (SC)	$S_1S_2-S_1S_2^{sp}$ (SI) (SC)	$S_1S_3-S_1S_4-S_3S_2^{sp}-S_4S_2^{sp}$ (SI) (SI) (SC) (SC)
$S_1S_2^{sp} \times S_1S_2$	$S_1S_2^{sp} \times S_1S_1$	$S_1S_2^{sp} \times S_2S_2$	$S_1S_2^{sp} \times S_3S_4$
↓	↓	↓	↓
incompatible (no progenies)	incompatible (no progenies)	incompatible (no progenies)	$S_1S_3-S_1S_4-S_3S_2^{sp}-S_4S_2^{sp}$

the presence of two different S specificities in the pollen grain would establish a state of competitive interaction perhaps analogous to co-suppression in diploid pistils (Sect. 4.2.3.1). This would enable the pollen tube to grow through otherwise incompatible stylar tissue. There would be no competitive interaction in S-homozygous pollen (Sx pollen with a centric fragment also bearing an Sx allele), because promotion, transcription and translation at the S of the fragment and at the S site of the S-bearing chromosome would all contribute to the production of the same Sx protein (Sect. 4.2.6). By extension, on this basis, one could assume that SC mutants that produce SC pollen without centric fragments have integrated a duplication of the S locus on the S-bearing chromosome. This confers SC to the pollen grain each time the two S alleles it carries are different. The competitive-interaction hypothesis also explains why the hetero-allelic diploid pollen produced by most tetraploids is self-compatible (Sect. 4.2.6).

Lewis (1961) challenged the view that competitive interaction should be considered a substitute for the mutation theory for the origin of function loss of the pollen S determinant ("pollen activity part" in the terminology of Lewis). He stressed the fact that involvement of centric fragments in the manifestation of SI was restricted to only a few species of the Solanaceae and could not be demonstrated in genera (such as *Oenothera* or *Prunus*) that also clearly display gametophytic incompatibility and yield PPMs after mutagenic treatment.

4.2.2.1.2 Complementation. Pandey (1965, 1967, 1969) endeavoured to settle the controversy and established that pollen SC mutations with centric fragments occur regularly (~50% of all SC pollen mutants induced in the progenies of irradiated plants of *N. alata*; Van Gastel and de Nettancourt 1975) in *N. alata* and *P. inflata*. He showed, through the identification (by test crosses) of the stylar S genotypes of the pollen mutants, that the pollen with a centric fragment could carry two different S alleles. The discovery by Pandey of a pollen mutant without a centric fragment that displayed three different specificities in the style was taken as proof that S duplications could be incorporated into the pollen genome. However, through the identification of the stylar genotypes, test crosses of certain pollen-part mutants associated with a centric fragment revealed that these plants were S homozygotes and could not produce self-compatible pollen resulting from competitive interaction. On the strength of these observations, Pandey (1967) confirmed the competitive-interaction hypothesis but concluded that, in some cases, the function of the centric fragment could also be the complementation of an otherwise lethal function loss of the pollen S determinant in the pollen genome by a healthy S allele on the fragment. Furthermore, if the S allele on the fragment differs from the complemented allele, pollen compatibility could result from both the complementation effect and the competitive interaction this effect renders possible. If the S allele on the fragment differs from the complemented allele, pollen compatibility could then result from both the complementation effect and competitive interaction. At an earlier date, Pandey (1965) conceded that the fragment did not necessarily complement a "lethal" S allele but alternatively, the damage to essential elements very closely linked to the S locus.

4.2.2.1.3 Restitution. After "competitive interaction" and "complementation", a third explanation of the role of centric fragments, "restitution", has been proposed by de Nettancourt et al. (1975). On the basis of electron-microscopy observations of a circular configuration of the endoplasmic reticulum (CER) in incompatible pollen tubes (Fig. 3.17), they postulated that the main consequence of the SI reaction is a general cessation of protein synthesis. This conclusion was supported at the time by the finding (Dereuddre 1971; Shih and Rappaport 1971) that a CER characterizes the cells of dormant potato tubers and resting *Betula* buds. Shih and Rappaport (1971) also demonstrated that applications of giberellin A3 to potato tubers simultaneously induced metabolic activity and the disappearance of the CER. The hypothesis implies that the fragment originates from a duplication, observed in certain SC pollen-part mutants of *N. alata* (Fig. 4.1 a, b, d), of the satellited region of chromosome 3, which provides an additional nucleolar organizer region to the incompatible pollen grain and enables it to re-establish protein synthesis. The "restitution" hypothesis does not require that the fragment carry an S locus. In fact, it predicts that the fragment does not carry an S locus, because chromosome 3 is not the S-bearing chromosome in *N. alata* (Chap. 2; de Nettancourt et al. 1975). The "restitution" hypothesis does not explain the origin of a centromere on the fragment.

The "restitution" theory receives some support from the discoveries that the stylar S protein in the Solanaceae is a ribonuclease (Sect. 3.5.5) and that the growth inhibition of incompatible pollen tubes is not necessarily irreversible (Sects. 3.5.7.2, 3.5.7.3). It could perhaps be tested by analyses of the morphology and evolution of the endoplasmic reticulum in pollen tubes with centric fragments or through the precise identification and characterization of the origin and nature of centric fragments.

4.2.2.1.4 Likelihood of the Three Hypotheses. For each of the three postulated causes of SC, Van Gastel (1976) predicted the S genotypes and segregation ratios of the offsprings obtained through selfing and different test crosses from self-compatible pollen mutants bearing a centric fragment. A comparison between expectations and observations revealed that, of the three explanations, "complementation" was by far the most likely in a majority of cases, and "restitution" was the least probable. Van Gastel also confirmed the report by Pandey that at least some of the centric fragments identified in SC pollen mutants carry an S allele.

4.2.2.1.5 Origin of the Centric Fragment in "Pollen-Part" SC Mutants of *N. alata*. Centric fragments that carry the S locus certainly originate from the longest un-satellited chromosome, as it was indirectly identified by Pandey (1965, 1967) and van Gastel (1976) and formally localized by Labroche et al. (1983). Fragments not carrying an S allele may originate from chromosome 3, as seen above (Fig. 4.1).

4.2.2.1.6 Current Approaches to the Biomolecular Study of PPMs Associated with Additional Chromosomal Material. The very first efforts in this direction were those of Thompson et al. (1991), who tried, with S-RNase gene probes and gels of stylar proteins, to find evidence of the duplication of the S locus in pollen-part mutants of di-haploid lines of *Solanum tuberosum*. They concluded that either a modifier locus unlinked to S was involved in the mutation or that S duplications, if they were responsible for PPMs, covered only part of the S locus. More recently, Golz et al. (1999 a, 1999 b) showed, through a combination of DNA blot analyses and cytogenetic research, that most PPMs (8 out of 11 analyzed) induced by γ-rays in *N. alata* are associated with the pres-

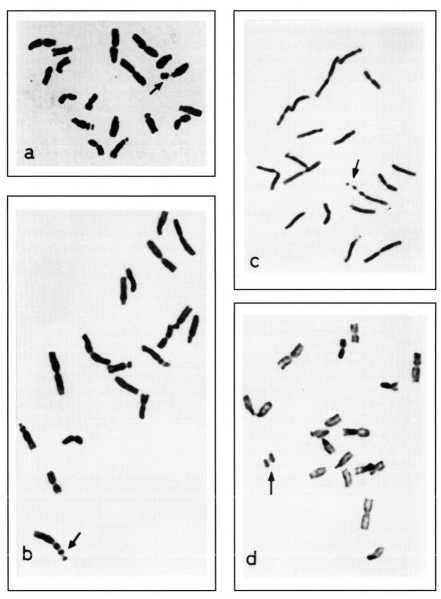

Fig. 4.1. Karyotypic modifications in a phenotypically unstable leaf-propagated clone of *Nicotiana alata* and some of its self-compatible progenies. **a–c** What appeared to be duplications of satellited regions of chromosome 3 were observed in all members of the leaf-propagated clone OL S2S3 and in two self-compatible offsprings (a "pollen-part" mutant and a "stylar-part" mutant) found in advanced progenies of the clone (de Nettancourt et al. 1975). **d** A centric fragment in a pollen-part mutant detected in the advanced progenies of the same clone. (Carluccio et al. 1974)

ence of an additional chromosome. In four of seven mutants, the PPM was coupled to an extra S allele (which, in three of these plants, resided on the additional chromosome). The pollination behavior of the pollen-part mutant was consistent with the hypothesis that their SC character results from competitive interaction between pollen-part S sequences closely associated with the S-RNase gene or forming part of it. However, in the (rare) cases of pollen-part mutants which did not display a complete additional S allele, it is presumed (Sect. 4.2.2.2) that either complementation or restitution could be involved. Golz and co-workers noted, furthermore, that the phenotype of PPM resulting from a duplicated S allele can be explained by either of the two models proposed as a molecular basis for SI in the Solanaceae (Figs. 3.21, 3.22). The predictions on the mutability of the pollen S determinant are different for each model. According to the receptor model, PPM can arise from either deletions or duplications of an S allele (Golz et al. 1999a, 1999b); in the inhibitor model, they can only result from S duplications. The data currently available, particularly that of Pandey (1967) and Van Gastel (1976), indicate that both interaction and complementation occur and support the receptor model (Sect. 4.2.2.2).

4.2.2.2 Function Loss of the Pollen Determinant Not Associated with the Presence of a Centric Fragment

In many species with GSI – for instance, *Prunus avium, Trifolium pratense, N. alata* or *Petunia hybrida* in the case of mono-factorial stylar GSI or *Oenothera organensis* (mono-factorial SI) and *Phalaris coerulescens* (bi-factorial SI) for stigmatic GSI – function losses of the pollen determinant that are not associated with a centric fragment can be induced. The possibility that these mutations consist of an integrated S duplication that presumably leads to competitive interaction between two different S alleles has been verified in *N. alata* by Pandey (1965, 1967) and Golz et al. (1999a).

With radiation and chemicals, Van Gastel (1976) obtained a considerable number of pollen mutants of *N. alata* without a centric fragment; he was thus able, for this species, to accurately compare the expected and observed segregation ratio for SC pollen mutations "without a fragment". He did this on the basis of the various hypotheses (competitive interaction, complementation, restitution, deletion/inactivation) used to explain the origin of the mutation. Again, as in the case of S mutations "with centric fragments", the comparison (Van Gastel 1976) between the expected values and the observed segregation ratios indicated that the most frequent mechanism (possibly the only one) through which a pollen mutant "without a centric fragment" expresses the self-compatible character is "complementation". To summarize the views of Pandey, complementation is the intervention (in the pollen grain) of an additional pollen determinant, whether or not it is integrated within the S-bearing chromosome. This intervention restores certain vital functions damaged within or beside the S locus at the time an SC mutation of the pollen determinant was induced.

The work of Van Gastel was confined to *N. alata*, and there is no evidence that his conclusions apply to species outside the Solanaceae. However, it is interesting to recall the finding by Lewis (1960) that the function-loss of a pollen S determinant was restored, and the original S specificity reinstated, in diploid pollen known to carry a functional pollen determinant ("pollen activity part"). In this case, complementation did not lead to the rescue of an otherwise lethal association of lesions and the expression of SC but to a reversion towards wild-type activity of a mutated S allele.

In parallel with research to understand the origin and molecular basis (presumably, in most cases, some form of co-suppression) of "pollen-part" SC mutants with and without centric fragments, related efforts should also be carried out to explain the SC character of the diploid hetero-allelic pollen grains produced by the tetraploid relatives of most SI species (Sect. 4.2.6). The work would be greatly facilitated by the identification of the pollen S gene in the Solanaceae.

4.2.2.3 The Frequency of S Mutations Leading to the Loss of SI Function in Pollen Grains

Since there is no screening system available for the large-scale scoring of S mutations in sporophytic systems or of SC-pistil mutants, the only data sufficiently important for a reliable assessment of the frequency of S function-loss mutation are those that originate from the detection of SC-pollen mutants in gametophytic systems. The scoring procedures are often difficult despite the availability of the attractive pistil sieve, which renders the screening of very large populations of pollen grains and the selective transfer of SC pollen mutations to the following generation possible in GSI species. Germinating pollen grains and growing pollen tubes are particularly sensitive to environmental changes and to discreet variations in experimental conditions. Interference from revertible mutations, pseudo-compatibility and occasional diploid-compatible pollen (de Nettancourt 1969, 1972) may modify the research results, particularly if the endpoint is pollen-tube growth to the ovary rather than seed set and the testing of putative SC mutants in the subsequent generation. Many factors need to be considered and controlled, including the age of the flowers, stigma size and receptivity, the moisture content of the pollen, the number of grains applied per stigma, and the percentages of pollen grains and pollen tubes that germinate and reach the incompatibility region of the pistil. Furthermore, in many species (such as *N. alata* or *L. peruvianum*; de Nettancourt 1972), floral abscission after incompatible pollination practically rules out the chance that an isolated compatible pollen tube will reach the ovary, unless special hormones (α-naphtalene acetic acid, for instance) have been applied around the calyx to avoid flower dropping.

Nevertheless, reliable data regarding mutation rates have been obtained in several laboratories. Analyses conducted by Lewis (1948, 1949b), Pandey

(1967), Van Gastel and de Nettancourt (1974, 1975) suggest that the spontaneous mutation rate ranges from 1.7 per million pollen grains to 4.3 per million pollen grains in different clones of *Oenothera*, from 0.2 per million pollen grains to 2.3 per million pollen grains in *P. avium* and from 0.2 per million pollen grains to 0.4 per million pollen grains in *N. alata*. Mutagenic treatments (the data available essentially originate from work with ionizing radiations) applied to PMCs undergoing meiosis [the ultimate stage for induced mutations to be expressed at the following pollen generation, according to Lewis (1949b)] strongly increase such mutation rates. After acute treatment with X-rays at dose rates exceeding 20 rad/min, the mutation rate in *N. alata* can reach an upper limit of approximately 15 times the spontaneous frequency (Pandey 1967; Van Gastel and de Nettancourt 1975). The ratio of "fragment" to "no-fragment" S-function loss in the pollen approximates 50:50; this could indicate that similar types of lesions are involved in the two classes of mutations. Fast neutrons are more effective than X-rays, and their relative biological efficiency, although it varies with dose, is between 2 and 4. For *O. organensis*, Lewis (1956) calculated (from data obtained at high doses) that the contribution of every Roentgen of X-radiation to the induced frequency of stable PPMs almost corresponds to the spontaneous mutation rate. This means that an approximately 100-fold increase in the spontaneous rate of transmissible mutations can be induced by X-rays on the "pollen-activity part" of the S locus in *O. organensis* if the dose–effect relationship is linear.

Mutation rates are much lower in the case of chronic irradiation. In *N. alata*, chronic exposure to γ-rays at dose rates not exceeding 14 rad/h appears to be almost ineffective, possibly because repair processes are particularly active during continuous exposure at relatively low dose rates.

4.2.2.4 Production of Cultivars with Modified Breeding Regimes: Examples of Traditional and Molecular Approaches

4.2.2.4.1 Cherry Stella. This "pollen-only" self-compatible sweet-cherry cultivar originated from a cross between cultivar "Lambert" and John Innes seedling 2420, which itself was selected from the progeny of "Emperor Francis", pollinated by Professor D. Lewis with irradiated pollen of "Bigarreau Napoleon". "The cultivar Cherry Stella was raised by Dr K. Lapins of the Canada Department of Agriculture at Summerland, British Columbia. The fruits are large, heart shaped, black skinned...; the tree is fairly upright and vigorous..." (Brown 1983).

4.2.2.4.2 Elstar. Self-compatible lines of apples that express low S-ribonuclease levels were obtained by Van Nerum et al. (1999) after transformation with either sense or an anti-sense copies of one of the endogenous S alleles. An expert review of SC in apple and almond was prepared by Feijó et al. (1999).

4.2.2.5 The Use of "Pollen-Part" Mutations for the Production of F1 Hybrid Seed

A simple scheme for the production of hybrid seed through the use of induced self-compatible S-heterozygous "pollen-part" mutations at the S locus has been proposed in the past (Van Gastel and de Nettancourt 1975). Such plants, as shown in Fig. 4.2, need to be selfed repeatedly during several generations and their progenies sorted for S heterozygosity at each generation until the desired number of inbreeding generations has been reached. Selected S heterozygotes are then submitted to a last generation of selfing (through bud pollination), and their self-incompatible homozygous progenies are screened for inbred lines with a superior ability to combine with other lines homozygous for a different S allele. Of course, modern techniques for site-directed mutagenesis, gene insertion or gene silencing (Sect. 4.2.4.2) should be used (rather than fast neutrons in the Van Gastel/ de Nettancourt scheme) as soon as the pollen S determinant is identified in the SI crop species to be involved in hybrid seed production. If the SC character originates from a competition effect mediated by a centric fragment (as it often does in the Solanaceae after the exposure of PMCs to fast neutrons), the inbred progenies will consist of self-compatible S heterozygotes only (Van Gastel and de Nettancourt 1975). The inbreeding procedure is facilitated.

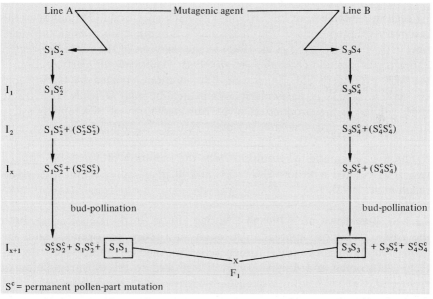

Fig. 4.2. Inbreeding scheme of self-compatible S heterozygotes and the production of self-compatible S homozygotes for the constitution of hybrid seeds. An Sc permanent mutation conferring compatibility to the pollen that carries it. If the self-compatibility character originates from a competition effect mediated by a free centric fragment, the inbred progenies will consist of self-compatible S heterozygotes only. (Van Gastel and de Nettancourt 1975)

The method could perhaps be extended to other families (see Wexelsen 1945 and Rowlands 1958 for early views on heterosis in the Poaceae) and systems other than mono-factorial GSI, such as homorphic SSI. It could be used particularly in the Brassicaceae, where wild-type S alleles have been utilized for F1 hybrid seed production (Odland and Noll 1950; Thompson 1957; Wiering 1958; Reimann-Phillipp 1965; Denna 1971; Frankel 1973; England 1974) and where modern techniques of genetic engineering are available. The causes of variations in the proportion of selfed seeds present among F1-hybrid seed lots have been established in the case of Brussels sprouts (Johnson 1972).

Possibilities also appear promising in the case of grasses. England (1974) and Wricke (1984) discussed potential applications and established the theory for F1 seed production with a two-locus gametophytic system. Hayman and Richter, (1992) working with *P. coerulescens*, assembled more than 150 SC mutants at the S or Z locus; their collection includes at least 16 PPMs.

4.2.3 Loss of S Function in the Stigmas of Species with SSI

4.2.3.1 Utility of SC Stylar Mutations and of Silencing Studies for the Understanding and Exploitation of SI

As seen in Chapter 3, in stigmatic SC *Brassica* species and SC mutants, the discovery of alterations of the SRK gene – ranging from a single BP deletion in *B. napus* (Goring et al. 1993) to the deletion of the promoter region and receptor domain (Nasrallah et al. 1994) – confirms the participation of the SRK protein in SSI. The single BP mutation reported by Goring and co-workers (1993) is particularly interesting, because it leads to a frame shift, results in the production of a truncated form of SRK and possibly originates from unequal crossing-over. Not all SC mutants, however, display damage to the SRK protein. A case has been reported where the manifestation of stigmatic SC resulting from the mutation of a gene not linked to S is associated with low SLG levels and normal expression of SRK (Nasrallah et al. 1992).

4.2.3.1.1 Breakdown of SI though Silencing Effects. In the Brassicaceae, both the loss-of-function approach (through anti-sense transgenes) and the gain-of-function approach (through sense transgenes) induce silencing effects (Baulcombe and English 1996). These lead to the downregulation of endogenous genes having sequence similarities with the S transgene. This last phenomenon was observed in *Brassica* after the transfer of wild and mutant forms of the SRK gene that induce, through co-suppression, the breakdown of SI in the stigma in SI plants (Stein et al. 1991; Conner et al. 1997; Stahl et al. 1998). Although the induction of S-gene silencing and SI by sense SRK transgenes is often considered a nuisance, it can serve useful purposes in certain forms of basic or applied research. The procedure involves, as in the case of the method outlined in Section 4.2.4.2 for the transfer of anti-sense DNA in GSI systems, the mediation of *Agrobacterium tumefaciens*, the use of powerful promoters and the utilization of leaf disks for transformation and plant regeneration.

Similar "silencing" damage to SI expression was also observed after transfer of the SLG gene. For instance, Toryama et al. (1991b) reported the induction of SC in *B. oleracea* through a co-suppression effect resulting from the introduction of a construct carrying a functional SLG gene from *B. campestris*. The results, complicated by inter-specific SI, suggested that S specificity may have been changed in the pistil of one recipient strain and in the pollen of another. Conner et al. (1997), in their analysis of the effects of S-gene silencing on the expression of SI in *Brassica*, observed that transformants harboring wild or mutated forms of the SLG gene also failed to express stigmatic SI. Interestingly, such "nuisance" results appear to confirm the involvement of SLG in SI (however, see other considerations on the role of SLG in section 4.2.3.1.3).

4.2.3.1.2 Why Gene Silencing Occurs. As defined by Vaucheret et al. (1998), transgene-induced gene silencing (TGS) by sense DNA essentially occurs when multiple repeats of a transgene are inserted in the genome of the transgenic plant. It can take place at transcriptional or post-transcriptional level and involves either the expression of transgenes at a single locus (cis-inactivation) or the silencing effect of one locus on another (trans-inactivation). Post-transcriptional trans-inactivation is better kown as co-suppression, which Vaucheret and co-workers define as "a reciprocal and synergistic phenomenon where host genes and transgenes can cooperate to produce an aberrant RNA and/or cDNA above the level that activates the RNA degradation machinery". TGS has been attributed to different effects, such as the disruption of transcription by methylation of promoter sequences, changes in the structure of chromatin flanking the silenced gene or, as seen in the case of co-suppression, alterations of post-transcriptional events (Flavell 1994; Conner et al. 1997; Vaucheret et al. 1998; Jakowitsch et al. 1999). Whether or not competitive interaction between S alleles, such as has been suggested to occur in heterogenic di-allelic pollen grains (Sect. 4.2.2.1), can accompany a case of co-suppression (as defined above) is not known.

4.2.3.1.3 Scientific Interest of S-Gene Silencing. The observations on the consequences of silencing SRK and SLG demonstrate the importance of the SLG/SRK tandem in SI. In particular, they confirm the observations (reported above; Nasrallah et al. 1992) that SLG levels are low in self-compatible mutants of *B. campestris*, and the finding (Shiba et al. 1995) that an anti-sense SLG construct leading to the suppression of SLG in *B. napus* induced the occurrence of SC. They do not, however, necessarily contradict the conclusion by Gaude and co-workers (1993, 1995) that SLG, found in high quantities in SC mutants and at low levels in certain SI individuals, is not an essential instrument of the SI reaction. SLG transcripts are usually so closely related to SRK transcripts from the same haplotype that SLG transgenes may be able to suppress both endogenous SLG and endogenous SRK expression (Franklin et al. 1996). In this case, the breakdown of stigmatic SI would result not from the silencing of the SLG gene but from the suppression of SRK. Because a mutant SRK transgene silences not only endogenous SRK and SLG but also SLR 1 (Stahl et al. 1998), it is possible that the entire S-gene family is not protected against collective suppression (Flavell 1994). However, the fact that the silencing of the stigmatic S genes has no effect on the expression of the pollen S phenotype in the sporophytic system of *Brassica* is additional evidence (Chap. 3) that SRK and SLG do not contribute to the specification of the pollen determinant.

4.2.3.1.4 The Importance of Specific S-Function Losses for Basic and Applied Research. One of many examples (Nasrallah et al. 1991; Hinata et al. 1994; Sect. 4.2.4 for the situation in GSI systems) of the utilization of function-loss mutations for the identification of S-stigmatic genes involved in SSI is provided by Franklin et al. (1996). They used an anti-sense construct to ablate the SLR1 gene in self-incompatible *B. oleracea* and self-compatible *B. napus*. They showed that the absence of the SLR1 protein did not affect the usual outcome of compatible and incompatible pollinations. In the field of plant breeding, the relative facility with which stable function-loss stigmatic mutations can now be obtained in the Brassicaceae may create interest in the creation of new autogamous cultivars or in hybrid seed production. Mutants as simple as the single BP-deletion frame shift observed in *B. napus* by Goring et al. in 1993 (above) may be used for the production of parental pure lines needed for hybrid production if their inbreds can easily be reverted to the wild type.

4.2.4 Loss and Gain of S-Function Approaches in the Stigma or Style of Species with GSI

4.2.4.1 SC Mutants Arising Spontaneously or from Conventional Mutagenic Treatment

Clear-cut cases of stylar-part mutations leading to the non-functioning of a specific S allele are rare but have been reported by some authors in monofactorial *Oenothera* (Lewis and Crowe 1954), *Nicotiana* (Van Gastel 1976), *Trifolium* (Pandey 1956a, 1956b), *Lycopersicum* (Royo et al. 1994) and *Pyrus* (Sassa et al. 1997) and in bi-factorial grasses (Hayman and Richter 1992). The great majority of these stigmatic or stylar mutants produce pollen that maintains and expresses an active S determinant. Of the 508 possible mutants affecting SI which were isolated by Hayman and Richter (1992) in *P. coerulescens*, 152 were self-compatible. In a sample of 37 such mutants, 16 (including pollen-only mutants and one complete mutant) were mutated at the S locus, six were pollen-only mutants at the Z locus and 16 were mutants at another locus (T).

These mutations, spontaneous or produced with conventional mutagens, contributed to current knowledge regarding the structure and function of the S locus. It was, for instance, through the study of an "S-complete" mutant from the Hayman-Richter collection that Li et al. (1996) were able to suggest a possible role for an S-linked thioredoxin gene in the SI reaction of *Phalaris* (Sect. 3.4.2). Similarly, the fact that an entire deletion (>4 kb) of the chromosome region bearing the stigmatic S determinant in Japanese pear (*P. serotina*) does not affect S function in the pollen led to the conclusion that different genes (linked, but not allelic) control S specificity in pollen and pistil (Sassa et al. 1997). Furthermore, it was the accessions of Professor Rick at the University of California that originated the self-compatible form of *L. peruvianum* (LA2157), which enabled Royo et al. (1994) to attribute the SC character to S proteins lacking ribonuclease activity and to confirm the evidence that active S ribonucleases are essential for the expression of GSI.

4.2.4.2 Genetic Constructs and Ablations of S-Gene Products Leading to SC and their Importance for SI Research

4.2.4.2.1 How to Induce Function Loss Through the Use of Anti-Sense DNA. It is possible to utilize messenger RNA (mRNA) technology to prevent the expression of a gene. In short, the method consists of constructing a chimeric gene containing anti-sense DNA (complementary to the DNA of the gene, or part of its sequence that one wants to silence), and a powerful promoter. The promoter/complementary-DNA fusion product is then introduced, through gene transfer procedures (usually involving *Agrobacterium*-mediated transformation, in the case of SI research), into the material to be transformed (for instance, leaf disks regenerated shoots from which will then be used to establish a population of transformed plants). In the tissue of these transformants, where the gene under analysis should normally function, the anti-sense mRNA produced by the inserted construct hybridizes with the sense mRNA originating from the target gene and prevents translation. Because *N. alata* regenerates poorly in tissue cultures and lacks promoter regions working efficiently in transgenic plants, Murfett et al. (1995) constructed a transgenic hybrid between *N. plumbaginifolia* and *N. alata*. This hybrid is not only vigorous and produces large flowers, it is also easily transformable.

In a series of transformation experiments with this hybrid, Murfett and McClure (1998) were able to cause suppression of the activity of an S allele from *N. alata* with three different gene constructs (two driven by RNA-polymerase-II-transcribed promoters and a third, driven by RNA polymerase III, containing a truncated soybean transfer-RNA met-i gene). They concluded that the cauliflower mosaic virus 35S promoter was more effective for anti-sense suppression than the tissue-specific tomato ChiP promoter. Nevertheless, it was with a hybrid between *N. langsdorfii* and *N. alata* as a transgene recipient and the ChiP promoter that Murfett et al. (1994) showed for the first time that the S proteins of *N. alata* were responsible for the S-specific recognition and rejection of self-pollen. For further information on anti-sense gene suppression and a detailed description of techniques, see also Xoconostle et al. (1993), Lee et al. (1994) and Murfett et al. (1994, 1995).

4.2.4.2.2 Loss of Function and Gain of Function Approaches Are Complementary. Anti-sense genetic constructs were also exploited by Lee et al. (1994 ; Sect. 3.5.6; Table 1) and Murfett et al. (1994; Sect. 3.5.6.2) to demonstrate (through loss-of-function approaches confirmed by evidence from gain-of-function experiments) that S proteins control the rejection of incompatible pollen in *Petunia inflata* (Sect. 3.5.6.1; Table 4.2). Furthermore, it was through a gain-of-function approach, and not with anti-sense genetic constructs, that rigorous proof that ribonucleases participate in the rejection of pollen tubes in species with GSI was provided during the same year by Huang et al. (1994). They showed (Sect. 3.5.6.3; Table 4.2) that S1S2 styles acquired the capacity to produce S3 ribonucleases and to reject S3 pollen after transformation with a wild-type S3 allele but failed to express such properties if transformed with a mutant S3 allele that produced a mutant protein with no detectable ribonuclease activity.

Another example (among many others) of the use of genetic engineering to assess S structures and S functions was provided by Karunanandaa et al. (1994). They constructed a non-glycosylated S3 mutant protein and, through the fact that the mutant form remained functional, demonstrated that the carbohydrate moiety of the protein is not required for SI interactions between the pollen and pistil (Sect. 3.5.7.5).

Table 4.2. Induction of loss and gain of function in self-incompatible plants as a tool for demonstrating the involvement of S-specific proteins (ribonucleases) in mono-factorial stylar gametophytic self-incompatibility

S genotype engineered	S proteins in the style	Pollen rejected	Source
S1S2+S1, S2 anti-sense DNA	None	None	Lee et al. (1994)
S1S2+S3 allele	S1, S2, S3 proteins	S1, S2, S3	Lee et al. (1994)
S1S2+S3 allele	S1, S2, S3 ribonucleases	S1, S2, S3	Huang et al. (1994)
S1S2+mutated S3 allele[a]	S1, S2 and modified S3 ribonucleases	S1, S2	Huang et al. (1994)
S2S3+mutated S3 allele[a]	S2 and modified S3 ribonucleases	S2	McCubbin et al. (1997)

[a] Replacement of His93 with Arg93

4.2.4.2.3 Competition Effects Occuring in the Styles of *Petunia* Plants with a Tri-Allelic S2–S3–S3* Genotype.

In *P. inflata*, McCubbin et al. (1997) engineered a mutant characterized by a normal S2S3 genotype to which was added a modified S3* gene encoding an S3 ribonuclease lacking ribonuclease activity (Table 4.2). The mutation results from the single amino acid (AA) replacement His-93 to Arg-93, which is not likely to affect the S3 recognition function of the allele unless folding configurations have been modified by only one substitution. McCubbin and his co-workers found that the mutant could not reject S3 pollen and attributed this incapacity to competition between the wild-type S3 allele and S3* (for binding to a common molecule, presumably the product of the pollen S3 allele).

4.2.4.3 Presence and Expression of S Ribonucleases in Self-Compatible Lines

It shall be seen in Section 4.2.5 that autogamy derives, in many instances, from SI. An illustration of such conversion can be found in a population (LA2157) of *L. peruvianum*, normally a very strict SI species, discovered in northern Peru (in the province of Cajamarca) by Rick (1982); this species splits into SI and SC individuals. In SC individuals, analyses revealed the presence of a protein associated with an SC allele of the S locus; it co-segregates in the progenies of SC×SI plants with wild-type S alleles (Rivers and Bernatzky 1994). This protein was identified as an inactive S ribonuclease (Kowyama et al. 1994) lacking a histidine residue (replaced by arginine at position 33) at its active site (Royo et al. 1994). As neither the self-incompatible founders of LA2157 nor the functional S allele from which the SC allele is derived are known, the possibility that changes other than the substitution of histidine were responsible for the loss of S function cannot be excluded. In the absence of knowledge regarding the specificity of the parental allele, it is also difficult to establish with certainty whether SC results in LA2157 from the production of an inactive ribonuclease (by far the most obvious explanation), from a PPM (which could suggest an involvement of S-RNases in pollen S specificity) or from a combination of the two. A situation somewhat analogous occurs with two forms of *L. hirsutum*: one form (*hirsutum*) is self-incompatible, and a second form (*glabratum*) is self-compatible. There is, however, no evidence for a product of an S gene in form *glabratum*, although the analysis of F2 progenies from the self-fertile hybrid *hirsutum*×*glabratum* revealed the presence of some self-incompatible individuals (Rick and Chetelat 1991; Bernatsky and Miller 1994).

Another example of the derived condition of SC has been provided by Ai et al. (1991), who screened the progenies of an SC cultivar of *P. inflata* and an S2S2 SI plant and discovered defective (So) and functional (Sx) S alleles that appear to have been transmitted by the SC parent. In self-compatible *Nicotiana sylvestris*, a relic S ribonuclease was found to be expressed in styles and could be purified (Golz et al. 1998). The protein is biochemically similar to *N. alata* S ribonuclease, except that an N-glycan chain in the *N. sylvestris* ribonuclease is attached to one of the hyper-variable (HV) domains. Golz and co-workers propose transgenic approaches for testing the hypothesis that the presence of an N-glycan chain in this region blocks an interaction between the pollen determinant and the HV domain and prevents *N. sylvestris* styles from rejecting self pollen.

4.2.4.4 Loss and Gain of Function in the Pollen of Plants with Mono-Factorial Stylar GSI

A summary of the work of Dodds et al. (1999), which tested the hypothesis that S ribonucleases determine the SI phenotype of pollen grains in *L. peruvianum*, is presented in Section 3.5.8.2. A short outline of the strategy followed in this loss-of-function approach is provided in Table 4.3.

4.2.5 SC through Genetic Changes Occurring Outside the S Locus

Clark and Kao (1994), Mather (1943) and Martin (1961, 1967, 1968) were among the first to demonstrate that the introgression of functional S alleles from SI species into SC species is not sufficient to confer SI. Other genetic material, presumably in a homozygous condition in the SI species, contributes to the expression of SI and must also be transferred with the S locus.

Table 4.3. Loss- and gain-of-function approaches for testing the hypothesis that S ribonucleases determine the incompatibility phenotype in pollen grains of *Lycopersicum peruvianum*. (An incomplete summary from some of the work by Dodds et al. 1999)

S genotype engineered	Self-incompatible phenotypes of pollen grains produced by transformed plants	
	Expected if pollen-expressed S-RNase is involved in SI[a]	Observed
S1S3+S3 anti-sense DNA	S1[b] and Sc	Unchanged (S1 and S3)[d]
S2S3+S3 anti-sense DNA	S2[b] and Sc	Unchanged (S2 and S3)[d]
S1S2+sense S3 DNA	S1, S2 and Sc[c]	Unchanged (S1 and S2)

[a] *SI*, self-incompatibility

[b] Unless S3 anti-sense is also able to suppress the expression of S1 and/or S2 (when the sequence homology is approximately 80% or more)

[c] Resulting from the co-operative interaction of different S alleles in hetero-allelic pollen

[d] Actually, one plant (out of 28 that expressed the anti-sense S3 DNA transcript) was self-compatible. However, there is no apparent relationship between the Sc character expressed in the pollen and the anti-sense construct, because self-compatibility (SC) and the trans-gene were inherited independently in the selfed progeny of the "pollen-only" SC mutant

4.2.5.1 In Sporophytic Systems

The weakening of incompatibility, through selection for SC in *Primula* (Mather and De Winton 1941) or in buckwheat by radiation (Sharma and Boyes 1961), has been attributed to the effects of mutation or recombination on a polygenic complex assumed to balance the whole incompatibility system in heteromorphic species (Chap. 3). A similar reduction in the intensity of SI in the breeding behavior of progenies from irradiated SC and SI forms of *B. oleracea* has also been explained by changes at a modifier complex rather than mutations at the S locus (Nasrallah and Wallace 1968; Rao 1970). Such a complex may be analogous to a group of polygenes reported (Richards and Thurling 1973 b) to regulate the SI reaction in *B. campestris*. They measured, for different temperatures, the responses to one generation of selection for SC and found that the expression of SC among the progenies of self-incompatible lines is primarily determined by interactions among S genotypes and different combinations of polygenic modifiers. These modifiers are temperature sensitive and probably differ from one another in their temperature requirements.

Major genes outside the S locus also participate in the control of SI, either in the preparation of the recognition phase or downstream, during the rejection process [Thompson and Taylor (1966, 1971) in cabbage and kale; Ronald and Ascher (1975) in *Chrysanthenum morifolium*; Martin (1973) in sweet potatoes]. A di-allelic M locus, not linked to S, has been observed (Hinata et al. 1995) to segregate in the progeny of a cross between SI *B. campestris* and a self-compatible cultivar. Plants with *mm* genotypes were self-compatible. In the same species, Nasrallah et al. (1992) identified a gene, SCF1, that regulates S-locus function through the encoding of a "transacting" factor necessary for the active transcription of several genes (including SLG, SLR1 and SLR2) belonging to the S-gene family. Homozygous recessive *scf1–scf1* plants express stigmatic SC and a down-regulation (at the RNA level) of SLG and SLR expression. The fact that SRK is not affected by this spontaneous mutation appears to suggest an involvement of SLG in SI or the occurrence of essential but undetected damage to certain recognition or rejection components of the SI reaction other than SRK.

In the Brassicaceae, there are several other examples of major genes, unlinked to the S locus, which contribute to the functioning of the SI mechanism at one stage or another and lead, when deleted or inactivated by a mutation, to the modification or breakdown of the system (Nasrallah 1974; Hinata and Okazaki 1986; Nasrallah and Nasrallah 1989; Hinata et al. 1993). A clear case of such involvement, typical of post-recognitition events, is illustrated by the MOD gene, which is epistatic to the S locus and encodes a protein related to membrane channels (aquaporins) regulating the transport of water through biological membranes (Ikeda et al. 1997). Plants of *B. campestris* homozygous for a recessive *mod* mutation do not encode the aquaporin protein and are self-compatible (Ikeda et al.

1997; Nasrallah 1997; Chap. 3). Another illustration of the involvement of genes outside the S locus in the rejection phase of SI was provided by Stone et al. (1999), who induced the breakdown of SI in *Brassica* through the construction of transgenic plants bearing the anti-sense mRNA against ARC1, a putative downstream effector of the SRK gene (Sects. 3.2.3.7, 3.2.5.5).

4.2.5.2 In Gametophytic Systems

As in sporophytic systems, SI has been reported to cease after the occurrence of mutations at sites other than those occupied by SI genes or in certain inbred plants derived from obligate selfing. In both cases, the phenomenon can be attributed to modifications of the genetic environment necessary for the proper recognition and rejection of self-pollen. As shall be seen in Chapter 5, in many cases, it emphasizes the difficulty of distinguishing the barriers of inter-specific incompatibility that stimulate speciation from those leading to an incongruous relationship as a result of isolation and speciation.

4.2.5.2.1 Mutations of Major Genes. A clear example of SI breakdown resulting from mutations outside the S locus has been reported by Townsend (1969). In diploid alsike clover, he observed the allele (A1) of a locus (A) that suppresses the action of a number of different S alleles when in heterozygous condition (A1A2) and in certain genetic backgrounds. Similarly, in tomatoes, Martin (1968) discovered the presence of a dominant switch gene necessary for the expression of SI (Bernatzky et al. 1995; Chap. 5). Ai et al. (1991) reported that the expression of *Petunia* S alleles is clearly dependent on the genetic background. Working with *P. coerulescens* (stigmatic bi-factorial SI), Hayman and Richter (1992) selected "pollen-only" SC mutations arising at a locus (T) not closely linked to S and Z. Among several explanations, Hayman and Richter envisage the possibility that the two-locus system of the grasses evolved from a multi-locus mechanism that involved a multi-allelic T gene now fixed in *P. coerulescens* as a gene that has lost its poly-allelism but continues to perform a function in SI. The exact role of T, before and/or after the specification of S and Z gene products in the pollen, is unknown. However, Hayman and Richter suggest that it may be related to the processing of the S and/or the Z products or to the establishment of their interaction.

4.2.5.2.2 Action of Polygenes. The action of polygenes is revealed by the weakening of SI in inbred populations (Atwood 1942; Lundqvist 1961; Denward 1963; Pandey 1965; Martin 1968; de Nettancourt 1969; Melton 1970, Hogenboom 1972b, 1972c). The SC character promoted by inbreeding is usually not established permanently and gradually reverts towards SI after a few generations of inbreeding (Bianchi 1959; Pandey 1959; Melton 1970; Hogenboom 1972b, 1972c). Such a return to the original breeding habit suggests that a certain amount of heterozygosity is needed for the manifestation of SC or that a mechanism re-establishing SI and heterozygosity operates as soon as an inbred individual has reached a critical level of homozygosity.

4.2.5.2.3 S Alleles Trapped in Translocation Rings.
Reciprocal translocations involving several pairs of chromosomes have accumulated throughout the evolution of certain species of the genus *Oenothera*. At meiosis, these translocations cause rings of chromosomes composed in some populations of the entire diploid set of chromosomes. As outlined by Harte (1994) in her review of scientific achievements with the genus *Oenothera*, two rings of four chromosomes each imply differences between the parental sets of chromosomes. These differences result from two translocations involving two pairs of chromosomes, while a ring of six chromosomes also indicates a difference of two translocations, but between three chromosomes. Five translocations can give rise to a ring of twelve chromosomes plus a bi-valent, but the same number of translocation events can also produce two circles of either ten and four or eight and six chromosomes. Six is the minimum number of reciprocal translocations necessary to assemble 14 chromosomes in a circle.

As for ordinary translocations, the fertility of male and female gametes will depend, in a large part, on the segregation (alternate or adjacent) of the chromosomes forming each ring, the presence or absence of lethal genes in the translocation, or the lethal association of certain combination of genes. Some North American species (belonging particularly to the *biennis* group of *Oenothera*) are "constant heterozygotes" with monotypic egg cells and pollen grains that behave as mono-types upon selfing and as either mono-types or di-types in crosses to other species or populations within the species (Oehklers 1943 and Steiner 1956, cited by Harte 1994). In such species or populations, the egg cell can receive only one complex of genes, called the α complex, which carries an SI allele (for instance, S1). The pollen grain receives either the α complex (with the S1 allele) or another complex of genes (the β complex), which carries a self-compatible allele (Sf) instead of Sl. Because the S1 pollen is incompatible, only the β pollen will reach the ovary and cause the reconstitution of the α–β heterozygote on fertilization. The only chance that S1 will travel all the way through a pistil and fertilize an egg cell results from cross pollination with plants (used as pistillate parents) that belong to populations where the S allele present in the α complex expresses a specificity other than that of S1.

4.2.6 SC in Polyploids

4.2.6.1 Tetraploid Forms and Tetraploid Species Are Often Self-Compatible

The tetraploid relatives (or artificially induced tetraploids) of diploid dicotyledons with mono-factorial GSI usually display a self-compatible phenotype (Lewis 1943, 1947, 1949 b; Atwood and Brewbaker 1953; Brewbaker 1954, 1958; Sampson 1960). Their ability to accomplish self-fertilization (and, in many cases, to set seed on selfing) does not result from a change in the pistil, which maintains the function of identifying incompatible pollen and rejecting it. Instead, it results from a loss in the diploid pollen grains produced by the tetraploid plant of the incompatibility phenotype. The restriction of the effects of polyploidy to pollen was first reported in *Prunus* by Crane and Lawrence (1929), in *Petunia* by Stout and Chandler (1942) and in *Pyrus* by Crane and Lewis (1942). These authors observed clear-cut differences between diploid and polyploid forms in reciprocal crosses. In all cases, this meant that the pollen, and not the pistil, was the site of the breakdown.

4.2.6.2 Competitive Interaction in Diploid Hetero-Allelic Pollen

Working with *O. organensis*, Lewis (1947, 1949 a, 1949 b) concluded that the production of compatible pollen by auto-tetraploids is limited to S-heterozygous plants (SaSaSbSb) and that only grains carrying two different alleles (SaSb) are compatible. Preliminary evidence in favor of this hypothesis had been provided earlier by Stout and Chandler (1942), who implicitly showed that, with regard to cross-compatibility between sibs and to unilateral cross-incompatibility with the original diploid mother plant, the progeny of self-compatible tetraploids of *Petunia* displayed features that would be expected if compatibility depended on the capacity of the tetraploids to produce S-heteroallelic pollen. The interpretation was confirmed in the same year by Lewis and Modlibowska, who found two different types of pollen tubes on self-pollination in the tetraploid form of *Pyrus communis*. Some, presumably S homozygous, failed to grow through the style, and others, assumed to be S heterozygous, behave compatibly.

Lewis also established, from data obtained by Atwood (1944) in *Trifolium repens* and from his own observations on *O. organensis*, that a state of S heterozygosity in diploid pollen was not always a sufficient condition for the manifestation of SC. Certain S alleles are dominant over other alleles in the pollen grain, which remains self-incompatible and expresses the S specificity of the "dominant" allele. From his observations, Lewis concluded that interactions between different S alleles are the general rule in diploid hetero-allelic pollen of species with mono-factorial GSI, and that the alternative to a relationship of dominance between two S alleles is that of competitive interaction leading to a loss in expressivity for each of the two alleles involved.

The biochemical basis of a competitive interaction that does not occur between identical alleles but takes place between different ones is not known. The di-allelic, diploid pollen grain is perhaps unable to maintain steady-state levels of transcripts at the threshold amounts (Lee et al. 1994) necessary for the expression of the phenotype of each of the two different S alleles (Sect. 4.2.2.1). If this was the case, the dominance of one allele over the other in diploid pollen could be explained by the assumption that the dominant allele begins transcription before the recessive allele.

The possibility that the interaction leading to the loss of S function in di-allelic pollen results from homology-dependent gene silencing (Meyer and Saedler 1996) appears remote, because one would not expect homology-mediated inactivation to occur only between different S alleles.

The ability of heteroallelic pollen to become compatible does not always depend on the identity of the two alleles present in the grain; it also, to a large extent, depends on the genetic background of the pollen-producing tetraploid. This influence was observed in alsike clover (Brewbaker 1954), *N. alata* (Pandey 1968) and *L. peruvianum* (de Nettancourt et al. 1974b).

4.2.6.3 Effects of Polyploidy on SI in Monocots and Certain Primitive Dicots

Polyploidy does not eliminate SI in species with a multigenic complementary GSI system. The genes (two or several loci) that work in association to establish the incompatibility phenotypes of the pollen and pistil interact among themselves but not with their own alleles. There is no possibility, if the system is to operate at all at the diploid level, of interaction or dominance between alleles of a same gene.

Interestingly, the maintenance of incompatibility at the polyploid level, which occurs in monocots (and in certain primitive dicots) with multigenic SI, also characterizes related genera or related families where a one-locus GSI system has been established (Lundqvist 1991). It even occurs in a one-locus sporophytic system in the genus *Cerastium* (Caryophyllaceae). The pollen produced by tetraploid forms of these plants fails to display any relationship of interaction or competition. This was clearly shown in *Tradescantia paludosa* 32 years ago by Annerstedt and Lundqvist (1967), who tested 12 pairs of S-allele combinations in di-allelic diploid pollen. Furthermore, cases (for instance, that of *Trifolium medium*; Fryxell 1957 cited in Lundqvist 1993) of the persistence of SI in tetraploids that are not restricted to monocots and primitive dicots have also been reported.

Therefore, the range of SI species producing self-incompatible pollen in tetraploids is relatively large and covers, in addition to the Poaceae, species of the Bromeliaceae, Caryophyllaceae, Chenopodiaceae, Commelinaceae, Fabaceae, Liliaceaeae and Ranunculaceae (Lundqvist 1975, 1990a, 1991, 1993). For *Ranunculus repens*, Lundqvist (1994b; Sect. 2.4.3) calculated the minimum degrees of cross-incompatibility to be expected for three types of intercrosses (Ii×Ii; F1×F1; P×BC; Sect. 2.4.3) under specific assumptions regarding the genetics and cytology of this polyploid species. Fearon et al. (1984) have shown that the level of cross-compatibility that can be expected between any pair of plants in an auto-tetraploid population of *Lolium perenne* (two-locus gametophytic complementary system) is significantly lower than a in diploid populations if only a few (but equally frequent) alleles segregate at each locus. The differences become negligible if the number of segregating alleles at each locus exceeds 12.

4.2.7 The Generation of New SI Alleles

The relative ease with which SC mutations can be induced and detected in species with mono-factorial GSI contrasts with the apparent inability of artificial mutagens to construct specificity changes at the S locus. It is possible that the number of events (probably BP substitutions; Sect. 4.3) required for a new specificity to replace a parental specificity is such that the probability of their induction in the same PMC is very low. Even the smallest numbers of AA differences ever observed in nature between distinct S ribonucleases in the Solanaceae – 13 (Ai et al. 1990) and 22 (Clark et al. 1990) in *Petunia* and ten in *Solanum* (Saba-El-Leil et al. 1994) – are unlikely to have appeared simultaneously. It is possible, however, that the only differences in AA distribution that really matter are those that distinguish the HV regions of the S protein. This eventuality, outlined in Section 3.5.7.5, is presented in more detail below.

4.2.7.1 Conflicting Evidence Regarding the Role of HV Regions in the Solanaceae?

4.2.7.1.1 Only Four AAs Are Responsible for the Difference in Specificity between the S11 and S13 Alleles of *Solanum chacoense*. The two alleles S11 and S13, between which Saba-El-Leil and co-workers in Montreal detected a difference of only ten AAs (out of 190), were found in a small population. The inbred parental line where S11 was found showed erratic behavior characterized by cycles of spontaneous seed setting after self-pollination. Its S-intron sequence is potentially translatable and could produce a new S protein, translated from unspliced message, which would have a different (longer) HVa region (Saba-El-Leil et al. 1994). Four of the AA substitutions that differentiate S11 from S13 appear in the HV regions, one in HVb and three in HVa. In *S. chacoense*, the HVa region is hydrophilic and is considered to be accessible for binding and specific configurations with other SI components (Saba-El-Leil et al. 1994). Furthermore, the region contains the only intron of the entire gene sequence and could also express new S-specific variations through BP substitutions, leading to changes of the intron splice sites. Splicing is the mechanism by which intron sequences are removed from precursor RNA molecules and adjacent exon sequences are re-ligated.

The Montreal group next transformed a plant of *S. chacoense* (wild type, S12S14 in genotype) with a chimeric S11 gene in which the four residues

Table 4.4. The role of the hyper-variable regions of S ribonucleases in the determination of specificity in different solanaceaous species

Research group	Species studied	Allele modified	Nature of the modification	Pollen rejected by the modified allele	References
Montreal	*Solanum chacoense*	S11	Replacement of four AAs responsible for sequence differences between HV regions of S13 and S11	S13	Matton et al. (1997)
Pennsylvania	*Petunia inflata*	S1	Swapping of HV regions between S1 and S3. One chimera combined HVa and HVb of S1 and the others combined regions of S3. The second contained most of the S3 protein except HVb of S1	Both chimeric genes unable to reject S1 and S3 pollen	Kao and McCubbin (1996)
Missouri	*Nicotiana alata*	SA2, SC10	Swapping of regions. Nine constructs made which sampled the entire S-ribonuclease sequence	None of the chimeric genes able to reject SA2 or SC10 pollen	Zurek et al. (1997)

AA, amino acid; *HV*, hyper-variable

had been substituted with those present in the S13 ribonuclease (Matton et al. 1997). The fact that the transgenic plants acquired the S13 phenotype (i.e., the ability to reject S13 pollen) provided evidence that the HV domains of S ribonucleases determine S specificity (Table 4.4). Matton and his co-workers concluded that "one allelic form of the S-RNase molecule can be converted into another by modification of the HV domains alone and that allelic specificity can be determined by the HV domains alone".

4.2.7.1.2 In *Petunia* and *Nicotiana*, the Ribonuclease Sequences Responsible for Pollen Recognition Appear to be Scattered Throughout the Molecule. The *Solanum* data summarized above contrast with the finding that different SLGs of *B. rapa* (Kusaba et al. 1997; Kusaba and Nishio 1999a) and *B. oleraceae* (Kusaba and Nishio 1999a) have identical or nearly identical HV regions. It is also not supported by data from Pennsylvania State University (Kao and McCubbin 1996) and the University of Missouri at Columbia (Zurek et al. 1997). In the course of experiments designed for studying the effects of swapping domains between different S ribonucleases in *N. alata* and in *P. inflata*, these two institutions showed (Sect. 3.5.7.5; Table 4.4) that the recognition functions of S ribonucleases, which are particularly sensitive to

domain disruptions in *N. alata*, were not restricted to specific domains of the proteins. On the contrary, the sequences responsible for allelic recognition appeared to be scattered throughout the molecule. An exchange of views between the Montreal group and the two U.S. groups (Matton et al. 1998; Verica et al. 1998) allowed a comparison of the respective merits of the two approaches and attributed particular importance to the extent of divergences between the alleles compared, the fragility of the "recognition" structures, the possible relationship between the dispersion of recognition elements and protein folding and, most importantly, the need to identify the pollen component of the recognition reaction in the Solanaceae for further clarification and progress.

The need for further research on the role of HV regions in other S systems is also essential, because new data could clarify current doubts (Nasrallah 1997; Charlesworth and Awadalla 1998) regarding a relationship between the polymorphism of a sequence and its involvement in recognition. A high diversity between regions may also result from their minor importance, which enables them to vary without disturbing a basic function (and vice versa; Matsushita et al. 1996; Sect. 4.3.1.4). The diversity generated may be substantial if the regions are linked to selectively maintained sites elsewhere in the locus and if each functional class of allele, behaving in the absence of recombination as an isolated population, is of a size sufficiently small to allow even non-synonymous mutations to drift to high frequency or fixation within each class. Both Nasrallah and Charlesworth and Awadalla refer to two *Brassica* alleles (SLG8 and SLG46 of *B. campestris*) with different specificities and identical HV regions (Kusaba et al. 1997). They suggest in vitro mutagenesis or reconstruction experiments (of the type carried out in *Nicotiana*, *Petunia* and *Solanum*) between *Brassica* alleles of different specificities or an identification of regions in identical alleles that do not participate in S determination (Charlesworth and Awadalla 1998). In an analysis of the possible role of intragenic recombination of *Brassica* S alleles (Sect. 4.2.7.3), Awadalla and Charlesworth (1999) propose, from considerations regarding clusters of linkage dis-equilibrium within the SLG gene, that HV regions are selectively balanced and cannot be assimilated into regions of relaxed selective constraint.

Another set of observations suggesting that both variable and conserved AAs are involved in the recognition and inhibition of incompatible pollen results from the identification of residues in a hydrophilic loop (loop 6) of the *Papaver* S protein. This loop plays a crucial role in pollen recognition (Sect. 3.3.3.3; Kakeda et al. 1998). The Birmingham group demonstrated, through the use of site-directed mutagenesis, that certain AAs are essential for S-specific activity because some of the mutated proteins are unable to reject incompatible pollen. Both variable and conserved AAs in loop 6 were found to be required for the pollen–pistil interaction. At the same time, the group (Kurup et al. 1998) showed that AA-sequence variation between the S alleles of *P. rhoeas* and *P. nudicaule* is not found in HV blocks but appears throughout the S protein, interspersed with numerous, short, strictly conserved segments.

Similarly, an alignment of the primary structures of five S ribonucleases from *Pyrus pyrifolia* with those of other rosaceous S-RNases showed that the positions of AA substitutions between pairs of alleles were scattered throughout the entire sequence. In one instance, however, the difference between two alleles was restricted to the 21–90 portion of the sequence, which includes the HV region (Ishimizu et al. 1998a).

4.2.7.2 New S Alleles Appear in Inbred Populations

New S alleles have been reported to occur in inbred populations on several occasions. As can be seen from Table 4.5, seven cases not attributed to contamination by stray pollen (Sect. 3.3.2.2) were recorded in species with GSI (*T. pratense, L. peruvianum, Nicotiana bonariensis, S. chacoense*) and one in a species with SSI (*R. sativum*). In most instances, the detection of a new S specificity was incidental.

In one study (de Nettancourt et al. 1971), the search for newly generated S alleles had been planned and was undertaken systematically by selfing, back-crossing and sib-crossing the members of advanced generations of inbreeding plants initially derived from a single S1S2 individual. The results of that work showed that a new S specificity (and always the same one, referred to as S3) was present in two of the several hundred plants analyzed during the third generation of inbreeding. Further attempts to obtain new specificities from other inbreds derived from the same initial S1S2 clone or its progenies failed (Ramulu 1982a, 1982b). The new specificity was detected in the pistils of plants otherwise homozygous for the S2 specificity operating. Through test crosses, it was demonstrated to be transmitted as a single gametophytic factor allelic to the S locus and to become fully operative, in progenies, in both the pollen and pistil. When back crosses were made between inbred plants stabilized for the new specificity (S1S3 or S2S3), S3 occasionally reverted to S2.

Table 4.5. Self-incompatible species in which new alleles have been detected in inbred populations. (de Nettancourt 1999)

	System	Source
Trifolium pratense	Stylar GSI	Denward (1963)
		Anderson et al. (1974)
Lycopersicum peruvianum	Stylar GSI	de Nettancourt and Ecochard (1968)
		de Nettancourt et al. (1971)
		Hogenboom (1972a, 1972c)
Nicotiana bonariensis	Stylar GSI	Pandey (1970c)
Solanum chacoense	Stylar GSI	Saba-El-Leil et al. (1994)
Raphanus sativus	Stigmatic SSI	Lewis et al. (1988)

GSI, gametophytic self-incompatibility; *SSI*, sporophytic self-incompatibility

Several (but not all) of these features were found to characterize the specificities discovered by Lewis et al. (1988) in *R. sativus* and by Saba-El-Leil et al. (1994) in the non-stable S allele (S11) of *S. chacoense* (Sect. 4.2.7.1). It is remarkable that transgenic plants bearing different mutations in the HV regions of this S11 allele express different SI stylar phenotypes, ranging from the loss of S11 specificity to the loss of S13 (the parental allele) specificity or the ability to reject S11 and S13 pollen simultaneously (Matton et al. 1998, 1999; Sect. 4.2.7.4).

4.2.7.3 Origin of New Specificities

It shall be seen below (Sect. 4.3) that S alleles diverged (during evolution) through the occurrence of non-synonymous BP substitutions (substitutions that lead to the replacement of one AA by another). When induced at the proper site and inducing the proper replacement, these substitutions contributed to highly selective modifications of the specificity of S proteins. Intragenic recombination appears to be improbable, because there is no inter-allelic homology in the flanking regions and introns of *Petunia* and *Nicotiana* alleles (Ioerger et al. 1990; Coleman and Kao 1992; Matton et al. 1995) and because statistical observations failed to show that crossing-over ever takes place among 12 different S alleles of the Solanaceae. Boyes et al. (1997), who observed highly divergent and rearranged S sequences in *Brassica*, suggest that structural heteromorphism may, together with haplotype-specific sequences, suppress recombination within the S-locus complex. In the case of *L. peruvianum*, where new S alleles seem to be generated in S-homozygous plants, there is little reason to think that equal crossing-over (or gene conversion) within S regions plays a role in S polymorphism. However, Awadalla and Charlesworth (1999), through an analysis of linkage dis-equilibrium within the SLG gene of two *Brassica* species, found evidence that intragenic recombination occurred during the evolutionary history of the alleles studied. Their conclusion is supported by patterns of synonymous nucleotide diversity within the SLG and the SRK genes, and between SRK domains (Sect. 4.3.1.4). Saba-El-Leil and co-workers (1994) suspect that the sudden SI behavior of the plant bearing the unstable S11 allele they discovered in *Solanum* is due to the temporary read-through of the intron sequence, which lacks a stop signal (Sect. 4.2.7.1).

4.2.7.4 The Dual Specificity of New S Alleles May Play a Key Role in the Generation of New S Alleles

The prediction of Fisher (1961; Sect. 4.2.7.5) that a new S allele (which he believed originated from recombination within the S locus) would remain incompatible with each of the parental alleles was verified by Matton et al. (1999) in the course of their research on the role of the HV regions of the S locus in *S. chacoense* (Sect. 4.2.7.1). After finding that the substitution of only four AAs in the HV regions converted one S phenotype into a different S phenotype, they showed that the replacement of three of these four AAs resulted in a new S allele that simultaneously rejected two phenotypically and genotypically distinct pollen types (Matton et al. 1999). In other words, in the case studied, there is only a single BP difference between bi- and mono-specificity. The Montreal group, although it refrained from making assumptions and predictions regarding the nature, identity or mode of action of the pollen component in SI of the ribonuclease type, noted that the phenomenon occurs in other plant cell–cell recognition mechanisms and provides relevant examples of disease-resistance genes (in *Arabidopsis* and *Vicia*) that express such dual specificities.

4.2.7.5 The Role of the Genetic Background

The fact that all known cases of new S specificities have been recorded in highly inbred material is suggestive of an influence of the genetic background on the occurrence of genetic changes. It is possible (but not known) that certain homozygous combinations of modifiers or polygenes intensify the rate of BP substitutions or that selfing generates large differences in recombination frequencies in the S locus. For an analysis of the effects of selfing on selection for recombination between a pair of loci, see Charlesworth (1976) and Charlesworth et al. (1977). It is also possible that the genetic environment of the S locus has no special influence on the generation of new S alleles and that inbreeding is only necessary to bring together, in a same plant, S sequences that spontaneously accumulate BP substitutions and diverge (with or without a stage of dual specificity) to become different S alleles. As predicted by Fisher (Sect. 4.2.7.4), who visualized the S locus as a short strip of chromosome with approximately 20 antigenically active points, the mutation of one allele into another, fully cross-compatible allele does not take place in one step. It occurs through an intermediate bi-specific allele, incompatible with the parent allele and impossible to detect, in the pollen and style, with self-pollination tests. There is no relationship between rarity and adaptive value for these transient new alleles, which are very similar to mutated forms of the S11 allele of *S. chacoense* (Sects. 4.2.7.2, 4.2.7.4; Matton et al. 1998, 1999) and, to some extent, to the new S3 specificity expressed in S2S2 and S1S2 inbred plants of *L. peruvianum* (de Nettancourt et al. 1971). These alleles progressively become cross-compatible with the parental allele through a further accumulation of spontaneous BP substitutions.

4.2.7.6 Methods for a Rapid and Reliable Identification of S Alleles in Plant Breeding

The need for methods for rapid and reliable identification of S alleles in plant breeding arises each time the breeder uses the identification of S genotypes as a means to distinguish between plants with identical morphology (for instance, Brewer and Parlevliet 1969 sorted out different clones of *Pyrethrum* on the basis of their cross-compatibility relationships) or requires a final verification of S genotypes before launching F1-hybrid seed-production programs. Techniques recently elaborated for such purposes are outlined in Section 4.3.1.6.

■ 4.3 Evolution of SI

Some of the pistil S genes participating in homomorphic SI provide exceptional material with which to study the evolution of complex and very ancient mechanisms of cell–cell recognition in higher plants, because they are polymorphic, of known genetic sequence and are submitted to potent selective forces that strongly favor the establishment of rare (new) alleles, their access to equilibrium and a particularly long lifetime. It is the purpose of this section to outline and discuss the nature, extent and age of differences among alleles of a same S gene and to review current opinions regarding the origin (unique or multiple) of contemporary SI systems and of SC (ancestral or derived). The method of quantifying S-sequence divergences among individuals, populations and species and trying to determine the evolutionary history of SI has been established or improved at regular intervals during the last 10 years (Uyenoyama 1988, 1991, 1995, 1997; Ioerger et al. 1990; Clark and Kao 1991, 1994; Hinata et al. 1993; Richman et al. 1996a, 1997; Boyes et al. 1997; Kusaba et al. 1997; Charlesworth and Awadalla 1998; Ushijima et al. 1998a, 1998b; Kusaba and Nishio 1999a, 1999b).

Research on the origin of new S specificities and on the identification of sites (within the S allele) that participate in pollen–pistil recognition are summarized in Section 4.2.7 and in Chapter 3. Such work is essential for a meaningful analysis of the evolution of S alleles. As can be seen below, the approaches followed by the "evolutionists" have usually been based on the assumption that, at least in the pistil, the HV regions of the S gene participate in pollen–pistil recognition.

4.3.1 Allelic Diversity

For each allele studied, the assessment and interpretation of diversity between two or more alleles require knowledge of the alignment of nucleotide and AA sequences and of the "shared polymorphic sites" in the sequence of the S protein when two or more nucleotides segregate in the populations or species compared. The exchange of nucleotides that takes place at such sites may be synonymous (not leading to the replacement of an AA) or non-synonymous (resulting in the presence of a different AA in the protein sequence). It may occur by chance, as the consequence of random mutation, at a rate that is homogenous throughout the gene (verification of the null hypothesis), or it may suggest a common ancestor for the alleles under consideration (rejection of the null hypothesis). Statistical tests (Clark and Kao 1991; Ioerger et al. 1991; Charlesworth and Awadalla 1998; Sect. 4.3.1.1) are available for evaluating the significance (effects of drift, linkage and direct selection) of an excess of non-synonymous substitutions at shared polymorphic sites and for establishing the degree of clustering of polymorphic sites.

Models for studying the evolutionary dynamics of SSI alleles in monofactorial multiallelic systems through stochastic simulations have been prepared by Schierup et al. (1997). These models address co-dominance in both the pollen and style (behaving independently in GSI, but with a stronger selection intensity), dominance hierarchy in the pollen and style, and co-dominance in the style, associated with dominance in the pollen. The implications of the results are particularly important for the prediction of equilibrium frequencies, the effects of genetic drift, the life span of alleles and allelic turnover. Based on the numerical simulation of the expected distribution of ratios of divergence times among S alleles, Uyenoyama (1997) proposed a method for characterizing the structures of genealogies that regulate SI.

4.3.1.1 Origin, Distribution and Extent of Divergences among Functional S Alleles in the GSI System of the Solanaceae

4.3.1.1.1 Intragenic Crossing-Over or Accumulation of Single BP Changes? It has been suggested, on the basis of a comparison of different S-gene sequences, that intragenic crossing-over leading to gene conversion or standard recombination between S alleles is at the origin of S-allele diversity in both gametophytic and the sporophytic systems (Ebert et al. 1989). The conclusion was not supported by the sensitive tests (Stephens 1985; Sawyer 1989) that were applied by Clark and Kao (1991) to detect the role of intragenic recombination on the accumulation of differences between S ribonucleases in the Solanaceae. Further analysis (and other arguments based on the distribution of point mutations at sites that appear to determine S specificity and on the heterogeneity of the flanking sequence of the S gene that implies a recombination-suppression mechanism) were presented by Tsai et al. (1992). Clark and Kao (1994) noted, however, that S alleles are so old (Sect. 4.3.1.5) that recombinations may have occurred in the distant past; the tests would be unable to detect this among the numerous nucleotide substitutions that subsequently took place. The demonstration (Despréz et al. 1994) that each of the ten AA replacements that distinguishes S14 from the S13 allele of *S. chacoense* could be explained by point mutations leads to the conclusion that conversion and intragenic recombination are not necessarily involved in the diversity of S ribonucleases. Unless one assumes that a single BP substitution can cause a change in specificity, this implies that the first point mutations contributing to the emergence of a new S allele remain neutral until all the necessary changes have taken place

(but see Sect. 4.2.7.1). Clark and Kao (1994) have assessed the consequence of this temporary neutrality on the detection of selection effects through a comparison of synonymous and non-synonymous substitutions. McCubbin and Kao (1999), who consider that alleles in the transition phase could lose their specificity and be rapidly eliminated, propose that these transient forms express a dual specificity, a suggestion also made by Fisher (1961) in his recombination model for the generation of new S alleles and as seen in Section 4.2.7.4, verified by Matton et al. (1998, 1999).

4.3.1.1.2 Distribution and Variability of S Alleles. An application of existing methodology to the study of divergence among 12 S ribonucleases from *N. alata, P. inflata, S. chacoense* and *L. peruvianum* revealed a lack of significant heterogeneity in the synonymous substitution rate in the S gene (Clark and Kao 1991). The HV regions of the ribonucleases displayed an excess of non-synonymous substitutions, consistent with the strong diversifying selection assumed to operate on this locus. Furthermore, the test of clustered polymorphism led to the detection of very old alleles that shared silent polymorphism. In an earlier paper from Pennsylvania State University (Ioerger et al. 1990), a pairwise comparison among 11 S ribonucleases and a gene genealogy of these alleles revealed that the divergence of AA sequences resulted from the very ancient origin of S alleles and could be larger within species than between species (Sect. 4.3.1.5). The data now available indicate that allelic divergences may considerably vary, with AA identity within and between species of *Solanum, Petunia* and *Nicotiana* ranging from 38.8% to 80.1% (Ioerger et al. 1990; Lee et al. 1992). This also provides evidence, as in the case of *Physalis crassifolia*, that contemporary alleles with low levels of sequence divergence may derive from recent diversification after different turbulences (such as population bottlenecks) during the life span of the species (Richman et al. 1996b). Exceptional cases (Sect. 4.2.7; Ai et al. 1990; Saba-El-Leil et al. 1994; Matton et al. 1998) of variations of only 10 bp and 13 bp between functionally different S ribonucleases and of only 4 bp between their HV regions are known. Such variations may be accentuated or completed by their effects on protein folding and the S-allele specific formation of loops likely to modify recognition specificities (Kakeda et al. 1998; Sect. 3.3.3.3).

4.3.1.1.3 Inter-Species Variation in S-Allele Age and Number. Richman and Kohn (1999) propose, in the light of current theory, that such variation essentially results from changes in the effective population size of the species concerned.

4.3.1.2 S Alleles in Other Families with a Ribonuclease GSI System

4.3.1.2.1 Differences between the Scrophulariaceae and the Solanaceae. Four (C1, C2, C3 and C5) of the five conserved regions identified in the Solanaceae are found in *Antirrhinum* (Scrophulariaceae) where the regions corresponding to the two HV regions of the Solanaceae also reveal a high degree of variability (Xue et al. 1996). Analysis of AA sequences indicates that S alleles from the Scrophulariaceae are not shared with those of the Solanaceae but form a separate group. The *Anthirrinum* sequences have 40–55% identity with each other and have extensive homology with other members of the S-ribonuclease family.

4.3.1.2.2 Differences between the Rosaceae and the Solanaceae. Five conserved regions were ultimately detected in the four sub-families of the Rosaceae (Ushijima et al. 1998 a), instead of the four regions initially reported (Sassa et al. 1996). The additional region, RC4, occupies the position of C4 in the Solanaceae but shares no homology with it, a fact Ushijima and co-workers interpret as an indication that RC4 arose on the occurrence of a population bottleneck that accompanied the differentiation of the sub-families after the divergence of the Solanaceae. Differences between the Rosaceae and the Solanaceae also appear at the level of the HV regions. Only one such region, located at the position of HVa in the Solanaceae, occurs in the Rosaceae and presumably plays an essential role in recognition. However, the region corresponding to region HVb of the Solanaceae is conserved in the Rosaceae and does not appear to be involved in the determination of S specificity. The flanking regions of S-RNases from Japanese pear and apple (characterized for three different alleles; Ushijima et al. 1998 b) bear no similarities with those of solanaceous S ribonuclease, although the position and sequence of the putative TATA box were conserved.

4.3.1.2.3 S-RNase Polymorphism in the Rosaceae. Ishimizu et al. (1998 a) performed a detailed analysis of the primary structural features of the rosaceous S ribonucleases and showed that S-RNase polymorphism predated the divergence of *Pyrus* and *Malus*. They established the position of AA substitutions between different alleles that, for some pairs of alleles, were spread throughout the entire protein and, for others, were restricted to the HV region. The group (Ishimizu et al. 1998 b) also undertook a search for possible recognition sites in the S-RNases of the Rosaceae. Window analysis of the number of synonymous and non-synonymous substitutions allowed the detection of four regions with an excess of non-synonymous substitutions. These four regions correspond to two surface sites on the tertiary structure of fungal RNase Rh.

4.3.1.3 Homology or Convergence among S Ribonucleases?

From their phylogenetic analysis of the Rosaceae and their two distant relatives, Scrophulariaceae and Solanaceae, Xue and co-workers (1996) concluded that the ribonuclease SI system was present in the common ancestor of the three families. This hypothesis, challenged by the suggestion (Sassa et al. 1996) that one could also attribute the emergence of S ribonucleases in different families with GSI to convergent evolution, was tested by Richman et al. (1997) through a reconstruction of the genealogy of angiosperm ribonucleases. The basic objective was to determine, through the use of out-groups of non-S ribonucleases as reference points, whether the distinct grouping of S ribonucleases from the Solanaceae and the Rosaceae consisted of sister groups or reflected different origins.

Richman and co-workers compared protein-sequence data from 34 ribonuclease genes (three fungal ribonucleases, seven angiosperm non-S ribonucleases, 11 Rosaceae alleles, three Scrophulariaceae alleles and ten Solanaceae alleles). They then used the neighbor-joining method (which allows the construction of the genealogy tree from the pairwise distance between AA sequences) to determine whether the exploitation of S ribonucleases by the different families of angiosperms is the result of homology or convergence. Protein distances among pairs of sequences were produced in two ways: either using the percentage of AA divergence or the PAM 001 matrix of Dayhoff, which corrects for the inferred number of nucleotide substitutions necessary for various AA changes. The results showed that the S ribonucleases of the Solanaceae and the Scrophulariaceae, which both belong to the sub-class Asteridae, belong to the same group of ribonucleases as that of the Asteridae; therefore, they have a common origin. However, the relationship (homology or convergence) of the Rosaceae to the two other families remains obscure and should be clarified through the use of additional reference points (non-S ribonucleases from a wider range of angiosperms) and less distant out-group sequences with which to determine the genealogy (Richman et al. 1997; Sect. 4.3.2.2).

4.3.1.3.1 What Happens in legumes? According to Lundqvist (1993), the Fabaceae form a major exception to the rule that families are characterized by one and same system of SI. His views are based on the finding of complex, multi-factorial genetic control of SI in diploid and tetraploid species of the genus *Lotus* (Lundqvist 1993), diploid *Vicia faba* (Rowlands 1958), tetraploid *Medicago sativa* (Whithehead and Davis 1954) and polyploid *T. medium* (Fryxell 1957). The general conclusion of Lundqvist is that the Fabaceae family represents the result of an evolutionary scheme in which the one-locus system emerged from a primitive, multigenic basis (Sect. 4.3.2.2). The rate of emergence of the *Nicotiana* type of SI is probably slow, at least in some legumes, because Sims and co-workers (1999), using a variety of heterologous probes and polymerase chain reaction (PCR) primers, were not able to detect S ribonucleases in the one-locus stylar GSI system of *T. pratense*.

4.3.1.4 SLG and SRK Allelic Divergences in the Brassicaceae

4.3.1.4.1 Hyper-Mutability of the S Locus. The possibility of hyper-mutability of the S locus has been studied by Uyenoyama (1995), who found that the rates of substitutions at the S locus of *Brassica* were lower than those recorded at S-related loci. His finding indicates, as emphasized by Richman and Kohn (1996), that the large number of alleles and their current levels of sequence divergence must be essentially attributed to the long persistence of allelic lineages. Richman and Kohn also propose an alternative to the explanation (by Uyenoyama) that the apparent slowing of S-allele origination may result from the preferential replacement of parental alleles by their immediate descendants. They suggest that the slowing reflects constraints on the nature of changes that can generate new specificities. Richman et al. (1995, 1996b) discussed the importance of the life history of populations and the effects of population bottlenecks (Sect. 4.2.7.3 on a possible influence of consanguinity on the emergence of new S alleles through domain swapping or other means).

4.3.1.4.2 The S Locus is not a Hot Spot of Recombination. The variability of SLG DNA sequences, such as have been reported by Dwyer et al. (1991) and, more recently, by Kusaba et al. (1997), is generally interpreted as an indication that intragenic recombination contributed to the evolution of *Brassica* S proteins (Nasrallah and Nasrallah 1993; Nasrallah et al. 1994; Kusaba et al. 1997; Charlesworth and Awadalla 1998). The main evidence originates from the work by Kusaba and co-authors, who determined and compared 31 new sequences of *B. oleracea* and *B. campestris* and developed the statistical treatments necessary to establish the probability that divergences between the variable regions had occurred under conditions of free recombination or complete linkage. Their calculations show that intragenic recombination, together with point mutations, is responsible for the high level of sequence variation in SLG alleles. This conclusion, in the absence of estimates of recombination rates, is in opposition to the results of structural analyses by Boyes et al. (1997). Boyes et al. suggest that heteromorphism at the S locus, together with haplotype-specific sequences, may "suppress" or "limit" recombination within the S-locus complex and maintain the linkage of co-adapted allelic combinations of SRK and SLG genes. In support of their views, Boyes and co-workers reported the complete absence of recombination between SLG and SRK in 500 plants heterozygous at the S locus. Nishio et al. (1997) also failed to detect any recombination between SLG and SRK in more than 200 plants segregating for different S haplotypes.

Kusaba and Nishio (1999a), while they concede that the S locus "is not a hot spot of recombination", maintain the view that recombination (or related events) has been involved in the evolution of S haplotypes. They identified two haplotypes, S8 and S46; S8 presented a head to head SLG-SRK configuration. The two haplotypes shared identical sequences (presumably of recent divergence) between the 3′ half of SLG and the transmembrane kinase domain and untranslated 3′ region of SRK. The two haplotypes also displayed variability (suggesting an ancient diversification) between the 5′ half of SLG and the S domain of SRK. Kusaba and Nishio showed that S8 could have resulted from recombination between S46 and an unknown SX haplotype. They also referred to several other cases where it can be inferred (from an identical sequence in the S locus of different S haplotypes) that recombination or related events might have occurred.

4.3.1.4.3 Distribution and Extent of Variations between S Alleles. The data available (Dwyer et al. 1991; Watanabe et al. 1994; Hinata et al. 1995; Yamakawa et al. 1995; Kusaba et al. 1997; Charlesworth and Awadalla 1998) reveal an important diversity of S sequences (which is, however, lower than in the Solanaceae). A comparison (Hinata et al. 1995) of nucleotide sequences between two SRK alleles (SRK9 from *B. campestris* and SRK 6 from *B. oleracea*) revealed that the number of non-synonymous substitutions per site (NS) was constrained and was much smaller in the kinase domain (NS = 0.048) than in the receptor domain (NS = 1.20). In contrast, the number of synonymous substitutions (SS) was approximately the same for the two domains. The SLG NS and SS values, calculated by Hinata and co-workers from six different alleles of *B. campestris* and *B. oleracea*, were mostly comparable to the figures obtained for the receptor domain of SRK. Peaks for SS were particularly high at the signal peptide, in variable region 1 (VR1) and for the 3' terminal region; NS peaks were observed at sites 600 (VR1), 810 (VR2) and 930 (VR3).

The analyses showed, as in the case of GSI, that the differences among species do not exceed the differences within species. In fact, inter-species sequence similarity can be very high (Sect. 4.3.1.3). As noted by Charlesworth and Awadalla (1998) in their study of the distribution of diversity values expressed in different regions of 19 SLG and three SRK alleles, this inter-species similarity is also found at synonymous sites and in the most conserved regions.

Matsushita et al. (1996) compared the entire SLG sequences of a same haplotype (S24) sampled from two populations of *B. campestris* (in Japan and Turkey) that had been separated approximately 2000 years ago and had started to diverge morphologically. With the exception of the polyadenylation site and the length of the poly-A tail (the run of adenylic residues, possibly involved in mRNA stability, which is added to the 3' end of most eukaryotic mRNA after transcription), these sequences were absolutely identical. Matsushita and his co-workers believe that their observations, i.e., the absence of divergence between geographically isolated sequences, provide evidence that the entire SLG sequence of haplotype S24 participates in the determination of S specificity. However, they acknowledge the alternative explanation that the period surveyed is statistically too short to allow the occurrence of variations between the Japanese and Turkish S24 alleles. This would be the case if the base substitution rate per site per year was lower than the estimate of 1.9×10^{-7} used in their experiments and if it approximated the value of 10^{-9} per site per year estimated by Uyenoyama (1995) and Hinata et al. (1995). Considerable similarity among different haplotypes has also been reported by Kusaba et al. (1997) at the inter-species level (95.7% between SLG8 of *B. campestris* and SLG13 of *B. oleracea*) and within species (97.5% between SLG 8 and SLG40 of *B. campestris*).

4.3.1.4.4 Extent of the Divergence between SLG and SRK of a Same S Haplotype.

This question has recently been raised by Kusaba et al. (1997), who discovered that SLG and SRK in each haplotype of *B. oleracea* and *B. campestris* could considerably differ in their HV regions. Percentages of SLG–SRK identity between the three regions of haplotype 3, for instance, ranged from 45.5% in HV region I to 70.3% in region II and 63.6% in region III. Such differences suggest that the SLG and SRK components of a same haplotype do not necessarily need to have the same specificity. This supports the assumption that SLG and SRK bind to different sites of the same pollen ligand if SLG is involved in the recognition process (Chap. 3; Kusaba et al. 1998). Nasrallah (1997) notes, however, that the analysis of large numbers of SLG/SRK gene pairs might identify protein domains (presumably involved in recognition) that are conserved in an SLG/SRK gene pair but differ among haplotypes.

4.3.1.4.5 How is Sequence Similarity Usually Maintained between SRK and SLG in *Brassica* Haplotypes?

It shall be seen, in Section 4.3.2, that the duplication of the S domain in SLG and SRK occurred well before the *B. oleraceae/B. campestris* divergence of S alleles. Therefore, it is difficult to understand how homology between the SLG and SRK alleles present in each haplotype, often very high, but low in certain cases, has been maintained throughout the evolution of the Brassicaceae. Trick and Heizmann (1992) proposed that a process involving unequal crossing-over or an RNA intermediate regulates the concerted evolution of the two genes in each haplotype. To explain known cases (above) of SRK–SLG divergence within haplotypes, they propose that the mechanism is imperfect.

4.3.1.5 S Alleles Are Very Old

The conclusion that S alleles are very old is a leitmotiv in all discussions on the evolution of the S gene in mono-factorial, poly-allelic systems. It results from the consideration that divergences between S alleles of a given SI system are often larger within species than between species and, therefore, preceded speciation (such as can be dated from the fossil record; Sect. 4.3.1.5). It also proceeds from a direct estimation of the time needed for the synonymous substitutions to accumulate to their present state of diversity, on the basis of a given mutation rate. The calculations are complex and require evolutionary measurements and assumptions regarding both the value and constancy of the average mutation rate supposed to occur at the S locus and the neutrality of synonymous substitutions. Several interferences (hyper-mutability, linkage to non-neutral genes, genetic drift, saturation of substitutions, changes and bottlenecks in population size, modification of breeding habits...) must be identified and weighted as accurately as possible are likely.

4.3.1.5.1 S-Ribonuclease Polymorphism in the Solanaceae Arose before the Emergence of *Nicotiana, Petunia* and *Solanum.* S-ribonuclease polymorphism in the Solanaceae has been assessed by Ioerger et al. (1991). They compared the sequences of 11 S ribonucleases from *N. alata* (six alleles analyzed), *S. chacoense* (two alleles) and *P. inflata* (three alleles) and established a distance matrix based on AA similarities (Table 4.6) and a gene genealogy created from a neighbor-joining tree (Fig. 4.3). The neighbor-joining phylogeny revealed two major branches, one with alleles from *N. alata* and the three *Petunia* alleles and a second lineage carrying the four remaining alleles of *N. alata* and the two *S. chacoense* alleles. Such results clearly suggest that the extant polymorphism of the S ribonucleases arose before the three solanaceous species diverged. Furthermore, an ancestral allele for each branch of the tree was the most recent common ancestor for the alleles it carried.

In an attempt to specify further the old age of the S ribonucleases, Ioerger and co-workers compared the synonymous substitution (SS) values calculated among S alleles from the same species and among sequences of the small subunit of ribulose-1,5-biphosphate carboxylase (RuBCs) from each pair of the three genera. It was found that the average number of SSs per site among S alleles of a same species (0.82) exceeded the SS rate from inter-species comparisons of the RuBC sequences (0.47). Taking the SS rate estimated for RuBC sequences (6×10^{-9}) as a standard, the *Petunia/Nicotiana* split is estimated to have occurred 28 million years ago (compared with 36 million years for *Petunia/Solanum* and 27 million years for *Sola-*

Table 4.6. Pairwise amino-acid similarity between S alleles (Ioerger et al. 1990). These values represent percent amino-acid identity. Note the unusually low similarity, even between alleles from the same species. For example, S_z and S_{1nic} are only 45% similar. Also, note that some alleles from the same species are even less similar than alleles from different species (45% between S_z and S_{1nic} vs. 62.4% between S_z and S_{1per})

	Nicotiana alata						Petunia inflata			Solanum chacoense	
	S_{F11}	S_z	S_{1nic}	S_{2nic}	S_{3nic}	S_{6nic}	S_{1pet}	S_{2pet}	S_{3pet}	S_{2sol}	S_{3sol}
S_{F11}	–										
S_z	59.7	–									
S_{1nic}	44.7	45.0	–								
S_{2nic}	44.7	43.6	67.7	–							
S_{3nic}	44.8	45.4	68.9	63.3	–						
S_{6nic}	43.1	45.1	61.0	60.0	69.9	–					
S_{1pet}	53.5	62.4	42.5	39.5	43.3	42.6	–				
S_{2pet}	53.2	61.7	40.8	40.3	41.7	39.9	73.1	–			
S_{3pet}	52.5	66.3	41.5	41.0	42.4	41.1	73.5	80.1	–		
S_{2sol}	40.1	41.6	46.7	44.7	46.0	45.3	38.8	38.8	39.3	–	
S_{3sol}	42.4	44.6	50.0	48.5	49.0	47.5	47.0	43.8	42.6	41.5	–

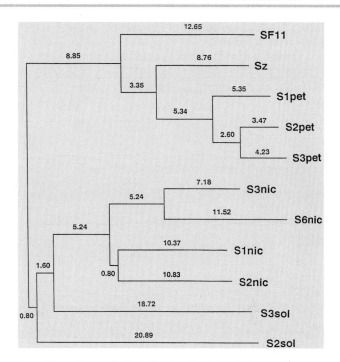

Fig. 4.3. S-locus gene genealogy. The authors note that "the tree indicates that some alleles are more related to alleles in other species than other alleles in the same species. For example, SZ and SF11 are clustered with the *P. inflata* alleles rather than with the other four *N. alata* alleles. Similarly, the other four *N. alata* alleles cluster with the two *S. chacoense* alleles rather than with Sz and SF11". The tree was constructed from a distance matrix by the neighbor-joining algorithm. The distances, labeled as branch lengths, were calculated by subtracting each pairwise similarity value from 100. (Ioerger et al. 1990)

num/*Nicotiana*). A more conservative test (resulting in a larger number of shared polymorphic sites by chance) and an analysis of shared AA polymorphism were also made in these experiments. These confirmed the conclusion that the excess shared polymorphism is due to common ancestry and that polymorphism pre-dated speciation. Similarly, the phylogenetic tree of rosaceous S ribonucleases reveals that S polymorphism predates the divergence of *Pyrus* and *Malus*.

4.3.1.5.2 The Origin of SLG and SRK. Several sets of data clearly show that SLG and SRK polymorphism pre-dated the emergence of the two species *B. oleracea* and *B. campestris*, which are usually compared for such purposes. The first indication came from Dwyer et al. (1991), who undertook (Table 4.7) pairwise comparisons of the number of synonymous substitutions among six different SLG alleles from *B. oleracea* (five alleles) and *B. campestris* (one allele). The degree of similarity observed between the *B. campestris* allele (SLG8) and the *B. oleracea* allele (SLG13) was obviously larger than the similarity among any of the remaining 14 pairs of alleles. This

Table 4.7. Pairwise comparisons of SLG sequences. (Dwyer et al. 1991, with kind permission from Kluwer Academic Publisher)

	$SLG-8_c$	$SLG-13_o$	$SLG-6_o$	$SLG-29_o$	$SLG-22_o$	$SLG-14_o$
Percentages of identical nucleotide residues						
$SLG-8_c$	–	92	88.8	88.2	86.3	84.6
$SLG-13_o$		–	89.7	89	86.6	86
$SLG-6_o$			–	89.5	89.7	88.5
$SLG-29_o$				–	87.5	86.5
$SLG-22_o$					–	88.5
$SLG-14_o$						–
Percentages of divergent amino acids						
$SLG-8_c$	–	10.3	17.2	18.9	20.9	23.6
$SLG-13_o$		–	15.4	17.5	18.8	21.7
$SLG-6_o$			–	17.9	17.2	20.6
$SLG-29_o$				–	21.1	23.1
$SLG-22_o$					–	19.7
$SLG-14_o$						–

similarity indicates that SLG8 and SLG13 originated from a common ancestral SLG gene that existed before the divergence of the two species.

The analysis of DNA variability among several class-I SLG alleles suggests that these alleles diverged more than 21 million years ago (Nasrallah and Nasrallah 1993). The calculation was based on an average inter-allelic nucleotide divergence of 21.4% and a substitution rate of 5×10^{-9} per site per year per lineage (a frequency representative of nuclear genes in plants; however, see Sect. 4.3.1.4). Boyes and co-workers (1997), in a study that included SLG8 and SLG13 of the Dwyer analyses, provided evidence that extensive re-structuring of the S locus (Sect. 3.2.3.2) preceded speciation in *Brassica*. Uyenoyama (1995) also confirmed that S lineages originated before the *B. oleracea/B. campestris* split. With the analysis of sequence data from S genes and from related non-S genes that do not co-segregate with the physiological expression of SI, he showed that it was possible to calibrate the entire phylogeny on the basis of lineages that are not subject to trans-species sharing. Through such calibration, Uyenoyama estimated that the coalescence time of all extant S alleles, both pollen dominant and recessive, exceeds the species-divergence time fourfold to fivefold, depending on the date of emergence of *Arabidopsis* receptor kinase relative to that of SRK. Accordingly, he calculates the age of SSI as approximately 40–50 million years. Kusaba et al. (1997) note that their own data indicate the same very ancient derivation of *Brassica* S haplotypes, especially since 95.7% identity was found between SLG14 (*B. oleracea*) and SLG25 (*B. campestris*) and 100% identity was found between their HV regions.

4.3.1.6 PCR Methods for Assessing Divergence among S Alleles

4.3.1.6.1 In the Crucifers. A molecular technique for identifying S alleles in *B. oleracea* has been developed by Horticulture Research International in Warwick (Brace et al. 1993, 1994). The technique is based on the PCR amplification of genomic DNA, followed by restriction analysis. A single pair of conserved primers are used (this can be a nuisance when confusion between different S genes is possible) to amplify S and S-like sequences. PCR products are then analyzed further by digestion with six restriction enzymes, followed by gel electrophoresis of the digestion products. The method requires little DNA as a starting template, and the procedure for restriction digestion does not involve hybridization. The method is simple and useful. It was used to amplify the sequences of the 48 homozygotes that constituted the reference collection of *B. oleracea* at Warwick. It not only characterized all different alleles, it was also unable to distinguish between certain samples that were later found (through pollination tests) to share the same specificity (they were presumably identical).

At the same time, a modification of the basic approach was developed by Nishio and co-workers (Nishio and Sakamoto 1993; Nishio et al. 1994, 1996). They designed highly specific primers that amplify only SLG and allow, from the determined nucleotide sequence of SLG PCR products, a precise calculation of the expected sizes of DNA fragments used as markers for the identification of S alleles.

4.3.1.6.2 In Solanaceous Species. The first applications of reverse-transcription PCR for the direct determination of the S-locus genotypes of plants sampled from natural populations of the Solanaceae was made by Richman et al. (1995) in *Solanum carolinense* and by Richman et al. (1996a) for *P. crassifolia*. The method, as outlined by Richman et al. (1996a) and Richman and Kohn (1996), exploits the abundance of S-locus mRNA in the styles of species with mono-factorial stylar GSI for the amplification of S alleles. A fragment of expected size from each individual plant analyzed is amplified by primer pairs designed to complement the conserved regions of published allele sequences from the Solanaceae. As could be anticipated from diploid populations with GSI where each plant carries S alleles with independent action in the pistil and is S heterozygous, restriction digests revealed the presence of two S alleles in each amplification product. These alleles could be cloned and sequenced from each amplification product. The presence of large quantities of S-allele RNA among the stylar RNA population prevents the detection of non-S ribonucleases and contributes to the high specificity of the test, which invariably indicates the presence of a maximum of two sequences. The reliability of the test was confirmed by the correspondence of its results and those arising from the study of fruit setting in the greenhouse. The advantages of PCR approaches for the study of S-allele variation (the collection of essential information on S sequences, avoidance of large crossing programs, the ability to compare alleles from different species or collected at different times...) have been listed by Richman and Kohn (1996).

4.3.2 The Multiple Origins of SI Systems

4.3.2.1 Early Views

4.3.2.1.1 SI is a Primitive Outbreeding Mechanism that Promoted the Expansion of Angiosperms. According to Whitehouse (1950), the key to the sudden rise of angiosperms during the Cretaceous period was the establishment of barriers that prevented self-fertilization and promoted efficient cross-pollination. The opinion is shared by Stebbins (1950), who agrees that the greater flexibility promoted by cross-fertilization must have been at the origin of the expansion of the angiosperms.

Because outbreeding mechanisms, such as dichogamy and dioecism, operate in both the gymnosperms and the angiosperms, Whitehouse suggested that the apparition of SI, essentially restricted to the angiosperms, must have coincided with the expansion of the phylum and, therefore, occurred as a primitive character during the mid-Cretaceous. Whitehouse states that 70% of the genera in angiosperms contain only hermaphroditic species, while no more than 5% are entirely restricted to dioecious species. Such a high proportion of hermaphrodites would not have emerged from competition with dioecious and dichogamous species if SI, the most likely mechanism for the promotion of outbreeding in hermaphroditic plants, had not strongly established itself in the angiosperms. As emphasized by Grant (1949), the rise of a very efficient outbreeding system, such as SI, was likely to occur in strict conjunction with the evolution of specialized pollinating insects (and, therefore, at a very early date during the history of angiosperms).

In addition to the fact that incompatibility systems are well scattered throughout most orders of angiosperms, another reason to consider SI as a primitive outbreeding mechanism is its prevalence at the centers of distribution zones. In his analysis of the relationships between dispersal and the prevalence of self-fertilization, Stebbins (1957) observed several examples where the occurrence of SI was restricted to the centers of origin in *Bromus*, *Hordeum* and *Secale* and where the manifestation of SC prevailed in peripheral zones. Stebbins noted many other cases illustrating the same phenomenon in the Plumbaginaceae (Baker 1948, 1953, 1955) and in the Primulaceae (Ernst 1953). A last argument formulated, at the time, by Whitehouse and Stebbins in support of the hypothesis of an early origin of SI during the evolution of angiosperms is that the requirements of the system for a close carpel and the evolution of styles and stigmas provided a very strong basis for the selection of characters contributing to the differentiation and complexity of the hermaphroditic flower.

4.3.2.1.2 Gametophytic Poly-Allelic Incompatibility Is the Ancestral System and Occurred Only Once. The proposition that gametophytic poly-allelic incompatibility is the ancestral system and occurred only once originated from Whitehouse (1950) and was examined and usually shared by many contemporary scientists (Brewbaker 1957, 1959; Pandey 1958, 1960; Crowe 1964; Koltin and Stamberg 1973). It attributes the profusion of SI systems now displayed by flowering plants to the emergence of mono-factorial poly-allelic GSI during the early history of angiosperms.

The arguments of Whitehouse in favor of the central role of GSI were essentially based on its apparent simplicity and its high outbreeding efficiency. Its early emergence and its spread coincide with the presence, in the primitive angiosperms, of characters now present in all flowering plants (the uniformity of male and female gametes, widespread occurrence of double fertilization, uniform basic plan of the stamens). Its unique origin is considered to result from the very rare combination of events necessary for the establishment and initial propagation of a self-rejection system.

However, the mechanism through which a functional SI system originated from late gymnosperms or early angiosperms was not (and still is not) understood. Charlesworth and Charlesworth (1979b) reviewed and discussed in depth the different models (East 1929; Mather 1943, 1944) proposed to explain the emergence of GSI. Some are based on the assumption that the mutation of a SC allele (if such an allele preceded incompatibility) into an active S allele is possible. Others are based on the hypothesis that a variety of allelic specificities, perhaps involved in disease resistance, were present in the primitive angiosperms without causing an incompatibility reaction between the pollen and pistil. In this case, a functional recognition-rejection system would be progressively established through the selection (against inbreeding depression) of mutations at other loci (polygenes, according to Mather; major loci according to Sect. 4.4.2.2). These loci co-ordinate the location and timing of S-gene expression in the pollen and pistil and the intensity of reactions among their products. The wide distribution of SI among the angiosperms has also been interpreted as evidence of its very early origin and ancestral nature; this view is contradicted (Charlesworth and Charlesworth 1979b) by the absence of SI in several supposedly primitive orders.

4.3.2.2 Current Thoughts

4.3.2.2.1 Towards a General Agreement Regarding the Multiple Origins of SI. Bateman (1952) did not accept the conclusions of Whitehouse and claimed that SI, like any other function, must have evolved progressively rather than from the coincidence of rare mutations. Furthermore, he thought that the highly dispersed distribution of SI throughout the evolutionary tree of the angiosperms was an indication that it had arisen de novo a large number of times and that its repeated occurrence among the angiosperms living today is much more probable than it was at the origin of the phylum. Finally, Bateman challenged the crucial importance of the style as a site of the manifestation and functioning of SI.

Some of the arguments presented by Bateman are particularly prophetic, because they anticipated (Bateman 1952) the discovery of incompatibility in gymnosperms (by Gustafsson) and in pteridophytes (by Wilkie) and hinted at a possible analogy between the origin of SI and that of immunological reactions. During the last 20 years, the thesis of multiple origins of SI systems has been considerably reinforced by the discovery that different proteins and different processes control the SI reaction in different systems [Chap. 3 and an illuminating discussion between Bell (1995) and Read et al. (1995)]. It is now usually agreed (by those who accept the principle of a multiple origin of SI systems) that identical or apparently closely related SI systems (or their essential features) appear in the species of very different families because their strict requirements are met by each family or subfamily through convergent evolution (Sect. 4.3.1.3).

At the same time, it must be remembered that essential information regarding the identity of recognition molecules, particularly in the pollen, is still needed and that, in several SI systems considered widely different (Cruciferae, Poaceae, Papaveraceae), information that may attribute a central role to the recognition and activation of kinase receptors by S-specific ligands is accumulating. It is possible that the recognition-activation mechanism of SI remains ancestral (at least in some systems) but that the choice of proteins submitted to SI-specific phosphorylation–dephosphorylation treatments during the rejection process is progressively adapted to the possibilities and requirements of individual families throughout their evolutionary development. If this is the case, the finding that the stylar ribonucleases from *N. alata* are phosphorylated in vitro by Ca^{2+}-dependent pollen kinases (Kunz et al. 1996; Sect. 3.5.8.3) could be meaningful. It must also be noted that the verification of the SC×SI rule (Chap. 5) in many plant families with different SI systems suggests relationships (possibly resulting from a common ancestry) among the S genes that operate in these SI systems.

4.3.2.2.2 Multiple Gene Systems as an Origin of Mono-Factorial GSI?

The possibility that multiple gene systems are an origin of mono-factorial GSI was raised long ago by Bateman (1952). He suggested that an SI mechanism could arise de novo several times from the selection of a few weakly active serological genes (and their specific modifiers) and their integration, through translocations and inversions, into single linkage groups (Sect. 4.3.3; Uyenoyama 1991; Newbigin 1966; McCubbin and Kao 1999). This integration would be followed by a selection of non-specific modifiers that contribute to the development, function and effectiveness of major S loci. Hayman (1956), who challenged the hypothesis that the complementary bifactorial system of *P. coerulescens* resulted from the duplication of an ancestral S locus, supported Bateman's views regarding the origin of SI. So did Lundqvist (1975) when he speculated that the complementary multi-locus system might be the primitive form of SI from which single-locus incompatibility was derived.

Pandey (1977, 1980) maintained his opinion, however, that the one-locus gametophytic mechanism is the basic ancestor of all SI systems. His first argument (complexity) was that an intricate super-structure, such as the S locus, was not likely to arise simultaneously and independently in different complementary copies unless one such locus evolved first. A second argument was that of protoplasmic limitations, which would not meet the physiological requirements of more than one independent S gene in pro-mono-dicotyledonous lines and, therefore, would favor the secondary evolution of duplicated S genes towards complementarity. Complementarity, in other words, is the logical attribute of a multi-gene system that evolved from duplications of a single ancestral gene.

Lundqvist, at that time and afterwards, was undertaking in-depth studies of the genetics and evolution of multigenic SI in families closely related to the primitive Magnoliales. The material under investigation included monocots (Liliales and one of their branches that led to the grasses) and, in the dicots, a cluster of derived branches leading successively to the Ranunculales (Osterbye 1975), the Cariophyllales and the Theligonales (Larsen 1977). Research results (Lundqvist et al. 1973; Larsen 1986; Lundqvist 1990a, 1990b, 1990c, 1994a, 1994b; Sect. 4.3.1.3) confirmed Lundqvist's conviction that multigenic SI was the origin of mono-factorial SI. He listed several selective advantages of multigenic SI (Lundqvist 1990a) and showed that the "complexity" argument of Pandey had more than one interpretation, because the transformation of a single independent locus into a team of complementary genes also requires considerable genetic engineering. Regarding the argument dealing with the protoplasmic limitations of pollen grains, Lundqvist observed that the upper limit for the number of S genes was not known and opposed examples of diploid (*Lilium martagon*) and tetraploid (grasses, sugar beet, *Ranunculus*) plants that carry several functional GSI loci but remain self-incompatible.

4.3.2.2.3 S Ribonucleases Could Be Operating in a Very Vast Majority of Species from the Dicot Families.

The possibility (Sect. 4.3.1.3), envisaged by Sassa et al. (1996) and tested in part by Richman et al. (1997), that the use of S ribonucleases by the Rosaceae and the Solanaceae constitutes a case of convergent evolution is very important, because the alternative (homology) implies that S ribonucleases are in exercise among the self-incompatible descendants of the most recent common ancestor of the Rosidae (the subclass to which the Rosaceae belong) and of the Asteridae (which include the Solanaceae and the Scrophulariaceae). Richman et al. (1997) calculated that this most recent common ancestor is shared by 80% of dicot families.

4.3.3 Origin of the Different Homomorphic SI Systems and their Relationships

The new evidence from molecular biology regarding the diversity of recognition and rejection mechanisms does not permit the characterization of current SI systems on the basis of their ancestry and of the conditions of their emergence (or re-emergence after population bottlenecks). However, significant approaches to the mathematical treatment and tentative interpretation of research data dealing with the divergence, selection and spread of S alleles have been made in recent years.

4.3.3.1 Origin of Stylar Mono-Factorial GSI and Properties of Allelic Genealogies at the GSI Locus

In the case of GSI, Uyenoyama (1991) presented a three-locus, stepwise model of the origin of GSI; it outlines the selective mechanism through which a genetic system of this kind evolves. The model represents the genetic determination of offspring viability by a locus (B) subject to symmetrically over-dominant selection. The specificity of the proto-S locus is encoded by an initially neutral antigen locus (A), and its expression is governed by an unlinked (enhancing) modifier (M). The evolution towards a complete SI reaction from weak incompatibility occurs through the intensification of the M modifier, which invades the population. Uyenoyama (1991) found that the (relatively) low levels of inbreeding depression generated by a single over-dominant gene can ensure the invasion of the modifier under conditions of sufficiently high exposure of the plants to self-pollen. In other words, as summarized by Clark and Kao (1994), the invasion of the population by the modifier can occur even when inbreeding depression is considerably lower than initially foreseen by Charlesworth and Charlesworth (1979b) if the mean fitness of the offspring is increased. SI is promoted by associations among the components of the system. Enhancement of heterozygosity at the initially neutral proto-S locus improves offspring viability through associative over-dominance (i.e., the apparent over-dominance conferred on a selectively neutral trait by selection elsewhere in the genome). The modifier that enhances the expression of SI develops a direct association with heterozygosity at the over-dominant viability locus.

The properties of gene and allele genealogies at a GSI locus were investigated analytically and checked against numerical simulations (Vekemans and Slatkin 1994) that revealed the presence (as with overdominant loci) of two distinct genealogical processes with markedly different time scales. As mentioned in Section 4.3.1, a method (Uyenoyama 1997) of characterizing the structure of genealogies among alleles that regulate SI was proposed. Generalized least-square estimates of scaled indices of divergence times were obtained for two solanaceous species from genealogies reconstructed from S-allele sequences. Comparison of observed indices and the expected distributions (generated by numerical simulation) indicated, for one of the species studied, that allelic genealogy was consistent with the symmetric balancing selection generated by SI. The genealogy of the second species exhibited long terminal branches (which suggested, however, the action of additional evolutionary processes). As noted by Sassa et al. (1996) and shown by Richman et al. (1997), the success of further research regarding the origin(s) and age(s) of the ribonuclease SI system(s) operating in different families will depend on our ability to distinguish between homology and convergence (Sect. 4.3.1.3).

4.3.3.2 The Origin of SSI

Uyenoyama introduced a generalized least-squares method to approximate estimates and standard errors of function-specific divergence rates and times of divergence in *Brassica* among S-allele sequences and sequences (used as calibrators) not co-segregating with SI and non-S sequences. The results (probably underestimates) do not contribute directly to current discussions of the multiple origins of SI systems; however, they increase the age of the SLG/SRK tandem to five times the age of the *B. oleracea/B. campestris* divergence. Even if an age of 120 million years is accepted as an upper bound, they post-date the monocot/dicot divergence, which is usually defined (Uyenoyama 1995) as more than 200 million years.

4.3.3.3 Evolution of Inbreeding Depression and its Importance for S-Allele Invasion

Lande and Schemske (1985) proposed a model of the joint evolution of inbreeding depression and self-fertilization. Steinbachs and Holsinger (1999) confirmed the central importance of inbreeding depression for the successful invasion of S alleles and showed that the more recessive the SI expression in heterozygous stigmas and the weaker the response induced, the easier it is for an SI allele to invade (Uyenoyama 1988). Clark and Kao (1994) emphasized the selective advantage of a weak and partial system of self-rejection, and Uyenoyama (1991) found that comparatively low levels of inbreeding depression generated by a single over-dominant locus can ensure the invasion of an enhancer of SI.

4.3.3.4 Transitions between GSI and SSI

The early theories or postulates that relate to the emergence of different types of SI were tested in the field. For instance, in 1958, Pandey predicted that a small change in the time of S-gene action at the onset of meiosis could account for evolutionary switches between gametophytic and sporophytic incompatibility. More than 30 years later, Lundqvist (1990b) actually verified the hypothesis when he found that the S genes were in a state of transition between gametophytic and sporophytic control of the pollen in *Cerastium arvense* ssp. *strictum* (Caryophyllaceae with mono-factorial SSI; Sect. 2.4.3). Lundqvist attributed the traces of gametophytic control observed in the small population studied to the effects of occasional successive cell-wall formation that restricted S-specificity determination to the individual microspore. Analogous observations were reported in *Stellaria holostea* (also a member of the Caryophyllaceae with a one-locus SSI system), where Lundqvist (1994a) observed that so-called "quick" alleles appeared to meet a barrier (early cell-wall formation) to the penetration of their S substances in the tetrad of microspores. In such cases, they behave as gametophytic genes that impart their inprint only to the microspore that contains them (Sect. 2.4.3). Finally, as seen in Section 4.3.2.2, elaborate genotypic and segregation analyses gave convincing support to the possibility that multigenic SI pre-dated mono-factorial homomorphic SI and was ancestral to some (or many) of the systems of self-rejection currently operating among angiosperms.

4.3.3.5 Co-Existence of SI and SC Alleles or Breakdown of the System?

Charlesworth and Charlesworth (1979a,b) studied the consequences of invasions of SI populations by different types of SC mutations that can arise via mutations at the S locus. There is always a critical number of active S alleles that ensure the elimination of PPM alleles; this number depends on the extent of inbreeding depression. If the inbreeding depression is small, alleles with no pistil activity can spread regardless of the number of active S alleles initially present.

4.3.4 The Origin of Heteromorphic Incompatibility

Heterostyly occurs and evolves independently in a few genera of 24 different families of angiosperms and is not present in the most primitive subclasses or in the most specialized families (Ganders 1979). It does not occur in flowers with elongated receptacles, apocarpous gynoecia and numerous, or spirally arranged carpels and stamens, which are characters among which the development of reciprocal herkogamy (a prerequisite of heterostylic incompatibility) is impossible (Dulberger 1992).

Considerable research efforts have been made to establish the origin of heteromorphic SI and to unravel the different events that led to the integration of heteromorphy and SI into an integrated and co-ordinated complex of the different genes managing this outbreeding mechanism. Such efforts and their results, outlined below, have been reviewed at regular intervals (Dulberger 1964, 1992; Charlesworth 1979, 1988; Charlesworth and Charlesworth 1979a, 1979b; Ganders 1979; Gibbs 1986; Barrett 1988, 1990; Lloyd and Webb 1992a, 1992b; Barrett and Cruzan 1994; Richards 1997).

4.3.4.1 Arguments against Evolutionary Relationships with Homomorphic SI

Arguments against evolutionary relationships with homomorphic SI (Barrett and Cruzan 1994) suggest the independent origin of heteromorphic SI.

4.3.4.1.1 Homomorphic SI Systems Probably have Multiple Origins. It was seen in Chapter 3 that the physiology and biochemistry of SI in the angiosperms are highly polymorphic, specific to one or few families, and not likely to derive from a common origin. At the same time, it must not be forgotten that current explorations have led to the identification of the pistil determinants of SI recognition and that there is only one system in which both the pollen and the pistil determinants have been characterized. The involvement in SI of protein kinases throughout the angiosperms is perhaps more generalized than anticipated.

4.3.4.1.2 Heteromorphic Incompatibility is Scattered among the Angiosperms and has Poly-Phyletic Origins. Unless evolutionary convergence has intervened, it is difficult to assume that the same physiological mechanisms are present in the 20 different points of origin that mark the distribution of heteromorphic SI in the family tree of angiosperms. In fact, there is evidence that this is not so and that the different patterns of heteromorphic incompatibility (Dulberger 1992; Sect. 3.1.1.2) reflect a diversity of physiological mechanisms. As emphasized by Ganders (1979) and Dulberger (1992), "the mechanisms based on polymorphism in pollen shape or exine sculpturing evolved independently from those based on pollen-size differences." At the end of her survey (1992) of floral polymorphisms and their functional significance in the heterostylous syndrome, Dulberger recognized that different constraints and selective forces participated in the evolvement of heterostyly among taxa. "Nevertheless," she concludes, "it seems that the same set of fundamental laws, presumably of developmental nature, must prevail in the repeated evolution of distylous and tristylous incompatibility".

4.3.4.1.3 There Are Basic Differences between Heteromorphic and Homomorphic SI. That there are basic differences between heteromorphic and homomorphic SI is particularly true of:
- Inhibition sites (usually restricted to one in each homomorphic SI system and several, often morph-specific, in the case of heterostylic species):
- The number of S alleles (usually high in polyallelic series in homomorphic species and usually two in heteromorphic systems):
- The completion of the rejection phase (immediate and, therefore, usually stigmatic in sporophytic homomorphic systems, and often postponed to the style or the ovary in sporophytic heteromorphic species):

Of course, this list of basic differences will remain very incomplete if more information regarding the molecular and cellular biology of SI in heteromorphic species does not become available.

4.3.4.2 Which Came First, Heteromorphy or Incompatibility?

One school of thought, perhaps the most likely for several experts, predicts that di-allelic incompatibility came first, with inbreeding depression as the selective agent (to avoid self-pollination), and was completed by reciprocal herkogamy, which promoted cross-pollination (Charlesworth 1979; Charlesworth and Charlesworth 1979a). Variations in style length appeared before differences in anther height. An important conclusion of the Charlesworths, emphasized by Ganders (1979), is that the evolution of di-allelic SI needs to pass through a functionally gyno-dioecious stage. Therefore, it is submitted to the same requirements and constraints (genetic and environmental) that apply to the evolution of gynodioecy. One weakness of the model is that, in heteromorphic families, there is no clear evidence of the occurrence of a monomorphic species with a two-mating-type incompatibility (Richards 1997).

A second model (Mather and De Winton 1941) suggests simultaneity in the emergence of heterostyly and incompatibility, because the two characters are physiologically associated. Dulberger (1992) notes, among other arguments, that the incompatibility reaction and morph-specific pollen size have not been found to be genetically separated. As seen in Chapter 2, Kurian and Richards (1977) were able to demonstrate independence between the genes for pollen size and male incompatibility.

In a third hypothesis close to the early views of Darwin, Lloyd and Webb (1992a, 1992b) consider that heteromorphic incompatibility was initiated by the evolution of reciprocal herkogamy from a herkogamous ancestor via a stigma-height polymorphism. Di-allelic SI occurred later, probably separately, in each morph. Furthermore, Lloyd and Webb reject the idea of a recognition system in heteromorphic species which would involve the identification of molecular specificities by pollen and pistil. They tend to associate heteromorphic incompatibility with incongruity, i.e., with the absence of a relationship between the pollen and pistil (Chap. 5).

The weakness of the Lloyd and Webb model appears to be the improbability that variations in stylar levels alone could contribute much to pollen exchange between morphs. However, a study by Stone and Thomson (unpublished, cited by Barrett and Cruzan 1994) with artificial flowers and captive bumble bees indicated significant levels of pollen transfer among flowers with different stigma heights.

Finally, a fourth proposal (the "Richards model") suggests the initial evolution of heterostyly (through thrum-linked recessive lethals) from a homostyle ancestor. Such thrum mutants would be established as heterozygotes without proceeding immediately to fixation. Subsequent mutations would cause homostyles to become "pin" and to acquire a self-rejection mechanism linked to the pin chromosome and affecting pollen hydration and germination. This mechanism would then evolve into a two-mating-type incompatibility system. The scheme proposed (Richard 1997) is complex.

Pros and cons of these models can be found in Charlesworth and Charlesworth (1979 a), Lloyd and Webb (1992 a, 1992 b), Dulberger (1992), Barrett and Cruzan (1994) and Richards (1997). A final choice will probably not be possible before the genetic sequences that control pollen–pistil incompatibility in heteromorphic SI species are identified and methods for making a rough guess of their ages and relationships are established (Lewis and Jones 1992). Meanwhile, some clues could be found by detecting (and analyzing) the SI character in "primary homostyles" (Chap. 2), the intramorph cross-compatibility of SI individuals in populations of *Narcissus* (Sect. 2.1.2.4) and the presence of poly-allelic series (Sect. 2.1.3).

4.3.4.3 Evolution of Tristyly

Little is known regarding the evolution of tristyly, for which Charlesworth (1979) tested a model based on many assumptions and relatively few facts. His main hypotheses (Ganders 1979; Barrett and Cruzan 1994) were that the genes governing stylar length and anther position have pleiotropic effects and that the syndrome is not governed by a master locus. In other words, the genes are integrated physiologically, not through linkage. In this case, very slight variations in the developmental pattern of the tristylic characters could be the origin (through different splicings of a same transcription unit or different post-translational treatments) of incompatible behavior or the expression of different pollen S phenotypes by a given S allele at different levels of the flower. However, in the absence of biochemical data, only inferences based on the distribution of tristyly among the angiosperms can be made regarding an evolutionary relationship between distyly and tristyly.

4.3.4.4 The Evolutionary Breakdown or Transformation of Heteromorphy

Ganders (1979) has identified the mechanisms through which distylous systems may be eliminated. They include the appearance of homostyly via recombination within the S supergene (Chap. 2; Sect. 3.1.2.2), the loss of SI after the loss of one morph, the action of modifier genes and the establishment of dioecy after the independent occurrence of female and male sterility. As a general rule, the breakdown of SI will increase under conditions that facilitate self-pollination, reduce pollen exchange and promote inbreeding. SC, as a derived condition of heteromorphic SI, will be more likely to spread after migration (as a single seed) to a favorable environment (Sect. 4.3.5.2). Cases where tristylic SI has evolved towards distylic SI or monomorphy are also known (Sect. 3.1.3.2). This generally occurs via the loss of the M gene (without which the Mid phenotype disappears) and/ or the S gene (which expresses the Short phenotype).

Ganders (1979) provides numerous examples of the evolutionary collapse or evolutionary transformation of tristylic and distylic SI. Richards (1997) reviewed the different means through which tristyly can be eliminated.

4.3.5 SC as the "Paradox of Evolution"

The presence, in most families, of a high proportion of long-established and well-adapted self-compatible species is a surprising fact, referred to as "the paradox of evolution" by Lewis and Crowe (1958). It appears to cast doubt on the ancestral nature and primary role of outbreeding as the driving force behind the spread of angiosperms since the Cretaceous period.

However, there are several indications, identified by Stebbins (1957) or resulting from contemporary research, that suggest a derived condition of SC in families known to host SI species. There are also a number of explanations of the unexpected success of contemporary self-fertilizers.

4.3.5.1 The Derived Condition of SC

The first argument advanced by Stebbins (1957) is that the morphological features of many self-fertilizing species are more specialized than in their cross relatives and, therefore, must have evolved recently. He refers to the elaborate specialized mechanisms for seed dispersal in SC populations of *Bromus*, *Hordeum*, *Medicago* and *Trifolium* and to their annual life cycle, thought to be derived from perennial growth habits (Jeffrey 1916; Ames 1939). A second line of evidence proposed by Stebbins is based on the observation that many self-compatible species clearly display floral features and structures (for instance, lodicles that contribute to pollen dissemination or floral structures specifically adapted to visits by cross-pollinating insects) typical of cross-pollinating populations. The general view of Stebbins is that these floral characters, once established in outbred populations, are essential for the reproductive activity of the plant and remain a permanent part of the SI system. A comparable assessment of the permanence of structures that outlive their function has been made by Ornduff. He found that heterostyly in *Jepsonia* (Saxifragaceae), although ill-adapted to the present pollination fauna and mostly inefficient for enhancing legitimate compatible pollination, is maintained in each population, because the genes governing floral dimorphism are firmly entrenched in the genetic architecture of the genus. The third reason proposed by Stebbins for considering that SC evolves from SI comes from the fact that a number of SC species are suspected to have originated from self-fertilizers during historical times. Two examples are provided: that of the cultivated snapdragon (*Antirrhinum majus*), which arose from unconscious selection for SC by man, and of *Primula vulgaris*, for which homostyly is rapidly replacing heterostyly under conditions that have ceased to be advantageous for heterostyled populations (Crosby 1949). Finally, Stebbins notes the frequent occurrence of mutations leading to the breakdown of SI (Sect. 4.2.2.3); as stressed by Baker (1955), these mutations are essential for the establishment of a single seed accidentally dispersed over a long distance.

4.3.5.1.1 More Recent Arguments. Some of the more recent arguments originate from the application of biochemical and phylogenetic techniques, which show that self-compatible species may harbor and express gene sequences (relic sequences) that belong to an S lineage representative of the SI system occurring elsewhere within the family. For instance, endogenous SLGs having high homology with either class-I or class-II SLG alleles were found to be expressed in self-compatible *B. napus* (Robert et al. 1994), and a relic S ribonuclease is expressed in the styles of self-compatible *N. sylvestris* (Golz et al. 1998; Sect. 4.2.4.3). S relics have also been found in self-compatible cultivars and self-compatible lines that originated from self-incompatible ancestors during historical times (Sect. 4.2.4.3).

A detailed comparison between the S region of the *Brassica* genome and the corresponding homologous region at the ethylene-response position on chromosome 1 (ETR1) of its self-compatible relative, *Arabidopsis*, was recently completed at Cornell University (Conner et al. 1998). The study showed that the sequences of the *Brassica* SLG and SRK genes were missing in *Arabidopsis*, probably as the result of a deletion from the ETR1 region that transformed *Arabidopsis* into a self-compatible genus. The alternative conclusion (that *Arabidopsis* is a born self-fertilizer) is not acceptable, due to the phylogenetic evidence available, particularly the high degree of microsynteny between the *Brassica* S-locus region and its homolog.

As outlined by Conner and co-workers, there are evidently many ways autogamy in a population of self-incompatible plants could emerge. Some are based on the deletion of the S locus, while others do not require the physical elimination of the gene but require its inactivation by major genes, modifiers (usually polygenes) or through genetic duplications leading to gene silencing or competitive interaction between gene products. In *Brassica*, *Petunia* and *Lycopersicum*, several cases (Sects. 4.2.3, 4.2.5 and 4.2.6) of SC plants that conserved relic sequences of an ancestral SI system have been found. Whether or not SI will reappear in such plants if the conditions listed in Section 4.3.5.2 cease to apply is not known, but evidence showing that silent S alleles may become active in modified genetic backgrounds has been found in *P. inflata* (Ai et al. 1991) and *L. peruvianum* (Rivers and Bernatzky 1994; Bernatzky et al. 1995). de Nettancourt et al. (1971) considered the possibility that the apparition of new S alleles in inbred populations of *L. peruvianum* resulted from the activation of silent S alleles.

4.3.5.2 Reasons for the Expansion of Self-Fertilizers

It is difficult to explain the success, as a derived condition, of a breeding system that appears far less advantageous than cross-breeding. In other words, to quote Stebbins (1957), the problem is to determine "why some species should have given up cross-fertilization, with its attendant advantages of population variability and genetic heterozygosity." Several answers are possible.

4.3.5.2.1 Outbreeding is not Always Essential Once the Environment has been Captured. Outbreeding, according to Lewis and Crowe (1958), is an absolute necessity for angiosperms only at the critical stage during which the principle of Gause (1934) – "that two or more forms with identical ecological requirements cannot coexist in the same environment" – is applicable. Past this period, the species, having captured its environment, can do without the benefits conferred by outbreeding and will tolerate autogamy if circumstances initiate the breakdown of SI.

4.3.5.2.2 Inbreeding–Outbreeding Alternations Provide Fertility Insurance. The second argument (Stebbins 1957) implies that a cross-fertilizer in a fluctuating environment will benefit from protection against accidents that affect the dissemination of pollen grains. Stebbins refers, in this connection, to the situation in *Bromus* (Harlan 1945a, 1945b) and *Myosurus* (Stone 1957), where inbreeding and outbreeding alternate from one season (or one set of experimental conditions) to the next. Several other cases dealing with pseudocompatibility and the tendency of many self-incompatible species to display a distinct capacity to set seed under extreme environmental changes can be found in the literature. Charlesworth and Charlesworth (1990) examined the stability of outbreeding populations in response to invasion by alleles that increase selfing.

4.3.5.2.3 SC Facilitates the Establishment of Colonies after the Long-Distance Dispersal of Single Seeds. Baker and Stebbins (Sect. 4.3.5.1) have provided examples of such dispersal; they show that accidental colony forming originating from a single seed associated with SC must have contributed to the formation of several SC species. Particularly if it results from a large deletion of genetic material at the S locus, the SC character in those species will practically never revert to they wild type, because mutations leading to the generation of new S alleles are very rare and because SI re-emergence requires the presence of several different S alleles. Furthermore, it is possible that unilateral incompatibility (Chap. 5) considerably complicates the relationships between isolated SI mutants and the population of SC plants from which they arise. Of course, the re-introduction of SI through pollen migration remains possible during the entire period preceding speciation.

4.3.5.2.4 Self-Fertilizers are Qualified Colonizers. Generation after generation, with regularity and uniformity, autogamous populations in a newly invested territory will propagate the well adapted trait(s) that made the initial colonizers successful. Stebbins (1957) provides several examples of such advantages but agrees that the first invaders must be few in number and that the invaded territories should remain favorable for the new colony for long periods of time.

4.3.5.2.5 Inbred Populations Display High Levels of Genetic Diversity. Although the evolution of SC in out-crossing populations is generally prevented by inbreeding depression (Lande and Schemske 1985), there is little doubt "that inbreeding is an integrated, evolutionary system of considerable importance and not the degenerate dead end previously assumed by the majority of investigators" (Allard 1965; Allard et al. 1968; Pandey 1970b, 1970c). Autogamous plants host a variety of mechanisms (such as cryptic or cyclic outbreeding, intragenic recombination or gene splicing, the storing of relic sequences and the management of heterozygous advantage) that explain why a number of species have been able to avoid self-incompatibility or to switch back to SC throughout evolution.

4.3.6 The Origin of Inter-Species Incompatibility

It was the thesis of Pandey (1979) that inter-species incompatibility (Chap. 5) is the primitive barrier and that intra-specific incompatibility developed later, as a superimposition on inter-specific incompatibility (Sect. 5.1.5.3). His argument was that a specificity that transcends the barriers of species and genera is likely to be more ancient than a specificity restricted to a single species. According to him, intra-species SI (in his opinion, it had to be mono-factorial GSI) developed very early during the evolution of angisoperms, whereas inter-species incompatibility evolved earlier, among the gymnosperms. Pandey's explanation relies on the fact that the protection of the "naked ovules" of the early anemophilous gymnosperms against distinct but related species was a more important priority than the prevention of selfing. As shall be seen in the next chapter, the opposite view tends to prevail for "incongruity" (the result and reinforcement of speciation rather than the initial cause of evolutionary divergence), which expands with biological diversification and complexity.

CHAPTER 5

Incompatibility and Incongruity Barriers Between Different Species

Pre-zygotic incompatibility and incongruity barriers are defined here as any of the relationships (or absence of relationships) between the pollen and pistil that impede the formation of hybrid zygotes between two fertile species. They occur through the failure of the pollen to germinate or failure of the pollen tubes to reach the ovules and discharge sperm nuclei able to fuse with the egg cells. The barriers prevent gene flow among species and [in opposition to self-incompatibility (SI), which restricts inbreeding] establish upper limits to outbreeding and panmixis. They contribute to the isolation of species and consequently favor speciation (or its reinforcement) and the gradual increase of polymorphism within the genus and the family.

It will be seen in this chapter that inter-species incompatibility, which is a starting point of divergence, operates between closely related species belonging to a same family or to a cluster of families hosting the same SI system. The barrier is usually unilateral, among SI species as pistillate parents and among SC species as staminate partners. In certain species of the genus *Nicotiana*, it has been shown (Sect. 5.1.4.3) to involve the intervention of pistil proteins encoded by the S gene active in the pistillate parent. Incongruity, on the contrary, is the consequence of divergence. It does not appear to require S-gene activity and often occurs between species that have been isolated for a long time and have differentiated considerably from one another.

▪ 5.1 Inter-Species Incompatibility Under the Control of the S Locus

5.1.1 The SI×SC Rule

The first suggestion that unilateral pre-zygotic barriers could contribute to speciation and the isolation of fertile species originated from the observation by Vargas Eyre and Smith (1916) that the homostyled self-compatible (SC) species *Linum usitatissimum* and *L. monogynum* accepted the pollen of different SI species but produced pollen that was rejected by heterostyled SI plants. In the Poaceae, Nordenskiold and Quincke (cited in Lewis and Crowe 1958) found examples of unilateral isolation among different species and different genera; this occurred only when the pistillate parent was SI and the staminate parent was SC. In *Lycopersicum*, McGuire and Rick (1954) noted that the barrier of incompatibility between *L. esculentum* and *L. peruvianum* only operated when *L. esculentum* was used as the pollen parent.

The relationship of dependence between the breeding habits of different species and the unilateral pre-zygotic barrier that separates them was first defined by Harrison and Darby (1955), who inter-crossed two SC and six SI species of *Antirrhinum* in all possible combinations. They found that the crosses prevented by a pre-zygotic barrier were those involving matings between SI species as pistillate partners and SC species as staminate parents.

5.1.1.1 Distribution of the Barrier

Three years after Harrison and Darby had characterized this unilateral relationship, Lewis and Crowe (1958) published a general review of the results of intensive crossing experiments that established the SI×SC rule that SI species (used as the pistillate parent) and SC species (used as the pollen source) are isolated by a pre-zygotic barrier. The task performed was considerable and covered the three known systems of SI (heteromorphic, homomorphic sporophytic and homomorphic gametophytic). In all, 43 species from the Cruciferae, Onagraceae and Solanaceae were studied experimentally, and additional data was obtained through a detailed survey of results published in the scientific literature.

5.1.1.1.1 In the Solanaceae. After completion of their analysis, Lewis and Crowe concluded that unilateral (SI×SC) incompatibility is the basic and most common (but not universal) manifestation of inter-species pre-zygotic isolation in the families studied, particularly in the Solanaceae and between that family and the Scrophulariaceae. Within the Solanaceae, four SI species (*Lycium chinense, L. peruvianum, L. hirsutum* and *Solanum jasminoides*) were reported to reject, in all combinations tested, the pollen of seven SC species (*Capsicum annuum, C. frutescens, L. esculentum, L. pimpinellifolium, S. capsicastrum, S. melongena* and *S. pseudocapsicum*). All other cross-combinations (SC×SC, SI×SI and SC×SI) were compatible, as expected from the rule.

5.1.1.1.2 In Sporophytic Systems. Three major studies of the occurrence of pre-zygotic barriers between species and genera of the Brassicaceae were made by Lewis and Crowe (1958), Sampson (1962) and Hiscock and Dickinson (1993). Exceptions were recorded, but the results in each analysis were always positive enough to exhibit a general conformity to the SI×SC rule. The exceptions outlined below in Sections 5.1.2.1 and 5.1.2.2 concerned the occurrence of a high frequency of incompatible crosses among

SI species and the acceptance, by SI pistils, of the pollen produced by SC species that appeared to have only recently evolved from SI.

5.1.1.1.3 The SI×SC Rule in the Grasses. Heslop-Harrison (1982) studied the relationship between the intensity of the SI reaction and the effectiveness of inter-generic pre-zygotic barriers among six different species of grasses (three SI species: *Gaudinia fragilis, Dactylis glomerata, Secale cereale*; one weakly SI species: *Alopecurus pratensis* and two SC species: *Anthoxanthum odoratum, Elymus arenarius*), as judged from the site of pollen inhibition. On the whole, the results showed clear differences between reciprocal crosses and a "general obedience to the rule that SI species tend to reject the pollen of self-incompatibles while producing pollen that functions on the stigmas of self-compatibles." At the same time, Heslop-Harrison insisted that final conclusions concerning the grasses could not be reached before the identification of the recognition factors involved in their intraspecies SI system. An interaction, if it occurs between SI and unilateral incompatible species at all, could involve the rejection mechanism without the participation of S specificities. He proposed two possible ways to explain cross-incompatibility. One implies a pollen–stigma recognition of differences through a possible intervention of the S and/or Z loci, followed by a rejection reaction and the arrest of the tube. Another path, typical of late-acting barriers between distantly related genera, are non-specific and result from disharmony or lack of co-adaptation between pollen and pistil.

Several authors also questioned the generalization of the SI×SC rule and/or the involvement of the S locus in the rejection of the pollen of SC species by SI pistils. Their arguments and concepts are presented in Sections 5.1.2, 5.1.3 and in 5.2, but it should be noted that none of these deny the widespread occurrence of inter-species, unilateral, pre-zygotic barriers.

5.1.2 Many Exceptions to the SI×SC Rule

5.1.2.1 They Occur Essentially in the Case of SI×SI Pollination

The exceptions to the SI×SC rule have been listed by Abdalla and Hermsen (1972) and Hogenboom (1972a, 1973, 1975, 1984). They show that unilateral pre-zygotic isolation may take place between SC species in genera such as *Antirrhinum* (Harrison and Darby 1955), *Lycopersicum* (Martin 1964, 1967; Mutschler and Liedl 1994) and *Nicotiana* (Pandey 1968, 1969; Kuboyama et al. 1994) and, more commonly, between populations of SI plants (Stout 1952; Pushkarnath 1953; Garde 1959; Dionne 1961; Martin 1961; Pandey 1962a, 1968; Sampson 1962; Ascher and Peloquin 1968). Cases where the barrier operates between SC forms as maternal parents and SI species as pollen sources are also known (Hardon 1967; Pandey 1968, 1969).

Hiscock and Dickinson (1993), in their re-examination of inter-species and inter-generic pre-zygotic relationships within the Crucifers, supported earlier results of Sampson (1962) and confirmed the existence of an SI×SC unilateral, pre-zygotic, inter-species barrier within the family. The results of the study, in conformity with the predictions of Lewis and Crowe, showed that a large majority of SI×SC crosses (93%) are incompatible and that the outcome of SC×SI and SC×SC crosses is fully compatible (usually) or partially compatible in more than 90% of the cases. The main deviation from the rule was SI×SI pollinations, which were found to be unilaterally incompatible or, rarely, reciprocally incompatible in 58% of the crosses studied.

5.1.2.2 SI×"Sc" Crosses

Lewis and Crowe were among the first to identify and recognize the outcomes of crosses that did not follow the SI×SC rule. In their conclusions (Sect. 5.1.1), they also noted that SC species of recent origin (i.e., species where self-compatibility has a high adaptive value, which do not display an inbreeding depression and which have SI relatives) produced "Sc" pollen generally accepted by the pistils of related SI species. The restriction is important, because it accounts for some of the SI×SC relationships that do not follow the rule. It also emphasizes the slow evolution of the inter-species pre-zygotic barrier and its secondary nature, which superimposes itself onto other isolation mechanisms.

On the basis of their observations, Lewis and Crowe postulated that the transition from SI to SC occurs in a series of mutational steps involving, in chronological order, the three following types of change (Table 5.1 A):

- SI into Sc, the condition of a recent SC species, such as *Antirrhinum majus*, which produces compatible pollen on the pistils of SI species and displays the same inhibition capacity as SI species in the pistil.
- Sc into Sc′, where the plant bearing this change does not inhibit the pollen of SC species and produces pollen compatible with SI pistils. This stage is purely hypothetical, and no example of a species expressing such a phenotype can be given.
- Sc′ into SC, the phenotype of a SC species that, like *L. esculentum*, evolved long ago from SI and sheds pollen inhibited by the pistils of SI species.

The transitional step (Sc′) appears to be an unnecessary complication, but Lewis and Crowe argued that a pollen grain carrying the newly arisen change would not find a compatible pistil. The explanation is debatable, because one could imagine that the mutation into SC occurs in an egg cell or in a microspore already determined for the SC phenotype.

In relation to the multiple forms of SC mutations predicted by Lewis and Crowe, reference must be made to the SC mutant LA2157, which was discovered by Rick (1982, 1986) in a population of SI *L. peruvianum*. Self-compatibility in this wild tomato mutant, the only one known to result from a change at the S locus, is due to the loss of a histidine residue at the

Table 5.1. Inter-species incompatibility under the direct control of the self-incompatibility locus. (de Nettancourt 1977)

	SI species	SC species		
A. S polymorphism for inter-species incompatibility in self-compatible species [a]				
Male and female	SI	S_c	S_c'	S_C
SI	+	+	+	–
S_c	+	+	+	–
$S_{c'}$	+	+	+	+
S_C	+	+	+	+
B. S polymorphism for inter-species incompatibility in self-incompatible species [b]				
Male and female	SI	SF	Sf	
SI	+	+	+	
SF	+	+	–	
Sf	+	+	+	

SC, self-compatible; SI, self-incompatible

[a] According to the scheme proposed by Lewis and Crowe (1958), different relationships between SI and SC species can be attributed to the occurrence of different forms of self-compatibility alleles

[b] As shown in *Petunia* (Bateman 1943; Mather 1943) and in *Nicotiana* (Anderson and de Winton 1931; Pandey 1964), different relationships between SI and SC species can also be attributed to the occurrence of different forms of self-incompatibility alleles

active site of the S-ribonuclease. The crossing behavior of LA2157 revealed no indication of a unilateral barrier with SI plants, and reciprocal SI×SC hybrids were easily produced (Bernatzky and Miller 1994). Progeny testing showed that the SC and SI alleles are co-dominant. Co-dominance was also observed in the offspring of hybrids between SI *L. hirsutum hirsutum* and SC *L. hirsutum glabratum*. In this case, however, although the exact nature of the SC character has not been determined, there is unilateral incompatibility between the sub-forms *hirsutum* and *glabratum*.

5.1.3 Differences between the SI Reaction and Inter-Species Rejection Processes

5.1.3.1 Situation in the Solanaceae

5.1.3.1.1 Observations with the Light Microscope. Lewis and Crowe (1958), de Nettancourt et al. (1973b, 1974a), Hogenboom (1972a–e) and Liedl et al. (1996) studied pollen growth in *Lycopersicum* and found that the inhibition reaction of SC pollen in SI pistils varies in intensity, time of occurrence and morphology with that of selfed SI tubes. In particular, very clear

differences between the growth of *L. esculentum* pollen in *L. pennellii* styles (which is inhibited in the first 3 mm of the style) and the performance of SI pollen [SI *L. pennellii* (selfed), which is able to reach the lower portion of the style at a very slow but continued speed] were observed by Liedl and co-workers. The finding is not surprising, given probable variations in the growth requirements and growth capacity of pollen tubes from different solanaceous species and the involvement (before or after the recognition phase) of a number of associated loci that are not necessarily identical to those participating in the SI reaction (Sect. 2.5; Martin 1961, 1964, 1968; Chetelat and DeVerna 1991; Liedl et al. 1996).

At the same time, however, it must not be forgotten that pollen-tube growth is irregular, particularly sensitive to external factors and likely to give rise to contradictory information. To illustrate this variability, Lush and Clarke (1997) noted that, in the case of *Nicotiana alata*, callose plugs are formed more frequently (Tupý 1959) or less frequently (Murfett et al. 1994) in compatible tubes and that inhibition occurs two-thirds of the way down the styles (Pandey 1979) or in the upper 5–10 mm (McClure et al. 1990). In their own work with *N. alata*, Lush and Clarke found that most SI tubes were less than 10 mm long 3 days after pollination, while many compatible tubes had traveled the 60 mm to the ovary. Ultimately, the ability of *N. alata* SI tubes to grow should depend on the concentration of specific S-ribonucleases in the style and, possibly, on de novo RNA synthesis by the incompatible tubes (Chap. 3). The growth of incongruous pollen in tobacco is also very diversified, as remarkable differences in the behavior, length and morphology of rejected tubes could be observed in *N. tabacum* styles after pollination by three different but closely related SC species of *Nicotiana* (Kuboyama et al. 1994).

5.1.3.1.2 Electron Microscopy. Interestingly, no differences between sections of *L. peruvianum* pollen tubes growing through compatible styles of *L. peruvianum* and those growing through styles of *L. esculentum* have been observed by de Nettancourt et al. (1973a,b, 1974a; compare Fig. 5.1 and Fig. 3.16). As far as can be inferred from pollen-tube morphology, the *L. peruvianum* tubes find an environment that is clearly adapted to the *peruvianum* requirements in *L. esculentum* styles. However, the inhibited tubes of *L. esculentum* are different from the incompatible tubes of *L. peruvianum* (Fig. 5.2a,b). Nevertheless, a relatively constant feature of the two types of rejection was the appearance of the endoplasmic reticulum as a whorl of concentric layers; this has been reported to be typical of a cessation of protein synthesis (de Nettancourt et al. 1973b, 1974a; Fig. 3.17).

5.1.3.2 In the Brassicaceae

The analyses of Hiscock and Dickinson mentioned in Section 5.1.2 concerned a series of inter-species and inter-generic crosses involving 13 SC species and 12 SI species from ten different tribes of the Brassicaceae. No

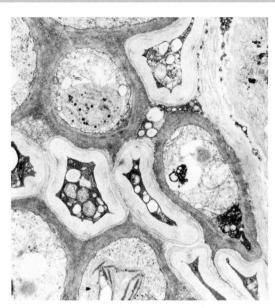

Fig. 5.1. Cross sections of pollen tubes from SI *L. peruvianum* growing in styles from SC *L. esculentum*. Note the similarities with sections of SI *L. peruvianum* tubes growing in cross-compatible styles of SI *L. peruvianum*. (Fig. 3.16; courtesy of Prof. M. Cresti, University of Siena)

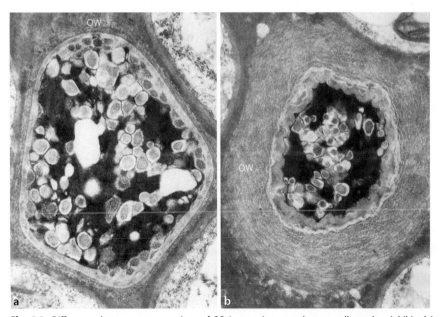

Fig. 5.2. Differences between cross sections of SC *Lycopersicum esculentum* pollen tubes inhibited in styles of SI *L. peruvianum* (**a**) and the pollen tubes of SI *L. peruvianum* rejected after self-pollination (**b**). Note the thin outer wall of the *L. esculentum* pollen tubes after rejection by the *L. peruvianum* styles and, in both cases, the presence of bi-partite particles in the tube cytoplasm. (de Nettancourt et al. 1974a)

significant differences were reported in the behavior of SI pollen after self-ing and the behavior of SC pollen in SI×SC crosses. Unilateral incompat-ibility (UI), like SI, was found to be overcome by bud pollination or by treating the stigma with the protein inhibitor cycloheximide. However, im-portant differences have been found between SI (clearly stigmatic) in *Bras-sica* and unilateral inter-generic incompatibility between *Brassica* and *Ara-bidopsis*, where rejection takes place in the ovary (Kandasamy et al. 1994).

5.1.4 The Involvement of the S Locus

The verification of the SI×SC rule in many combinations of inter-species crosses is certainly suggestive of participation of the S locus in the pre-zy-gotic isolation of related species with different breeding habits. The argu-ment is not decisive (Sect. 5.2) but other lines of evidence confirm, when materials and techniques are available, the role of the S locus in the separa-tion of species within certain families.

5.1.4.1 Unilateral Pre-Zygotic Isolation is an Active Process in *Brassica*

Support of the hypothesis of an involvement of S genes in inter-species in-compatibility was provided (Sect. 5.1.3.2) by a study of the effects of cyclo-heximide and bud pollination on pre-zygotic inter-species isolation in cab-bage (Hiscock and Dickinson 1993). The barrier obviously shares impor-tant features with the SI mechanism of *Brassica*: it is not established in flower buds and requires protein synthesis at the stigma surface.

Hiscock and Dickinson were also able to identify a positive correlation between the ability of an SI species to inhibit other SI pollen and the abil-ity of its own pollen to grow on other stigmas. They constructed a "cross-ability" index that expresses the combined ability of the species to protect its own germ plasm and to invade those of other species. It is unlikely that such a complex relationship could have evolved exclusively from the loss of a relationship between species.

5.1.4.2 SF, a Class of S Alleles that Clearly Display a Dual Function

SF alleles, first discovered in the SI species *Petunia violaceae* (Bateman 1943; Mather 1943) and *N. alata* (Anderson and De Winton 1931; Pandey 1964, 1967, 1969, 1973), are known to occur in at least five different SI spe-cies of *Nicotiana*. Strains that carry them are normally SI but have the ad-ditional property of rejecting the SC pollen of *Nicotiana langsdorfii* pollen and, for some of them, the SC pollen of other *Nicotiana* species (Table 5.1b). In other words, SF alleles behave as though they express two levels of specificity: one that functions within the species and the other (inde-pendent of it, allelic to normal S alleles or very closely linked to them), which controls inter-species incompatibility. In more than 12 000 plants from five SI species tested by Pandey (1969), not a single case could be found where the relationship between intra- and inter-species rejection processes, typical of each allele, could be broken. Nevertheless, inter-species incompatibility usually occurred in the stigma and intra-species SI occurred in the style (Pandey 1964). The discovery of SF seems to provide a parallel, for SI species, of the situation postulated to occur in SC pollen, where Lewis and Crowe (Sect. 5.1.2.2) predicted the existence of Sc and Sc' alleles. It would be interesting to sequence the SF allele and its product and to use the approaches successfully followed (Sect. 5.1.4.3) by Murfett et al. (1996) to introduce loss or gain of SF function in the backgrounds of SI and SC species.

5.1.4.3 Unilateral Pre-Zygotic Isolation Requires the Action of S-Ribonucleases in *Nicotiana*

The decisive proof that the S gene participates in inter-species isolation, at least in the genus *Nicotiana*, was presented by Murfett et al. (1996). The team from the University of Missouri at Columbia (Fig. 5.3) used sense and anti-sense constructs to manipulate the expression of S-ribonuclease alleles (SA2 and SC10) from SI *N. alata* in different genetic backgrounds. The breeding behavior of transformed plants was tested with pollen from three SC species (*N. plumbaginifolia*, *N. glutinosa* and *N. tabacum*). These three species exhibit UI with *N. alata*, but only *N. plumbaginifolia* follows the SI×SC rule, as the pollen of *N. glutinosa* and *N. tabacum* is rejected by both the SC and SI lines of *N. alata*. The difference cannot be attributed to a recent origin of SC in *N. alata*, because this accession was utilized only as a pistillate partner and because, even if an age effect could be exerted on the compatibility performances of the pistil, the *L plumbaginifolia* pollen was rejected by SI *N. alata* but not by SC *N. alata*.

The results showed (Fig. 5.3) that pollen from all three SC species was consistently rejected by transgenic plants expressing S-RNases. However, there was a very strong interaction between the S-RNase transgenes and the genetic background. Pollen from *N. plumbaginifolia* was only rejected when S-RNase was expressed in conjunction with other factors from the *N. alata* background, but pollen from *N. glutinosa* or *N. tabacum* could be rejected in the absence of such factors (Fig. 5.3). For example, transgenic *N. plumbaginifolia* plants expressing SA2-RNase reject pollen from *N. glutinosa* and *N. tabacum* but remain compatible with pollen from non-transformed *N. plumbaginifolia*. However, when such plants are crossed with SC *N. alata*, the resulting transgenic (*N. plumbaginifolia*×SC *N. alata*) hybrids have the ability to reject *N. plumbaginifolia* pollen. Thus, on the basis of the different interactions between the S-RNase and the genetic background of the transgenic plants, there are at least two S-RNase-dependent mechanisms for rejecting pollen from SC species.

Interestingly, these experiments also revealed at least one S-RNase-independent pollen-rejection mechanism: pollen from *N. glutinosa* and *N. taba-*

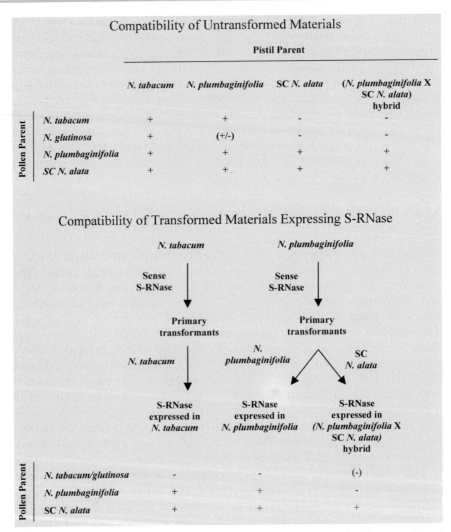

Fig. 5.3. Behavior of the pollen from four different SC species in transformed and non-transformed styles (with or without S ribonuclease) in different genetic backgrounds. (Courtesy of Dr. B.A. McClure, University of Missouri at Columbia)

cum is rejected by SC *N. alata* that expresses no S-RNase. Pollen from *N. plumbaginifolia* is not rejected by this S-RNase independent mechanism. Such results highlight differences between UI mechanisms that follow the SI×SC rule and those that do not. The rejection of pollen from *N. plumbaginifolia* requires S-RNase and other factors that are probably needed for SI. Therefore, a mutation that leads to a loss of style-part function in SI species will also lead to loss of the ability to reject pollen from SC species, such as *N. plumbaginifolia*. However, because the genetic mechanisms for rejecting pollen from species such as *N. glutinosa* and *N. tabacum* are different and do not depend on the same factors as SI pollen rejection, it is possible that a mutation leading to a loss of SI will not result in a change in the UI relationship. To add to this complexity, so typical of pollen–pistil interactions, multiple, redundant mechanisms may contribute to the rejection of pollen from some species (i.e., pollen from *N. glutinosa* and *N. tabacum* is rejected by both S-RNase-dependent and S-RNase-independent mechanisms). In such cases, a breakdown in SI is not expected to be associated with a change in a UI relationship.

Murfett et al. (1996) also examined the morphologies and sites of pollen-tube inhibition associated with these different pollen-rejection mechanisms. Rejection of *N. plumbaginifolia* pollen occurred on the stigmas of transgenic (*N. plumbaginifolia*×SC *N. alata*) hybrids, in conformity with earlier results suggesting a tendency towards an earlier site of action for UI mechanisms. However, *N. plumbaginifolia* pollen rejection appears to be mechanistically very similar to SI in *N. alata*. In contrast, rejection of *N. tabacum* pollen occurred in the styles of transgenic plants expressing S-RNase and was similar to pollen rejection in SI even though the genetic mechanisms are clearly different. Overall, the conclusion of the study was that the morphology and site of pollen-tube inhibition are not good indicators of the underlying rejection mechanism.

In separate experiments, Beecher et al. (1998) tried to find out whether a non S-RNase (i.e., carrying no S specificity) could substitute for S-RNase in *N. plumbaginifolia* pollen rejection. The results were negative: S-RNases, regardless of their specificity, are required for the rejection of pollen from SC species. McClure's group is currently testing whether chimeric S-RNases that have lost the capacity for S-allele-specific pollen rejection are nonetheless still active in rejecting inter-species pollen (McClure, personal communication).

The results from the McClure group clearly implicate the S locus in unilateral pre-zygotic isolation (i.e., unilateral inter-species incompatibility) and demonstrate that, at least in *Nicotiana*, these pollination barriers result from active barriers and not from the mere absence of a relationship. It now remains to examine other genera among the Solanaceae, Scrophularicaceae and Rosaceae and families with very different SI systems, such as the Brassicaceae and the Poaceae, for which ribonucleases are not implicated.

5.1.5 S-Recognition Structures Participating in Inter-Species UI

A discussion is probably premature, because the results of the *Nicotiana* experiments have to be confirmed with other plant material and because the identities of the determinants that participate in the recognition phase of SI are not known. However, now that there is evidence (to some extent in *Brassica* and certainly in *Nicotiana*) of a participation of the S locus in both SI and inter-species UI, it seems appropriate to briefly review past and current thinking regarding the structures that could regulate this dual function.

5.1.5.1 The Antigen–Antibody Model

In the antigen–antibody model (Lewis and Crowe 1958), elaborated more than 40 years ago, it is assumed that every S allele of an SI species produces a non-specific pollen enzyme (PE) and a pollen protein (PPS), specific to the recognition sequence of the S allele, which binds itself to PE to form a protein–enzyme complex (PE×PPS). The same S allele produces two different proteins (antibodies in the terminology of Lewis and Crowe) in the pistil: a primitive one (PA) able to combine with the unbound pollen enzyme and a more recent one (RB), S-specific and possibly identical to PPS. The latter protein combines upon selfing with the S-specific protein of the enzyme–protein complex present in the pollen. The SI reaction is initiated as soon as the S-specific protein of the pollen PE×PPS complex (acting, for instance, as an usher of self-ribonucleases into the pollen tube or as a ligand of SRK) recognizes the pistil-specific antibody (RB) upon selfing.

To explain incompatibility between SI and SC species, Lewis and Crowe propose that the S locus in SC species lacks the specific recognition sequence (deleted or silent) and only codes for the non-specific PE. In an SC×SI cross, the pollen or pollen tubes of the SI parent do not find any of the non-specific (PA) or S-specific (RB) proteins able to recognize PE or PPS in the SC pistil; consequently, they germinate or grow normally towards the ovary. The pollination is cross-compatible. In the reciprocal combination (SI×SC), PE remains unbound, because SC pollen does not produce PPS and is recognized by PA, the non-S-specific protein. The two proteins combine to form a complex able to promote the rejection of SC pollen.

5.1.5.1.1 SI×SI Crosses.
On the contrary, inter-species or intra-species crosses between plants with different S alleles remain compatible, in theory, because the non-specific pistil antibody (PA) matching the non-specific PE does not bind to PE if PE is established in a complex with PPS. Because the specific stylar RB protein fails to recognize the PPS element of the pollen complex when different S alleles are present in the pollen and pistil, the cross is compatible.

5.1.5.1.2 SI×Sc Crosses and Sc×SC Crosses.
Lewis and Crowe explain cross-compatibility between SI species as pistillate parents and SC species of recent origin (Sc) by making the assumption that Sc plants still produce the PA and RB antibodies in the style but encode a pollen sequence (PE) not recognized by PA. Therefore, the pollen of Sc species is not inhibited by SI plants. Its pistil, as suggested by rare cases of cross-incompatibility among certain SC species, rejects SC pollen.

5.1.5.1.3 Actuality of the Model. The model implies the intervention of several S-specific and non-specific sequences that emphasize the complexity of the S locus (as it exists today) and the possible origin of monofactorial SI via linkage, gene loss or gene silencing through allele erosion. The dispersion and independent evolution, possibly in many species, of unlinked non-specific sequences, such as PA and PE, could explain some of the exceptions to the SI×SC rule and the erratic breeding behavior of inter-species hybrids or their progenies. Irregularities of this kind usually do not appear in inheritance studies of SI, because the pollen and pistil determinants of S-specificity need to be either allelic or very closely linked if the system is to be maintained. Whether or not the various determinants imagined by Lewis and Crowe are real will not be ascertained until far more information is available regarding the molecular biology of UI. In the meantime, it is possible to transpose the model to cases of inter-species incompatibility involving GSI systems of the *Nicotiana* type or SSI typical of *Brassica*. In genera characterized by stylar monofactorial GSI, the PA×PE complex formed after an SI×SC cross would have the "gatekeeper property" identical to that of RB×PPS in a self-pollinated SI style [i.e., the ability to liberate active (or activated) stylar S-ribonucleases in SE pollen tubes]. In *Brassica*, the PA×PE complex would behave as a kind of universal S ligand, possibly designed to activate SRK.

5.1.5.2 The "Area Hypothesis"

According to Sampson (1962), a state of compatibility after self- or cross-pollination cannot be achieved (at least in the Cruciferae) if a substance in the pollen fails to recognize a corresponding substance in the pistil. Sampson postulates that each of the two substances displays two attachment sites: the "species" area and the "S-allele" area. One attachment between pollen and stigmatic substances in the species area is required for inter-species compatibility. A second attachment in the S-allele area results in intra-species incompatibility. Compatibility within the species is the outcome of species-area combination in the absence of S-allele-area combination. In other words, as summarized by Heslop-Harrison (1982) in his attempts to distinguish a relationship between SI and inter-species incompatibility in the grasses, one of the process (SI) is based on the recognition of genetic similarities, while the second, when it involves a specific recognition exercise, is based on the recognition of genetic differences. The model does not seem to conform to the SI×SC rule, because no difference is anticipated in the outcome of inter-species reciprocal crosses between SI and SC species. It explains, however, why inter-generic SI×SI crosses fail for lack of species-area attachment. It is interesting, given the current debate regarding incongruity (Sect. 5.2), that Heslop-Harrison (1982) identified cases of inter-generic pre-zygotic isolation which, on the basis of the intensity of responses, could have resulted from physiological maladjustment of the pollen and pistil.

5.1.5.3 The S Locus as a Cluster of Primary and Secondary Specificities

Pandey (1962b) suggested that the physiology of the S gene is controlled by four components: growth substance, primary specificity, protective substance and secondary specificity, each with corresponding pollen and stylar units attached in the same order, the component for secondary specificity being added last. The components of primary and secondary specificities protect the growth substance in the pollen against primary and secondary specificities in the pistil, which themselves protect the growth substance of the pistil. If the two specificities are lost, this growth substance can still be prevented from inactivation by a separate protective substance present in the pollen and pistil.

In this scheme, the majority of SI species are tabulated as SI and are considered to possess all four components. The SI pollen is thus fully protected against the pistil of SC and SI species in all circumstances. The only exceptions are SI pollinations where the primary and secondary specificities in the pollen are identical to those of the pistil and combine, after selfing, to form a tetramer peroxidase (Pandey 1967) having the ability to destroy the growth-promoting hormone. The S-specificity and involvement of this tetramer peroxidase in pollen rejection have not been demonstrated, because, in *N. alata*, Bredemeyer and Blaas (1980) were unable to find any relationship between the patterns of peroxidase isoenzymes of pollen or styles and different S genotypes.

Other species, whether SI (like the SF strain of *N. alata*; Sect. 5.1.4.2) or SC, lack one of the four components defined by Pandey in the pollen and/ or the pistil. The selfing and crossing behaviors of two given species depend on the presence of the components left undisturbed in the pollen and/or style and correspond (Pandey 1962b) to the situation currently met in nature. The SC species that accept all pollen and produce pollen rejected by SI styles are thus considered to lack primary and secondary specificities in both the pollen and style. In these species, the growth substance of the pollen is completely unprotected and exposed in any pistil containing a secondary specificity. On the contrary, the Sc species (Sects. 5.1.2, 5.1.5.1) known to be reciprocally cross-compatible with SI plants, such as *Nicotiana langsdorfii*, are assumed to possess (in the pollen) the protective substance and the primary specificity necessary to protect the growth substance and to display (in the pistil) the four components that constitute the phenotype of regular SI species. Because there is no primary specificity in the pollen of the *N. langsdorfii* plants to match the primary specificity in the pistil, the plants are all SC and would only reject the pollen of SI individuals carrying the same primary specificity as that of the *langsdorfii* pistil.

In the same manner, Pandey has explained the compatibility phenotypes of unusual strains or species of *Nicotiana* or *Solanum* by assuming that these exceptional plants lack one or several of the components that usually make up the S phenotype of SI plants. In fact, the strengths and weaknesses of Pandey's model are in its complexity, which requires not less than eight sub-components of the S gene. The model can explain all exceptions to the SI×SC rule, but it is extremely speculative and is not supported by current knowledge regarding the molecular biology of SI. As far as stylar GSI is concerned, any correspondences between the hypothetical units of Pandey's model and the various processes [both involving and not involving the S-specific ribonucleases active in SI plants (Sect. 5.1.4.3) that participate in the different rejection systems responsible for inter-species incompatibility] will be established from work of the type performed by Murfett et al. (1996).

5.1.5.4 Why SI×SI Crosses Often Fail to Follow the SI×SC Rule

Hiscock and Dickinson (1993) have proposed a model inspired by the "area hypothesis" of Sampson (1962). Their model provides an interpretation for the high frequency (>50%) of crosses between SI species of the Brassicaceae that are unilaterally or, rarely, reciprocally incompatible. Their general assumption, in this model, that barriers to pollen tubes only occur in SI pistils readily accounts for the compatibility of SC×SC and SC×SI crosses and the incompatibility of SI×SC pollinations. The unpredictable outcome of SI×SI crosses is attributed to the presence (in the pollen of SI species) of a specific (key) molecule of a size that may vary between different SI species but that always matches the dimensions of the lock (pistil barrier) within the species. After an intra-species pollination, the key opens the lock, and the pollen is accepted, provided that its specific S phenotype does not antagonize the S phenotype of the pistil (SI). In the case of reciprocal crosses between SI species, the outcome only depends on the ability of the key to enter the lock. Three different situations can occur: UI when there are large inter-species differences in key sizes, and either reciprocal cross-compatibility or reciprocal cross-incompatibility if the lock and key have approximately the same size and match each other in some or many of the cases.

The model of Hiscock and Dickinson is simple and particularly useful, because it accounts for the most important deviations from the SI×SC rule. The molecules that act as keys and locks in SI species must be identified, and the mechanism involved in the opening and closing of the inter-species pistil barrier must be explained. It is possible (Stephenson et al. 1997) that a molecule involved in inter-species incompatibility belongs to the class of basic, cysteine-rich proteins (PCP-A) located in the pollen coating. Such proteins are considered to provide the male determinant of SI in *Brassica oleracea* (Doughty et al. 1999; Schopfer et al. 1999; Sect. 3.2.4.4). This hypothesis is supported by the ability of PCP-Al to bind products of the S locus and, possibly, by certain features of bioassay data (Stephenson et al. 1997).

5.1.6 Other Genetic Loci Also Participate in Inter-Species Incompatibility

The analysis of F1 hybrids originating from crosses between SC and SI species often reveals the presence of major genes and polygenes that not only control the expression of S alleles in SI or incompatibility between species but also participate, through their activity or the transformation (for instance, phosphorylation) of their products, to the rejection phases of reactions of self- and inter-species incompatibility. These genes are present in the parental species and segregate in the advanced progenies of inter-species hybrids to produce individuals with different compatibility and incompatibility phenotypes. The complexity of the genetic situation and of interactions among gene products is such that it is very difficult to study the inheritance of inter-species UI. It will be seen, however, that the genetic factors involved in the control of inter-species incompatibility are often the same elements that participate in the rejection of self-pollen.

5.1.6.1 The R Locus of *Solanum*

The control of S genes by other genes was first suspected by Pushkarnath (1953), who discovered UI in *Solanum* and noted that, in the cross *S. subtilis*×*S. aracc. papa*, all F1 hybrids are SI, are reciprocally incompatible among themselves and with the male parent, and are unilaterally compatible with the female parent. Pushkarnath explained this behavior by postulating that *S. arrac. papa* contributes the dominant sporophytic allele of a locus R, which inhibits the growth of any pollen tube containing an SI allele. Garde (1959) discovered similar relationships of UI in *Solanum* hybrids, which he also attributed to an interaction between S and R alleles.

5.1.6.2 A switch gene in *Lycopersicum*

Martin (1961, 1964, 1968) constructed hybrids between SI *L. chilense* and SC *L. hirsutum f. glabratum*, SC *L. esculentum* and SI Cajamarca (*L. hirsutum f. hirsutum*), and between SC *L. esculentum* and different SI accessions from Cajamarca and from the SI hybrid *L. esculentum*×*L. peruvianum* hybrids. Through detailed analyses of F2 and back-cross progenies from each of the combinations, Martin found that unilateral inter-species incompatibility and SI were regulated by the same dominant sporophytic switch gene(s), one in some cases and two in others, transferred from SI parents. When the dominant switch gene(s) and an S allele are present together in the same plant, UI and SI are expressed. If either the switch gene or the S allele is lacking, the incompatibility character is not displayed. According to the interpretations of Martin, the genotypes of SI and SC species and their progenies are represented, with the specification of their breeding behavior, in Fig. 5.4.

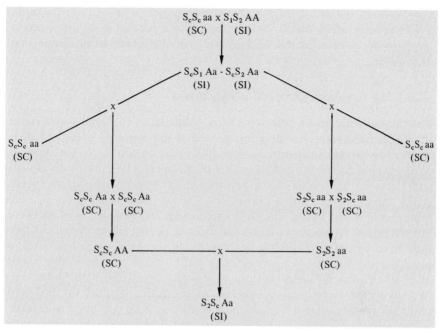

Fig. 5.4. In certain tomato strains, SI and unilateral inter-species incompatibility are governed by the S locus and a sporophytic dominant switch gene present (AA) in SI species and absent (aa) in Sc forms. The implications of the presence of this switch gene were almost entirely verified by Martin (1961, 1968), who showed, with material where incompatibility was regulated by two dominant loci (A and B), that compatible back-crosses of the Sc×SI hybrid progenies to the SC parent, used as the pistillate partner, lead to the occurrence of self-compatible forms (ScSc AA and SxSx aa) that inter-cross to produce self-incompatible hybrids. (de Nettancourt 1977)

The results of Martin have many implications. For instance, in material where incompatibility was regulated by two dominant loci (A and B), he found that the hybrid between two SC forms expressed an SI phenotype. This result, which can easily be explained by the assumption that the SC forms had genotypes SvSwAAbb and SxSyaaBB, shows that the hope of Larsen et al. (1973) – that an SI system could be reconstructed through crosses between distantly related SC forms – is justified. Another important contribution of Martin was his discovery that the S alleles present in F2 and F3 segregants, although they maintain their functions in the style, progressively lose the specific property that permits pollen containing them to grow in styles with different S alleles. Thus, it is possible to explain the occurrence of UI between two SI forms on the basis of such a partial alteration of the activity of S alleles in certain genetic backgrounds.

The results of Martin were convincingly confirmed by Bernatzky et al. (1995), who showed that S proteins could be recombined with self-compatibility in back-cross progenies of SI (*L. hirsutum*)×SC (*L. esculentum*) inter-species hybrids. This result and other data indicate that the presence of

S proteins is not sufficient for the expression of SI in SC *L. esculentum*. Evidence suggesting that *L. esculentum* hosts a silent S gene that becomes active in the proper genetic background was also obtained in inter-species derivatives.

5.1.6.3 The Two-Power Competition Hypothesis

The two-power competition hypothesis (Abdalla 1970; Abdalla and Hermsen 1972) takes into consideration several of the elements presented above and the discovery (Grun and Radlow 1961) that certain forms ("acceptors") of SI species accept the pollen grains of SC species. According to Abdalla and Hermsen, UI is due to the presence in standard SI species of specific genes, the UI genes, which prevent any pollen grain bearing an SC allele from reaching the ovary. The authors consider the emergence of an SC allele within an SI species as a specific disease against which each plant protects itself by developing UI genes specific to the newly arisen Sc allele. A pistil bearing such UI genes rejects any pollen grain containing this specific SC allele. The UI genes are not necessarily associated with SI and may be established directly in SC species as protection against the pollen of other SC species or may remain in species that have not been able to develop resistance against certain Sc alleles and have become SC. There is no known limit to the number of different SC alleles and, if molecular biology permits, there should be no limit to the number of different UI genes. The hypothesis accounts for all possible combinations of unilateral or reciprocal incompatibility among species. It resembles early views of Fisher (1961), who visualized the S locus as a short strip of chromosome with 10–20 antigenically active points that could provide, on recombination, 1000 or more new combinations. To explain variations between the properties of different S alleles, Pandey (1964, 1968) also assumed that the S gene could consist of complex assemblies of genetic segments able to evolve independently. We now know that such representations of S alleles are incorrect, at least in the main plant families studied to date, and that there is no evidence, if one excepts the Sc and Sc′ variants of SC postulated by Lewis and Crowe (1958), that supports the hypothesis of extensive polymorphism among SC alleles. Nevertheless, the analogies between SC pollen and parasites remain attractive.

5.1.6.4 Non-Functional and "Conditionally" Functional S Alleles in an SC Cultivar of *Petunia*

An SC cultivar of *Petunia inflata* hosts two S alleles: So, which encodes a 31-kDa protein, and Sx, which encodes a 24-kDa protein (Ai et al. 1991). Both proteins were found to have ribonuclease activity comparable to that of *P. inflata* S proteins. While inter-species hybrids between the cultivar and SI *P. hybrida* that inherited So were SI, only 50% of the hybrids with the two alleles S2 (from *P. hybrida*) and Sx (from the SC cultivar) were SI.

It is the conclusion of Ai and co-workers (Clark and Kao 1994) that So is a non-functional S allele that cannot operate in certain genetic backgrounds lacking activator genes or containing suppressor genes.

5.1.6.5 An Interaction Between the S Locus and Major Genes in *Lycopersicum pennellii*?

The probability of interaction between the S Locus and Major Genes in *L. pennellii* was suggested by Chetelat and DeVerna (1991; Sect. 5.3.1) after segregation studies that involved a search for linkage between chromosome markers and genes involved in the control of inter-species bridging ability (unilateral inter-species compatibility, if one assumes that an active recognition process is involved). The study centered on the ability of SC forms of the wild tomato species (*L. pennellii*) to cross (as staminate parents) with diploid (LS) and sesquidiploid (LLS) hybrids between SC *L. esculentum*×SI *S. lycopersicoides*. The markers chosen for identifying the chromosomes involved in the inheritance of inter-species bridging ability were restriction-fragment length polymorphism (RFLP) and isozymes. The character for bridging ability was back-crossed from *L. pennellii* to *L. esculentum* for the development of bridging lines used to obtain progeny from the inter-generic hybrids and to study the inheritance of bridging ability.

L. esculentum×*S. lycopersicoides* diploid hybrids were crossed to these bridging lines as pistillate parents. In the resulting progeny, preferential transmission of *pennellii* alleles was observed by Chetelat and DeVerna (1991) for isozymes and RFLP markers on chromosomes 1 (the carrier of the S locus), 6 and 10. The data indicated skewing of the segregations in favor of the *pennellii* alleles for certain regions of the genome, particularly a region flanked by two isozymes (Prx-1 and Skdh-1) close to the S gene on chromosome 1 (Tanksley and Loaiza-Figeroa 1985; Quiros et al. 1986; Gadish and Zamir 1987). Chetelat and DeVerna interpreted this as evidence that a gene at or very near the S locus was involved in bridging ability. Observations of pollen-tube growth also showed that compatibility with the pistils occurred only in bridging lines at least heterozygous for the *pennellii* markers on chromosomes 1, 6 and 10. These requirements were reduced to chromosome 1 and 6 when the sesquidiploid hybrid was used as the female parent. In an F2 population of *L. esculentum*×*L. pennellii*, preferential transmission of *L. pennellii* alleles was observed for the same markers on chromosomes 1 and 10, as were other markers on chromosomes 3, 11 and 12 (but not 6). Chetelat and DeVerna conclude that their linkage data support the hypothesis that inter-species UI is governed by an interaction between expression at the S locus on chromosome 1 and other genes on chromosomes 6 and 10.

The results of Chetelat and DeVerna have been examined in depth by Mutschler and Liedl (1994), who argue that they do not necessarily support a model of unilateral inter-species isolation based on the involvement of the S locus. In crosses combining E and L genomes as the female parent and E and P genomes as the pollen source, the skewing of segregations in favor of the *L. pennellii* alleles would also occur if SC pollen (derived from *L. esculentum*) was inhibited through a regular SI function of the S gene of *L. pennellii*. In other words, Mutschler and Liedl (Liedl et al. 1996) suggest that pre-zygotic isolation between *L. pennellii* and *L. esculentum* and isolation involving the crossing behaviors of their inter-species hybrids may be governed by different barriers. One barrier involves SI function and a second, based on incongruity only, is a barrier between SI plants or their progenies (as males) and *L. esculentum* (as the pistillate parent). Liedl et al. (1996) "believe that the data of Chetelat and DeVerna (1991) are insufficient to support a relationship between their barrier and SI due to the single time point used to observe pollen tube growth, the existence of aberrant segregation of markers in the genomic regions analyzed and the complex tri-species population used." The controversy will probably not be settled without studies of the type performed by Murfett and co-workers (1996), which attempt the elucidation of the function(s) of incompatibility genes.

■ 5.2 Incongruity Between the Pollen and Pistil

5.2.1 What Is Incongruity?

It is the thesis of Hogenboom (1972 d, 1972 e, 1973, 1975) that inter-species incompatibility, termed incongruity by this author, is a process completely distinct from SI and foreign to the S locus. According to Hogenboom, the non-functioning of the pistil–pollen relationship may be due to two separate mechanisms. One is incompatibility, a mechanism that "prevents or disturbs the functioning of the relationship" and regulates inbreeding and outbreeding at the intra-species level. The second principle for non-functioning is incongruity, "the incompleteness of the relationship". The partners, usually different species, do not fully fit together, and the matching of the gene system is not complete. In contrast to incompatibility, which is an evolutionary solution through a precise and specific reaction to the negative effects of inbreeding, incongruity is considered by Hogenboom to be a by-product of evolutionary divergence that may affect any part of the relationship between the pistil and pollen. Thus, it can be of a very different nature in each case.

5.2.1.1 When Does It Start?

Therefore, it is important, when defining incongruity, to separate it from processes that initiate the emergence of reproductive barriers and are at the origin of speciation between sympatric populations. These processes, presumably derived from existing rejection mechanisms, such as SI, or new mating strategies involving, for instance, the pre-potency of cross pollen (Barrett 1998) or the intervention of "crossability" genes (Thorn 1991), are based on rare mutations likely to increase outbreeding capacity and its maintenance from generation to generation. They are active and governed by new forms of genes or new alleles able to express modified functions and to propagate rapidly in their environment.

It is not necessarily in this manner, nor at this early stage of the evolution of the future species, that incongruity contributes to reproductive isolation. Under allopatric conditions, incongruity interferes non-specifically, through the progressive loss of co-adaptation of the pollen and stigma. The result is what Heslop-Harrison (1982) identified as the "physiological maladjustment" of the pollen or pollen tube and the pistil. A recent example of the incompleteness of a pollen–pistil relationship between different families of plants was supplied by Wolters-Art (1998), who found that lily exudates could only satisfy some of the requirements of the tobacco flower (Sects. 3.5.1, 4.1.1.1). As demonstrated in *Eucalyptus*, for which Ellis et al. (1991) analyzed the outcome of three intra-species, 57 inter-species and six inter-generic crosses, the severity of the abnormalities and the probability of pollen-tube arrest in the pistil are proportional to the taxonomic distance between the parent species. Incongruity, in other words, does not seem to be an active process but is the consequence of genetic divergences resulting from independent evolution in allopatric species or from selection pressures favoring the complete interruption of gene flow between sympatric populations.

The end point of incongruity, which causes chaos in the pistil for visiting pollen from incongruous species, is not necessarily very different from that of SI. The alterations in mono-factorial stigmatic SI that the activation of protein kinase is likely to induce in proteins responsible for compatible pollen-tube growth are equally dramatic. They are even worse in mono-factorial stylar GSI, where stylar S-ribonucleases are thought (but see Sect. 3.5.7) to cause the destruction of the pollen-tube RNA. However, the significant difference with incongruity is that the final inhibition of the incompatible pollen grain or pollen tube is initiated by an exact, precisely controlled and active recognition process involving the recognition of two or a few specific pollen and pistil molecules encoded by the same alleles of a single gene. Through small changes in the specificity or activity of these molecules, or due to interference from other loci, such a mechanism is adapted for further evolution towards new functions.

5.2.1.2 When Does It End?

No provisions appear to have been made by Hogenboom to take into account the incompleteness of relationships that may occur between species or populations at the post-zygotic level. Nevertheless, it seems obvious that incongruity barriers, the existence of which no one denies (how else would one explain pre-zygotic isolation between *Prunus* and *Triticum* or between the field poppy and a cauliflower?), are not restricted to the inhibition of pollen germination or pollen-tube growth. All reproductive barriers between sympatric species, including embryo abortion and hybrid sterility, may be indications of incongruity (Hermsen and Sawicka 1979). Over long periods of time, divergences between species that no longer exchange germ plasm will continue to expand and will obviously not be limited to pollen–pistil relationships; they will extend to all phenotypic traits submitted to natural selection. If the circulation of information is maintained between the parent and its daughter species, leaky pre-zygotic and post-zygotic incongruity barriers may be specifically consolidated through the strength, if any, of selection pressure.

5.2.2 Genetic Basis of Incongruity

Hogenboom describes the barrier capacity (b) as the sum total of the pistil characters that are relevant to pollen-tube growth and fertilization. In the same manner, the sum total of information in the pollen used for normal functioning in the pistil is called penetration capacity (p). The barrier capacity is considered to be governed by dominant genes, while the corresponding penetration genes may be dominant or recessive. In his model, Hogenboom designates the b–p complex of genes by letters from A to Z. For example (b:AA) refers to the barrier capacity genes AA to ZZ and (b:BB) designates a case where AA is lacking; similarly, (p:AA) indicates the presence of all penetrating genes, while (p:AACC) means a penetration capacity from AA to ZZ where BB genes are lacking. Two plants with notations such as (b:AA, p:AA) and (b:AA, p:AA), which match one another, are reciprocally "congruous". Because only pollen bearing all the penetration genes that correspond to the barrier capacity in the pistil can function, the pistil of a plant (b:AA, p:AA) will reject the pollen of a plant (b:BB; p:BB). There will be no incongruity, however, in the reciprocal cross, because all pistil barrier genes in (b:BB, p:BB) are matched by the penetration genes in the pollen (Table 5.2). It is clear that a model of this kind explains all possible types of cross-incompatibility, whether reciprocal or unilateral, between SC and/or SI species. At the limit, the model could also account for SI itself, because any plant that, through the action of repressors, gene splicing or of certain mutations, acts as (b:AA, p:BB) will reject its own pollen.

Table 5.2. Examples of congruous and incongruous relationships (reciprocal or non-reciprocal) in the incongruity hypothesis of Hogenboom (1973). A pollen grain is congruous if it displays the penetration capacities corresponding to the barrier capacities manifested in the foreign pistil, or if the pistil fails to express a barrier capacity corresponding to the penetration capacity lacking in the pollen. For instance, two species lacking the same barrier capacity in the pistil (BCP) and lacking the same penetration capacity in the pollen (PCP) and that both express, as in the table, AA – CC DD, are reciprocally congruous (Rc). Two species with different penetration and barrier capacities (for instance, AA BB – DD in one species and AA – CC DD in the second species) are reciprocally incongruous (Ri). Other combinations lead to a non-reciprocal relationship (NRc or NRi)

Congruity				Incongruity			
Rc		NRc		Ri		NRi	
BCP	PCP	BCP	PCP	PCP	PCP	BCP	PCP
AA	AA	AA	AA	AA	AA	AA	AA
–	–	–	BB	BB	–	BB	–
CC	CC	CC	CC	–	CC	CC	–
DD	DD	DD	DD	DD	DD	–	–

5.2.3 Origin of Tissues and Genomes Involved in Incongruity

Apical meristems at the tip of each stem consist of differentiated outer layers (L) which, in most angiosperms, constitute the outer tunica (L1), the inner tunica (L2) and the corpus (L3), from which originate the aerial parts of the plant. Each layer is responsible for the formation of some of the specific tissues that compose these parts. Periclinal chimeras combining the layers of different species exist in nature or can be constructed.

Liedl et al. (1996) used a series of periclinal chimeras consisting of layers originating from two SC tomato species (*L. pennellii* and *L. esculentum*, which are unilaterally incongruous when *L. pennellii* is the pistillate parent) and their Fl hybrid to identify the layers involved in the male and female components of incongruity. For a list of tissues produced by the different layers or for further reference on the function of apical meristems and the construction of the inter-species chimeras, see Liedl et al. (1996). Four types of inter-species chimeric plants in which the apical meristems therefore originated from SC *L. esculentum* (E), an SC accession of *L. pennellii* (P) and the F1 hybrid (F) between E and P were constructed. The four chimeras had the following L1, L2, L3 layer constitution: PPE, PEE, FFE or FEE.

Through the detailed analyses of pollen-tube growth after cross-pollinations involving all possible combinations between the parental species, the hybrid and the four types of chimeric plants, Liedl and co-workers attempted to determine the tissues involved in unilateral incongruity (for more details and a different interpretation of the crossing behavior of *L. pennellii*, see Sect. 5.1.6.3). The results of crosses showed that the pistillate parent must be *L. pennellii* at either L1 alone or both L1 and L2, and the staminate parent must be *L esculentum* at either L2 alone or both L1 and L2 to observe a rejection reaction (UI, in the terminology of Liedl and co-workers) similar to that expressed in the *L. pennellii*×*L. esculentum* cross. The experiment did not settle matters dealing with possible participation of the S locus in unilateral inter-species rejection resulting from crosses between different tomato species. However, it constitutes an original approach for attempting to subdivide populations of inhibited pollen tubes on the basis of their meristemic origin.

The authors intended to continue their work with the help of new chimeras, such as EPP, and to extend current research (Liedl et al. 1996) involving the use of a GUS reporter gene for the identification of the different populations of pollen tubes growing in vivo in a same pistil after mixed pollination.

5.2.4 Justification of the Hypothesis

Hogenboom justifies the incongruity hypothesis and the distinction between incongruity and incompatibility on the basis of anatomical (Bateman 1943; McGuire and Rick 1954; Lewis and Crowe 1958; Rick 1960; Ascher 1986; Ascher and Peloquin 1968; Pandey 1968, 1969; Ascher and Drewlow 1971), physiological (Ascher and Peloquin 1968; Newton et al. 1970; Günther and Jüttersonke 1971) and genetic studies (Grun and Aubertin 1966; Hogenboom 1972e) that suggest basic differences between the two processes. Considerable additional information, accumulated during the last 25 years, has been reviewed (Heslop-Harrison 1982; Hiscock and Dickinson 1993; Mutschler and Liedl 1994) or obtained through research (Liedl et al. 1996; Murfett et al. 1996).

5.2.4.1 Arguments in Support of the Hypothesis

It is widely agreed that selection pressure in favor of inter-species breeding barriers and the independent evolution of pollen–pistil relationships in isolated species initiate and reinforce physiological maladjustments between pollen and pistil that are typical of pre-zygotic incongruity. Current knowledge regarding the important and complex mechanisms that control, for instance, the hydration of compatible pollen (Roberts et al. 1980; Sarker et al. 1988; Ikeda et al. 1997) in *Brassica* or the navigation of pollen tubes towards the ovary (Wolters-Arts et al. 1998), shows how the loss of some of the numerous, normally interactive, pollen and pistil proteins can endanger the success of pollination. Similar observations have been made in *Arabidopsis thaliana* (Hülskamp et al. 1995), where several genes (Cer1, Cer3 and Cer6) are required for pollen–stigma recognition and the regulation of water transfer from the stigma to pollen. Pollen grains from non-cruciferous species are not recognized and have presumably become incongruous through the loss of these Cer genes. The importance and complexity of pollen–stigma interactions explain why SI, with its lethal ability to destroy pollen-tube RNA (stlylar GSI) or to phosphorylate and de-phosphorylate proteins essential for pollen hydration and pollen germination (stigmatic SSI), is such an efficient rejection mechanism. They also show how incongruity and its barriers to pollen penetration can be progressively established and maintained to build up or increase the integrity of species.

5.2.4.1.1 Pollen–Pistil Barriers between SC Species. The strongest argument for the incongruity hypothesis appears to be the (relatively few) cases of pre-zygotic cross-incompatibility between SC species, such as those reported in *Antirrhinum, Lycopersicum* and *Nicotiana* (Sect. 5.1.2.1). One of the recent descriptions of the inhibition of SC pollen tubes in SC styles was made by Kuboyama et al. (1994), who observed the specific rejection profiles of the pollen tubes of *N. rustica, N. trigonophylla* and *N. repanda* in styles of *Nicotiana tabacum* (Sect. 5.1.3.1). There is a possibility that the strain of *N. tabacum* used in these experiments contained an active relic S-ribonuclease, which may be involved in the rejection of SC pollen. This thesis is not supported by the recent finding of relic S-ribonucleases genes in *N. tabacum* and *N. sylvestris*, which are 98% identical at the nucleic acid level, because Golz and co-workers (1998) found that this ribonuclease gene is not expressed in *N. tabacum*.

5.2.4.1.2 The Conditions That Overcome SI Often Have No Effect on Incongruity. The argument, wrongly based on the work by Vantuyl et al. (1982), who reported, on the contrary, that both SI and incongruity respond to mentor pollen and to the inactivating effects of irradiation, is supported by the results of Ascher (1986), who observed that sensitivity to inhibitors was different in SI and inter-species pre-zygotic isolation. It is also confirmed by the work of Davies and Wall (1961) who, in the course of large-scale experiments involving several different plant families as research material, were unable to induce mutations leading to the removal of pre-zygotic barriers between species (Sect. 5.3.2). It is probable that incongruity (which, in some of the cases studied by Davies and Wall, complemented an UI barrier) either does not mutate (as one would expect from an absence of function) or is governed by numerous unlinked genes that only very rarely mutate simultaneously in a given target cell.

5.2.4.1.3 The Genetics of Acceptance and Non-Acceptance. The genetic arguments considered by Hogenboom in support of the incongruity hypothesis essentially stem from work performed using different species of *Solanum* (Grun 1961; Grun and Radlow 1961; Grun and Aubertin 1966) and *Lycopersicum* (Hogenboom 1972 c–e, 1973, 1975). In *Solanum*, it was possible to show that certain SI forms accept the pollen of the SC species *S. verrucosum*, while others reject it (Grun and Radlow 1961). Because similar differences also appeared within populations, Grun and Aubertin (1966) were able to determine the genetics of acceptance and non-acceptance at the in-tra-species level. They discovered that UI was governed by four independent dominant genes. Any plant homozygous recessive for these four genes accepts foreign pollen. Hogenboom (1972 d) performed extensive selection work among inbred progenies of the SI species *L. peruvianum* and obtained a number of plants that accepted the normally incompatible pollen of the SC species *L. esculentum*. In addition, Hogenboom observed that different patterns of rejection characterized the incompatibility behavior of the inbred segregants that still rejected *L. esculentum* pollen. In some cases, pollen-tube inhibition was uniform and took place after penetration of approximately one-third of the style length. In others, the incompatible tubes assembled in thinning bundles at the base of the style or did not appear to be impeded at all during their growth through the style. On the basis of these data and various test-crosses between progenies and their parental lines, Hogenboom concluded that the growth inhibition of *L. esculentum* pollen tubes in *L. peruvianum* styles is governed by a number of independent dominant genes, active in *L. peruvianum* styles and ovaries, which each probably govern a distinct step of the barrier to UI.

These results and those obtained in *Solanum* by Grun and Radlow (1961) and by Grun and Aubertin (1966) showed that genes other than the S locus may participate in the rejection of SC pollen by SI pistils. The conclusion is clearly confirmed by the recent work of Murfett and co-workers with *Nicotiana* (Sect. 5.1.4.3).

5.2.4.2 Questions Remain

Certain features or implications of the hypothesis are not obvious. The site (stigmatic or stylar) where incongruity occurs is often, but not always, the place at which SI is expressed in different species of the family. This relationship between two supposedly independent mechanisms was perhaps not expected. However, despite this common finding, incongruity is often manifested at an earlier stage in the upper portion of the pistil, and is manifested somewhat more vigorously than SI (Lewis and Crowe 1958; Marcucci and Visser 1987).

Also, one might expect that, after the initial growth period fueled by the pollen grain reserves, a number of incongruous barriers to pollen-tube penetration take place at the lower portion of the style or in the ovary, where a diversity of sites are also available for interactions between the tube and the female sporophytic tissues. However, the involvement of several levels throughout the stigma and the upper-style pistil for acceptance or non-acceptance of SC pollen has been observed by Hogenboom in SI *L. peruvianum* (Sect. 5.2.4.1) and by Kuboyama et al. (1994) in *N. tabacum* (Sect. 5.1.3.1). Perhaps late-acting inhibition after inter-species pollination occurs more frequently than presently estimated.

Furthermore, the finding in *Solanum* and in *Lycopersicum* that recessive genes control the acceptance of SC pollen by SI pistils (Sect. 5.2.4.1) is not a decisive argument in favor of the incongruity hypothesis. All loci, including the S gene, can be submitted to the action of other genes (modifiers). This is certainly true of the SI gene itself, for which several authors (Chap. 4) have observed a weakening or an inhibition of function in modified genetic backgrounds. Also, it is now established (Sect. 2.5; Figs. 2.9, 3.12) that many genes are implied in the rejection reaction as an immediate consequence (through activation, inactivation, differential processing and phosphorylation/de-phosphorylation events) of recognition between incompatible pollen and pistils.

The different models proposed in the past as illustrations of the SI×SC rule (Sect. 5.1.5) can all accommodate the mutability of genes responsible for the promotion of S-coding sequences or the processing of their products. Recently, Chetelat and DeVerna (1991) showed how the expression of UI in *Lycopersicum pennellii* could be ascribed to the joint action of the S locus on chromosome 1 and other major loci on chromosomes 6 and 10 (for a different interpretation, see Liedl et al. 1996; Sect. 5.1.5.4). Thus, it seems premature to conclude, from the fact that dominant loci govern the strength, morphology or occurrence in the pistil of inter-species incompatibility, that the process is performed in the absence of S-gene action. The control of inter-species incompatibility by SF is dominant (Sect. 5.1.4.2). In addition, one must remember the warning from Pandey (1979) that incongruity, recognized by Hogenboom to be haphazard and a natural by-product of evolutionary divergence, cannot be expected to result in the "intricate but systematic pattern of incompatibility relationships discovered between disjunct flora."

Finally, and despite evidence demonstrating the reality of incongruity, particularly in the case of SC×SC crosses, one must not forget that the involvement of the S locus in SI×SC UI has now been clearly demonstrated in *Nicotiana* (Murfett et al. 1996; Sect. 5.1.4.3). It is possible that, each time genetic engineering permits it, the S locus will be shown to operate in the vast majority of cases where SC pollen is rejected by SI pistils. Mather (1975) anticipated this development when he stated "…one is not surprised that the incompatibility reaction, having come into being to regulate the occurrence of inbreeding with the species, should be pressed into service to regulate the occurrence of over-wide crossing with other species."

▪ 5.3 The Removal of Pollen–Pistil Barriers Between Species

Inter-species incompatibility is generally far more genetically complex than SI; incongruity is even more so. The reasons are quite understandable, because the efficiency requirements of SI (prevention of selfing and maximum outbreeding capacity within the species) are clearly different from those of an inter-species barrier, which essentially needs to be present in the portions of habitat exposed to gene invasion and to become as impermeable as possible. This barrier, even if it begins with a simple inter-vention of an S locus with a dual function, may progressively involve a large number of different genetic elements and accumulate several unre-lated rejection mechanisms that reinforce the protection (and independent divergence towards incongruity) of the species. Such differences, added to the fact that only function losses through deletions or frame shifts are needed in order to change SI into SC, explain why it is much easier, with mutagens, to induce the breakdown of SI (Chap. 4) than that of pre-zygotic inter-species isolation. Post-zygotic barriers, needless to say, are usually equally complex and are rarely removable through mutation induction at single targets.

In the medium to long term, germ-plasm improvement of cultivated spe-cies will essentially result from the recently acquired capacity of man to transfer, almost at will, the genes isolated from one species into the cyto-plasm or the chromosomes of another species. In most cases, provided that the transferred genes are not silenced and are expressed at the right site and the right moment in the recipient plant, this technique of genetic sur-gery ought to allow numerous applications in plant-breeding (Sect. 5.6). Nevertheless, there are examples of past and current work intended to re-move pollen–pistil barriers between species or to introduce, through genet-ic engineering, the S locus of SI plants into the genome of autogamous spe-cies, which could contribute to crop improvement.

5.3.1 Intra-Species Inbreeding

One way to produce inter-species hybrids between SI and SC species is the selection, within the SI species, of exceptional individuals that accept SC pollen (Sect. 5.2.4.1). Such plants can be of great potential value for the inter-species transfer of germ plasm and the production of cultivars com-bining the features of different species. The acceptors of SC pollen, such as those discovered in *Solanum* (Grun and Radlow 1961; Grun and Aubertin 1966) and in *Lycopersicum* (Hogenboom 1972e; Chetelat and DeVerna 1991), are homozygous recessive at several unlinked loci. Thus, it appears that the best way to obtain them is to screen (as did Hogenboom) highly inbred progenies for the loss of their ability to reject SC pollen. Systematic crossability predictions have also been made by Takashina and co-workers (1997), who exploited the "ovule-selection method" (OSM), successfully established by Imanishi and co-workers (Imanishi et al. 1996; Imanishi 1988), to bypass the post-zygotic barriers that prevent the development of inter-species hybrids between *L. esculentum* and *L. peruvianum*. The Taka-shina team used OSM to evaluate the cross-incompatibility of different lines of the "*peruvianum* complex" with *L. esculentum*. Cross-incompatibil-ity, measured by the number of fertilized ovules per fruit "germinating" in vitro (GOF), which is itself correlated to the production of F1 or BC1 seed-lings, was strongest in lines from the *L. peruvianum* var. *humifusum* group.

It was intermediate in most of the other *L. peruvianum* lines and lowest in two lines from *L. chilense*.

A far more complex approach, also involving the production of bridging lines, is the one followed by Chetelat and DeVerna (1991) to transfer traits from wild nightshade (*S. lycopersicoides*) to cultivated tomato. In order to overcome the male sterility and UI of these hybrids, Chetelat and DeVerna made use of the compatibility of the pollen of an SC accession of *L. pennellii* with the pistils of inter-generic hybrids. They produced bridging lines of *L. esculentum* by back-crossing *L. pennellii* to *L. esculentum*. Pollen of bridging lines at least heterozygous for major loci on chromosome 1 (which carries the S locus), chromosome 6 and chromosome 10 was accepted by the inter-species hybrids, and seed was produced (Sect. 5.1.6.5). None of the bridging lines was able to overcome the stronger incompatibility barrier presented by "pure" *S. lycopersicoides*. Different populations of SC species considered to have recently acquired their SC phenotype (according to Lewis and Crowe 1958), which produce Sc pollen accepted by SI pistils, could also be searched.

5.3.2 Induced Mutations

The earliest and most detailed analysis of the effects of artificial mutagens (ionizing rays) on barriers to hybridization is certainly the one performed by Davies and Wall (1961) with the aim of overcoming incompatibility and post-zygotic isolation among certain species of the genera *Trifolium*, *Brassica*, *Vicia*, *Hordeum*, *Anthirrinum* and *Lycopersicum*. From each genus, two species were chosen which were inter-pollinated in reciprocal directions after radiation exposure of the male or female gametes at all possible developmental stages. In the three cases (*L. peruvianum*×*L. esculentum*; *A. majus*×*A. orontium*; *T. medium*×*T. pratense*) where the barrier was clearly pre-zygotic, Davies and Wall failed to record any significant action of radiation on crossability. The only instance of a marked effect of the treatment was found in the cross *B. oleracea*×*B. nigra*, where reproductive isolation is post-zygotic and results from embryo abortion.

Tomato hybrids between *L. peruvianum*×*L. esculentum* were produced after radiation treatment of the parental species by Günther and Jüttersonke (1971), Yamakawa (1971), who used chronic γ irradiation, and Hogenboom (1972b), but without any evidence that damage had been induced at the S locus or anywhere else in the parental genomes. A somewhat different approach was used by Schibilla and Louant (1975), who tried to mutate the S gene (and/or the other loci involved in the inter-species barrier) not in *L. peruvianum* and *L. esculentum* but in the SI inter-species hybrid itself. All the results were negative.

At approximately the same time, more promising results were obtained by Pandey (1974), who reported that X-irradiation (500 rad) of the anthers of SI *Nicotiana glauca* enabled the pollen of this species to be accepted (70 hybrid seedlings obtained after the pollination of 257 flowers) by the otherwise incompatible pistil of *N. forgetiana*. The same approach failed when anthers of SC *N. langsdorfii* were irradiated and used as pollen source in crosses with SI *N. alata* or SI *N. forgetiana*. It is not possible to conclude from such isolated experiments that the positive effects of irradiation on the pollen of *N. glauca* are associated with the SI character of this species.

5.3.3 Effects of Mentor Pollen

The use of mentor pollen to overcome inter-species isolation has been exploited in a number of crop plants and ornamentals, such as black cottonwood (Stettler 1968; Stettler and Guries 1976), poplars (Knox et al. 1972; Stettler et al. 1980), apples and pears (Visser 1981), *Cosmos* (Howlett et al. 1973), *Lillium* (Vantuyl et al. 1982), *Cucumis* (Stettler and Ager 1984; Pickersgill 1993) and *Manihot* (Nassar et al. 1996). The most significant and reproducible results were obtained in *Populus*, where high numbers of inter-species hybrids (essentially between *P. nigra* and *P. deltoides*) of important agricultural value could be produced. With other material, the main advantage of the mentor effect, when it occurs, is the production of an amount of seed per pollinated flower several times (20, in the case of *Lilium*) the yield normally obtained without mentor pollen.

5.3.3.1 Can Mentor Effects on Self Pollen Consolidate Reproductive Barriers between Species?

The possibility that mentor effects on self pollen might consolidate reproductive barriers between species was envisaged by Rieseberg et al. (1995) and Desrochers and Rieseberg (1998) during a study of mentor effects on selfing in wild species of *Helianthus* (Asteraceae). They found that the mentor effects did not play an important role in the reproductive isolation of the two species studied (*H. annuus* and *H. petiolaris*) after pollination with different pollen mixtures that either included or excluded self-pollen. It was noted, however, in conformity with predictions by McCarthy et al. (1995), that even a modest increase in self-fertilization in hybrid populations, such as that exerted by mentor pollen, could reduce hybridization frequencies and enhance the rate of hybrid-species establishment.

5.3.3.2 Nature of the Mentor Effects

An important preliminary step towards our understanding of mentor effects was made when Knox and co-workers (1972) extracted the rapidly leachable proteins from the pollen of *Populus deltoides* and mixed lyophilized extracts with equal volumes of viable pollen of *P. nigra*. This pollen was then, at least to a certain extent, accepted by the normally incompatible pistils of *P. deltoides*. The average seed set per capsule (~2.7) was relatively low, but the significance of the treatment cannot be challenged, because all progenies were hybrid.

5.3.4 Bud Pollination and the Action of Protein Inhibitors

The range of possible pollen–pistil barriers between species is wide and practically unpredictable, particularly if one includes incongruity (the incompleteness of a relationship) as an important speciation force. Isolation may occur at every one of the different phases that follow the landing of pollen on the stigma (adhesion, communication, interaction, germination, wall loosening, penetration, orientation, growth, entry in the ovary and ovule, integumentary growth, delivery of sperm nuclei...). Each of these phases may break down if any of the numerous gene products essential to each step is altered, deficient or produced at the wrong moment, at the wrong place or in insufficient quantities.

In comparison, SI seems to be simple, because it depends, initially, on a single recognition reaction that takes place between a small number of gene products, possibly only two: one from the pollen and one from the pistil. It does not result, as in incongruity, from the absence of a relationship but from the presence, in the pollen and style, of a specific information.

This basic distinction between SI and incongruity is important, because it indicates that bud pollination and protein inhibitors should have very little effect, if any, on incongruity barriers. However, this is not what occurs in the Brassicaceae, where Hiscock and Dickinson (1993) demonstrated that both SI and inter-species incompatibility were overcome by bud pollination or the application of cycloheximide to stigmas (Chap. 4). In this case, there appears to be a mechanism involving the participation of the S locus (or of an active gene with somewhat similar functions), which controls inter-species incompatibility. The conclusion does not imply that certain inter-species barriers could be easily removed through the use of simple techniques designed to prevent or avoid specific gene action, because several different loci are probably involved in their control. However, together with the results of Murfett et al. (1996) regarding the participation of S-ribonucleases in inter-species incompatibility in *Nicotiana*, it shows that not all mechanisms of pollen–pistil rejection between closely related species result from the absence or incompleteness of a relationship.

▪ 5.4 Transfer of the S Gene to Autogamous Species

The introduction of a functional S locus in an autogamous crop species should constitute a spectacular application of gene technology to plant breeding; it could be of considerable interest for the possible emergence and exploitation of hybrid vigor in traditional self-fertilizers. Most efforts to transform SC plants with SI genes attempt to acquire basic knowledge regarding the requirements and site of S expression in a genetic background where there is, at least in theory, no interference from endogenous S genes (Sects. 4.2.3, 4.2.4). The current work centers essentially on the use of *Agrobacterium*-mediated gene-transfer methods in readily transformable plants, such as *N. tabacum*, and some diploid inter-species SC hybrids and their diploid S relatives in the genus *Nicotiana*.

5.4.1 Introduction of the *Brassica* SLG and SRK Genes in SC Species

Most attempts to transform SC species with *Brassica* S genes have been confined to the amphi-diploid *B. napus*, *A. thaliana* and *N. tabacum*. The research, as seen earlier, is not expected to contribute directly to crop improvement (SI strains of *B. napus* occur in nature and are available). *Arabidopsis* has no agricultural value, and tobacco plants might benefit to a greater extent, if they are ever exploited as a cross-fertilizer, with an S-ribonuclease rejection mechanism. However, the research may contribute to the development of the materials and methods necessary to provide S transgenes with the complex genetic environment necessary (as seen throughout this book) for the full expression of the SI character and its exploitation in agriculture.

5.4.1.1 Transfer of SLG to *B. napus*

The transfer and expression of SLG in *B. napus* did not confer SI to the transgenic plants (Nishio et al. 1992). Since transfer experiments were not undertaken with SRK alone or in association with SLG, it was not possible to attribute this absence of effects to gene silencing or to the non-involvement of SLG in the manifestation of SI. Before this work, Toriyama and co-workers (1991b) found that a partially SC accession of *B. oleracea* transformed with a functional S allele from *B. campestris* had become fully compatible. They observed similar silencing effects (Sect. 4.2.3; Conner et al. 1997) when completely SI plants were used as recipients in the transformation experiments.

5.4.1.2 Transfer of SLG and SRK to *A. thaliana* and *N. tabacum*

Several successful attempts have been made to insert SLG and SRK, with their promoters, in *Arabidopsis* or *Nicotiana* (Trick and Heizmann 1992; Conner et al. 1997; Sect. 3.2.3.5). In *Arabidopsis*, after insertion of an SLG:GUS construct, GUS activity was observed in the stigma papillae cells and, in rare instances, in the tapetal cells of anthers at the uni-nucleate microspore stage (Toriyama et al. 1991a). In tobacco, the transfer and expression analysis of the SRK gene (Stein et al. 1991, 1996; Susuki et al. 1996) and the SLG gene (Kandasamy et al. 1990; Moore and Nasrallah 1990; Thorsness et al. 1991; Susuki et al. 1996) showed that SLG is produced more abundantly by the style than by the stigma. The SRK protein, predominantly located in the stigma, was detected in non-negligible quantities in the style (Susuki et al. 1996). Stein and co-workers found that SRK (SRK6) was targeted to the plasma membrane. This observation, coupled with similar and other information directly obtained in *Brassica*, provided biochemical proof of the existence of a plasma-membrane-localized transmembrane kinase with a known cell–cell recognition function (Sect. 3.2.3). However, neither *Nicotiana* nor *Arabidopsis* transformants expressed SI in any of these experiments. From the data available at the time, Trick and Heizmann (1992) concluded that "*Brassica* S-expression appears to be a function of the degree of taxonomic affinity of the recipient genome, with S-transgenes being expressed gametophytically in *Nicotiana* and sporophytically in *Arabidopsis*".

5.4.2 Transfer and Expression of the S-Ribonuclease Gene in SC Species and SC Inter-Species Hybrids of *Nicotiana*

Jane Murfett performed essential research with SC species and SC inter-species hybrids of *Nicotiana*, working successively in Prof. Adrienne Clarke's lab at the Plant Cell Biology Research Centre in Melbourne, then in Dr. Bruce McClure's lab at the University of Missouri at Columbia. These contributions (Murfett et al. 1992, 1994, 1995, 1996; Murfett and McClure 1998) not only allowed a direct demonstration that S-RNases cause S-allele specific pollen rejection in *Nicotiana* (Sect. 3.5.6), they also demonstrated a role for S-RNases in UI (Sect. 5.1.4.3). This type of research may soon lead to the introduction of SI in certain solanaceous crop plants.

However, if such efforts are to succeed, it will be necessary to pay careful attention to genetic background effects. For example, Murfett et al. (1996) showed that expression of S-RNases using either the tomato Chi2;1 or the *Brassica* SLG13 promoter in purely SC backgrounds, such as *N. plumbaginifolia* or *N. tabacum*, does not cause S-allele-specific pollen rejection or the rejection of pollen from certain SC species such as *N. plumbaginifolia*. This is in marked contrast to the results recorded when the same S-RNases are expressed in hybrids (*N. plumbaginifolia*×SC *N. alata* or *N. langsdorfii*×SC *N. alata* hybrids) and S-allele-specific rejection occurs normally. Thus, the pollen-rejection phenotype is strongly dependent on the genetic background of the transgene recipient. Moreover, the genetic background has also been shown to affect S-RNase levels per se (Murfett and McClure 1998). It is also worth noting that S-RNases appear to be specifically adapted for their function in pollen rejection, because a non S-RNase (i.e., RNase I from *Escherichia coli*) cannot substitute for S-RNase in pollen rejection (Beecher et al. 1998).

5.4.3 Other Genes Should be Transferred with the S Gene

In addition to the need, emphasized in the preceding paragraph, to transfer the S locus as a whole (i.e., complete with pistil and pollen determinants), if autogamous species are to reject their own pollen, it is also necessary to modify, perhaps considerably, their genetic background. The conclusion arises not only from the work on *Nicotiana* presented above (Sect. 5.2.6.2). It also stems from the results of research performed with inter-species tomato hybrids and their offspring (Martin 1961, 1968; Chetelat and DeVerna 1991; Bernatzky et al. 1995), which clearly demonstrates the dependence of the S-RNase gene on its cellular environment. S alleles transferred from a related SI species do not function in the genetic background of SC *L. esculentum*. Bernatzky and co-workers (1995) even consider the possibility that *L. esculentum* normally contains a silent but potentially functional S allele that maps to the S-bearing chromosome 1 (Liedl et al. 1993). The implication – which certainly holds for *Lycopersicum* and *Nicotiana* and probably other Solanaceae, such as *Petunia* (Dana and Ascher 1986a, 1986b) and other systems (like that of the Brassicaceae, where the family of expressed S-related genes is very large; Sect. 3.2.3.7) – is obvious. Transfer experiments that attempt to introduce a functional SI breeding system in an autogamous species for scientific or breeding purposes must not be restricted to the S locus and appropriate promoters. Instead, they should involve some of the other genetic sequences believed to cooperate or interact with S alleles during the recognition and/or rejection phases. It is likely that the amount of germ plasm that should accompany the S locus in transgenic experiments will increase with taxonomic distance, but data is lacking. A rare example of the inter-generic transfer of the S-ribonuclease gene within the Solanaceae was provided by Chung and co-workers (1999), who introduced the S11 allele of *L. peruvianum* into SC *N. tabacum*. The styles of transgenic tobacco plants abundantly expressed the S11 ribonuclease at developmental stages analogous to those at which the S-ribonucleases are produced in *L. peruvianum*. The S11 transcript was also found, at low concentrations, in developing anthers and mature pollen of transgenic plants. However, pollination behavior was not modified in the transgenic plants, which remained SC.

▦ 5.5 Reconstruction of Multigenic SI in SC Species

Larsen and co-workers (1973) conclude that it is possible to regenerate a functional SI population from different SC forms in an SC species. Their working hypothesis involves three postulates and two corollaries.

- ▦ Postulate 1. Certain "self-compatible species are derived from self-incompatible ancestors" with a multigenic SI system.

▦ Postulate 2. "Mutations to loss of activity at any of the S loci result in self-compatibility."

▦ Postulate 3. "Self-compatible forms may be mutants at different S loci."

▦ Corrolary 1. "Crosses between different self-compatible forms may bring together in the same F1 plant non-mutated functional S genes."

▦ Corrolary 2. "Plants with a reconstituted SI system may segregate from crosses between different F1 combinations."

This hypothesis should be tested in genera of families (Poaceae, Chenopodiaceae and Ranunculaceae) where multigenic SI systems are known to occur and within which there is evidence that the silencing of S genes has progressively led (through homozygosity rather than by mutation) to the establishment of mono-factorial SI and, ultimately, self-compatibility.

▦ 5.6 Crop Improvement through the Transfer of Individual Genes

It is not the purpose of the present review to outline the methods available to completely bypass the pollen–pistil barrier between species through the transfer of the individual genes or groups of genes, nuclear or cytoplasmic, necessary for the improvement of a specific character in a given cultivar. Table 5.3 provides a list of publications, among many excellent ones, in which the reader will find further references to the state of the art and a review of the important scientific and technical problems that still have to be solved.

Table 5.3. Some basic references for the genetic engineering of flowering plants

General	
Molecular biology and crop improvement	Austin et al. (1986)
Gene cloning and identification	Xoconostle-Càzares et al. (1993)
Approaches to gene transfer and available techniques	Potrykus (1993)
The role of gene technology in plant breeding	Salamini and Motto (1993)
Introduction and expression of transgenes	Gelvin (1998)
Somatic hybridization	Pelletier (1993)
Control of gene expression in transgenes	Gallie (1998)
Examples of applications	
Gene transfer in cereals	Komari et al. (1998)
Plant cell walls as targets for biotechnology	Chapple and Carpita (1998)
Economical aspects and commercial markets	
Engineering new plant strains for commercial markets	Briggs and Koziel (1998)
Transgenic crops for new products	Hitz (1999)

CHAPTER **6** **Conclusions**

The aim of this monograph has been to overview more than a century of research on the genetics, physiology, biochemistry and evolution of the pollen–pistil rejection processes (especially incompatibility, but also incongruity) that operate within and between species to establish lower and upper limits of outbreeding in flowering plants. An attempt is made below to list some of the most significant results of this work and to comment briefly on possible priorities for further research efforts.

6.1 High-Quality Research and Abundance of Achievements

6.1.1 High-Quality Research

The proportion of research activities that have yielded results is clearly high. Probably because self-incompatibility (SI) involves unique systems of cell–cell communication, cell–cell recognition and cell-cell rejection and provides strong opportunities for grafting fundamental to applied research, it has continuously attracted laboratories of very high quality. In particular, the methods and techniques currently used to study, engineer and transfer S genes form part of the most advanced technology presently available in plant research.

6.1.2 An Abundance of Achievements: Classification, Distribution and Inheritance of Self-Incompatible Systems

Knowledge, especially in the case of SI, is now available regarding the classification and distribution of the pre-zygotic barriers (Sects. 1.5, 2.1, 2.2, 2.3, 5.1, 5.2) and, to some extent, their inheritance (Chaps. 2, 5).

6.1.3 Fine-Structure Studies of Pollen and Pollen Tubes in Compatible and Incompatible Surroundings

Such studies, involving scanning and electron microscopy of stigmas or styles after compatible and incompatible pollination, have been performed in detail for heteromorphic (Sect. 3.1.2) and homomorphic species (*Brassica*: Sects. 3.2.1, 3.2.2; *Papaver*: Sect. 3.3.1; grasses: Sect. 3.4.1; Solanaceae: Sects. 3.5.1, 3.5.2).

6.1.4 Identification of S Genes Active in the Pistil

Some S genes, such as the kinase receptor of *Brassica*, the stigmatic S gene of *Papaver rhoeas* and the ribonucleases of the Solanaceae, Scrophularia-ceae and Rosaceae, which determine the incompatibility phenotype of the pistil, have been identified, cloned and sequenced. In the case of SRK, the different domains and the occurrence of truncated forms were character-ized through their functions, sites, mutations and on the basis of the pro-cesses to which they are submitted at the time of expression (Sect. 3.2.3.3). With the specific ribonucleases typical of mono-factorial stylar gameto-phytic SI (GSI), decisive "loss- or gain-of-function" experiments could be conducted which clearly demonstrated the involvement of RNases in the re-cognition and rejection phases of SI and in the expression of certain inter-species barriers (Sect. 5.1.4.3).

6.1.5 Discovery of a Putative Pollen Determinant in the Cabbage Family

Just before the year 2000, Schopfer and co-workers at Ithaca reported the discovery of a putative pollen determinant in *Brassica campestris* (Sect. 3.2.4.4). Loss- and gain-of-function studies (Sect. 3.2.4.4) proved that the product of this gene (SCR) is necessary and sufficient for determining pol-len specificity, possibly by acting as a ligand of the stigmatic receptor. If it is confirmed, this result, the first identification of pollen S determinants, should permit considerable progress in our understanding of sporophytic homomorphic SI (the role of the tapetum, interaction of pollen determi-nants on the pollen coat, nature of the recognition process leading to SRK receptor activation, identification of the recognition molecules...).

6.1.6 Progress towards the Understanding of S Specificity

In the meantime, through different routes involving the assessment of the distribution of amino acid (AA) variations among S haplotypes (Sects. 4.3.1.1, 4.3.1.4), an evaluation of the specificity of folding and loop configurations in the S protein, the swapping of S-gene domains and the replacement of individual AAs (Sect. 4.2.7), preparations are being made for the identification of the S sequence that participates in the recognition process for the pistil. However, the first indications (Sects. 4.2.7.1 and 4.3) have not been clear, because observations differ (for instance, with regard to the distribution of variations among different alleles) and arguments are contradictory (as in the case of the significance of regional concentrations of high variability). Opinions also diverge regarding the role of recombination for the production, circulation and accumulation of S-allele diversity (Sects. 4.3.1.1 and 4.3.1.4). Recent evidence shows that recombination or related events has played a role in the evolution of SLG and SRK alleles. Awadalla and Charlesworth (1999) conclude that the hyper-variable regions of the SLG gene are under conditions of balancing selection and are not merely regions of relaxed selective constraints.

6.1.7 Advances in Cellular and Molecular Surgery

When gene silencing did not interfere (Sects. 4.2.3.1), techniques that allowed the follow-up of S genes transferred in other species or that improved their expression (and over-expression) in foreign tissues through the use of powerful promoters were developed or adapted (Murfett and McClure 1998). Chimeras of widely different kinds, combining portions of incompatible and compatible styles (Sects. 3.5.7.3, 4.1.5), different layers of apical meristems (Sect. 5.2.3), or individual regions of S alleles (Sect. 4.2.7.1), have been constructed to properly assign functions.

6.1.8 New Information Regarding the Evolution of SI Systems

The methodology necessary for the genealogical study of the S gene since mid-Cretaceous times (and, in particular, the assessment of allelic diversity among species), its nature and origin, and the age of S alleles, was successfully established (Sect. 4.3.1). Several different points of origin appear to characterize heteromorphic SI, which is often assumed to have resulted, in most cases, from the combination of a primitive di-allelic incompatibility mechanism (necessary for avoiding self-pollination) and reciprocal herkogamy, which promoted cross-pollination (Sect. 4.3.4). Current evidence, particularly that arising from molecular biology, also tends to attribute an independent origin, possibly multigenic, to most homomorphic SI systems (Sect. 4.3.2).

6.1.9 Bypassing Pre-Zygotic Inter-Species Barriers

Few (but spectacular) achievements, such as the use of mentor pollen to produce inter-species hybrids of poplars (Sect. 5.2.5.3) or the demonstration that unilateral SI×SC pre-zygotic isolation between different species of *Nicotiana* requires the action of S-ribonucleases (Sect. 5.1.4.3), have been made towards the bypassing of certain inter-species pre-zygotic barriers. Some progress has also been registered in understanding the main differences between inter-species incompatibility (a starting point, through the establishment of a pollen–pistil relationship, for the isolation of two species) and incongruity, not necessarily pre-zygotic in its manifestation and effects, which results from the consequences (loss of relationships) of the independent evolution of these two species.

■ 6.2 There Are Still Numerous Gaps in Our Knowledge and Skill

6.2.1 Unraveling the S-Gene Family

6.2.1.1 Genetic Control Is More Complex Than Expected

It is usually thought that, for practical purposes, sufficient knowledge regarding the inheritance of SI in most of the annual species that play a role in plant breeding has been accumulated. Furthermore, the discovery that S-ribonucleases govern the SI reaction in the Rosaceae greatly facilitated the identification and analysis of SI in several orchard species, such as cherry, apple, pear, almond and apricot. It gave rise, in particular, to rapid assays (Tobutt et al. 1999) allowing the identification of different bands which correlate with S-ribonuclease activity on electrophoretic gels. Methods of this kind (Sect. 4.3.1.6) are very useful for predicting the S genotypes of perennials, detecting new alleles and selecting for desirable traits.

However, the finding that multigenic, poly-allelic mechanisms participate in the determination of pollen and pistil specificities in certain plant families important for agriculture (Chenopodiaceae: Sect. 2.3.3; Fabaceae: Sect. 2.3.6; Brassicaceae and Compositae: Sect. 2.4.2) shows that many different gene products can be (and probably are) biochemically associated with the recognition phase of SI in these families. Inheritance studies in such cases remain very difficult and should be encouraged. Furthermore, there is a need for a better exploitation of SI to identify the genes involved or affected in pollen grains, stigmas, styles or ovaries by the pollen–pistil rejection phase (Sects. 2.5.1, 2.5.2, 2.3.7.3).

6.2.1.2 The S-Gene Family of *Brassica* is Surprisingly Large

In 1997, Boyes and co-workers estimated that several hundred kilobases are spanned by the highly polymorphic "master-recognition" S complex in *Brassica*. Indeed, dozens of sequences related to the SLG-SRK haplotype, expressed or silent, possibly recruited from vegetatively expressed genes with functions often unrelated to pollination (Dwyer et al. 1994), have been reported during recent years (Sect. 3.2.3.7; Susuki et al. 1999). An elucidation of the functions, largely unknown, of these genes has now become an important research objective (last paragraph of Sect. 3.2.4.3).

6.2.1.3 S-Locus Complexity Has Also Been Found in the Solanaceae

Unpublished results of McCubbin, Wang and Kao (cited by McCubbin and Kao 1999) suggest that more than a dozen pollen-expressed genes are tightly linked to the S locus of *Petunia inflata*. Here also, nothing is known regarding the functions of these genes (which, in contrast with S-ribonucleases, appear to display very little sequence diversity; Yu et al. 1996; Liu et al. 1997, also cited by McCubbin and Kao 1999; a review of current research is in Sect. 3.5.8.5).

6.2.1.4 The S Alleles of *Papaver* May Also Belong to a Large Family

Very little is known regarding the open reading frames with homology to the stigmatic S gene of *Papaver*, which Ride et al. (1999) discovered in *Arabidopsis* (Sect. 3.3.3.4). Research on their nature, origin and distribution may lead to the recognition of links between the S system of *Papaver* and those of other taxa.

6.2.2 Analysis of Recognition and Rejection

6.2.2.1 Identification and Function of S and S-Related Proteins in the Pistil

Considerable progress towards the identification and function of S and S-related proteins in the pistil has been made during the last 20 years, particularly with regard to the structure and function of pistil in systems of the *Brassica*, *Papaver* and *Nicotiana* types. Nevertheless, knowledge gaps remain regarding a wide diversity of issues, such as:

▩ the nature of haplotype dominance in sporophytic stigmatic SI (Sect. 2.2.2.1)

▩ the contribution, if any, of SLG during the recognition and rejection processes (Sects. 3.2.3.2, 3.2.5.4, 4.2.3.1) and the extent and significance of its haplotype co-evolution with the extra-cellular domain of SRK (Sect. 4.3.1.2)

▩ the function of region C4 (RC4 in the Rosaceae), which is apparently lacking in the Scrophulariaceae (Sects. 3.5.7.1, 4.3.1.2).

There is also a need to identify and characterize the product of the S gene in legumes, particularly in genera, such as *Trifolium*, that play an important role in agriculture.

6.2.2.2 The Search for Pollen Determinants

The search for pollen determinants is the current priority of several laboratories throughout the world (Sects. 3.2.4.3, 3.5.8.5), because our understanding of the nature, origin and evolution of SI and the domestication of the rejection process depends on detailed knowledge of the molecules that participate in the recognition phase in both the pollen and pistil. Now that a putative pollen determinant (SCR) has been identified in *Brassica* (Sect. 3.2.4.4) its function as a ligand of SRK (Fig. 3.11) must be confirmed, and the search for pollen S (and Z) genes in stigmatic and stylar GSI systems must be continued. Further progress in this area could permit the testing of some ancient and very remarkable hypotheses regarding the involvement of protein dimers in the SI reaction (Sect. 3.5.3.3) and the tapetal origin of pollen determinants in sporophytic SI (Sect. 3.2.4.2). Furthermore, the discovery of the sequence(s) that determine pollen specificity in some of the systems presently being studied should help to clarify, through the identification of pollen–pistil recognition complexes, the current discussion regarding the role of hyper-variable regions and protein folding in the specificity of pistil determinants (Sect. 4.2.7.1). In mono-factorial stylar GSI, the discovery of pollen S proteins could contribute to the understanding of pollen-part mutations and their frequent association with S-bearing centric chromosome fragments in the Solanaceae (Sect. 4.2.2.1). Reciprocally, research in the Solanaceae on the molecular nature of these mutations may lead to the identification of pollen determinants.

6.2.2.3 Identification of the Genes and Processes Affected by the Rejection Phase of SI

Some of the permanent and occasional symptoms of the rejection reaction in stigmatic and stylar homomorphic SI are known. However, in the case of stigmatic SI, numerous questions remain regarding the nature of signal-transduction pathways initiated by the recognition reaction, the phosphorylation/de-phosphorylation processes that follow, the different genes concerned in the pollen and stigma, and the involvement of programmed cell death as an end point of the rejection response. Also, it is not known to what extent the knowledge accumulated on *Brassica* (Sect. 3.2.5.5) can be extrapolated to *Papaver* (Sects. 3.3.4.2, 3.3.4.3) and the grasses (Sects. 3.4.3.2, 3.4.3.3). In mono-factorial GSI (Sect. 3.5.2), it would be helpful to gain information regarding the significance of certain symptoms of the rejection phase (expansion of the tube outer wall, degeneration of the tube cytoplasm, occasional bursting of the tube apex, concentric configuration of the endoplasmic reticulum, retardation of tube growth) and the extent of the degradation of ribosomal and messenger RNA by S-ribonucleases (Sect. 3.5.7.2). Similarly, the inability of the S-ribonucleases of the stigma to inhibit incompatible pollen tubes in stylar GSI (Sect. 3.5.7.4) should be explained.

6.2.3 The Molecular Biology of SI in Heteromorphic Species

In contrast with the considerable and successful research efforts (Sects. 2.1, 3.1) made for understanding the ecology and genetics of heterostyly only limited attempts have been undertaken to identify the products and function(s) of the S gene(s) which governs SI in distylic and tristylic species. There is definitely a need to continue, diversify and expand the research currently under way (Sect. 3.1.3) in this area as then, and only then, will it become possible to progress substantially in our knowledge of the evolution and functioning of SI in these species.

For instance, it is generally considered (Dulberger 1992) that the developmental relationship between incompatibility and floral polymorphism is so intimately integrated in tristyly that each of the two anther levels typical of each flower expresses a specificity of its own. As the mechanisms through which a given gene (or a given set of genes) generates different products in different tissues are now being clarified, it will be possible, once the molecules encoded by the M and S genes are characterized, to investigate the occurrence of differential splicing or differential post-translational processing (in the flowers) of tristylic species. Similarly, the S-supergene structure, typical of distyly, may constitute an interesting material for the introduction (through genetic engineering) of the complete outfit necessary to express SI in related autogamous plants. It carries not only the fixing necessary for the expression of SI (and reciprocal herkogamy) but, possibly, a valuable switch gene. Also, new knowledge regarding the molecular structure and possible diversity of the genes that govern heteromorphic SI may contribute to the testing of different theories regarding the origin of the system and the relationship between distyly and tristyly (for example, Sect. 4.6).

6.2.4 Barriers to the Expression of Transferred Genes

Induced mutagenesis and genetic engineering have played a major role in the understanding of SI (Sects. 4.2.2, 4.2.3, 4.2.4, 4.2.7) and, to a much smaller extent, in the improvement of crop plants (Sects. 4.2.2.4, 4.2.2.5). Unfortunately, work in this area can be prevented or hampered by silencing effects, regularly observed on endogenous SLG or SRK after the transfer of stigmatic genes of the *Brassica* S family (Sect. 4.2.3). Research activities and applications to agriculture can also be prevented by the fact that a number of species, such as *Nicotiana alata*, are refractory to genetic transformation (Sect. 3.5.6.2 for the way Murfett et al. 1994 were able to bypass this obstacle) or the fact that certain genes are sensitive to genetic surgery (Sect. 4.2.7.1 for the sensitiveness of S-ribonucleases to domain disruption). Research should be devoted to the detection of transformable material, the improvement of molecular tools and, in a general manner, to the control of the numerous factors (genetic and physiological) that determine the acceptance and expression of a foreign SI system in important crop plants.

6.2.5 Evolution of SI

6.2.5.1 The Multiple Origin of Homomorphic SI

The hypothesis of a multiple origin of homomorphic SI, justified by the differences observed among systems, is probably correct. One must not forget, however, that current knowledge regarding mechanisms is still fragmentary and that three (*Brassica*, *Papaver* and grasses) of the four most studied systems could function through the participation of kinase receptors. Perhaps the ligands for these receptors are S determinants of the recognition phase. The involvement of a kinase receptor in the *Nicotiana* system appears far more unlikely, but it must be noted that protein kinases of *N. alata* use S-ribonucleases as substrates in vitro (Sect. 3.5.8.3).

6.2.5.2 A Relationship Between GSI and Sporophytic SI?

A relationship between GSI and sporophytic SI would be plausible if, as postulated by Brewbaker (1957, 1958), Pandey (1958) and, in the case of the Caryophyllaceae, by Lundqvist (1990b), the transition between the two systems could be attributed to a shift in the timing of events at the stage of cell-wall formation within the tetrad of microspores. It would be less obvious if the S phenotypes of pollen grains in sporophytic SI were determined by substances received from the tapetum (Sect. 3.2.4.2) or from the mixing of different gametophytic products in pollen coats (Sect. 3.2.4.1).

6.2.5.3 A Multiple Gene System as the Starting Point for GSI

It seems that none of the arguments presented by Pandey in favor of mono-factorial GSI as the ancestral system of SI resists the evidence, collected by Lundqvist (1990a), which supports the hypothesis that a multigenic mechanism was the origin of mono-factorial GSI in at least several families of angiosperms (Sect. 4.3.2.2). It ought to be possible to adapt the three locus model proposed by Uyenoyama (Sect. 4.3.3) to this new requirement. Simultaneously, research regarding the relics of S-multigenic structures should be undertaken in families characterized by an S-ribonuclease SI system.

6.2.5.4 Is the Emergence of New S Alleles a Gradual Process?

Most questions concerning the origin of S alleles cannot be answered properly before the identification of the sites within the S genes that determine allelic specificity is made for each SI system, in both pollen and pistils. However, unless one assumes that a single BP substitution can cause a change in specificity, the most likely hypothesis – that individual base-pair (BP) substitutions contribute substantially to the generation of new alleles – implies that emergence occurs gradually, with transition phases during which the new allele remains neutral or displays two or several specificities (Sect. 4.2.7.4).

6.2.5.5 Nature of the Relationship (Homology or Convergence) Between the S Proteins of the Different Families of Plants That Share a Common SI System

Before attributing a common ancestry to a system of molecules, it is necessary to find out whether the relationship between these molecules reflects homology or convergence. This is what Richman and co-workers (1997) have done with ribonucleases, through a reconstruction of their genealogy within the angiosperms (which covered both S- and non-S-ribonucleases). Their results (Sects. 4.3.1.3, 4.3.2.2) showed that the S-ribonucleases of the Solanaceae and the Scrophulariacea have a common origin. The relationship (homology or convergence) of the ribonucleases of the Rosaceae to the ribonucleases of the two other families could not be identified, because the necessary reference points (non-S-ribonucleases from a wide range of angiosperms and less distant out-group sequences) were insufficient. Further research in this area is needed.

6.2.5.6 Evolution of Heteromorphic SI

Research on the molecular biology of heteromorphic SI should contribute significantly to the testing of current models and hypotheses (Sects. 4.3.4.2, 4.3.4.3, 4.3.4.4) of the origin, evolution and breakdown of distyly and tristyly. However, once the incompatibility genes are identified and cloned, progress could be seriously hampered by the absence of allelic diversity that is likely to characterize the di-allelic rejection mechanism operating in heteromorphic SI. Special attention could be directed towards the rare heterostylic species, such as *Anchusa officinalis* or *Narcissus tazetta*, which are suspected of poly-allelism at the S locus (Sect. 2.1.3).

6.2.6 Inter-Species Incompatibility and Incongruity

6.2.6.1 The Complexity of Pollen–Pistil Barriers Between Species

When they demonstrated the participation of the S locus in the pre-zygotic barrier between transformed and non-transformed SI and SC species of *Nicotiana*, Murfett and co-workers discovered the overall complexity of the inter-species pollen–pistil relationship. While S-ribonuclease was implicated in rejecting pollen from each of the SC species tested (Sect. 5.1.4.3), other cases of pollen rejection that implied an S-ribonuclease-independent mechanism or that required both the S-ribonucleases and other genetic factors present in the pistillate SI parent (*N. alata*) could be identified. In other words, in a self-incompatible species like *N. alata*, there is a redundancy of mechanisms, still not understood, that contributes to inter-species pollen rejection and protects the species against SC pollen in case of breakdown of the SI character.

6.2.6.2 The Need for Further Research on the Molecular Biology of Pre-Zygotic Barriers Between Species

Analyses of the involvement of S-gene products in inter-species barriers must continue. It would be helpful to know, for example, what the non-ribonuclease factors involved in pollen rejection in *Nicotiana* are, and to what extent they can be considered relevant to incongruity, "the absence of a relationship." There is also a need, if gene-transfer methods become fully available, to expand the research to other materials, such as the tuber-bearing species of *Solanum* (Xu et al. 1990a, 1990b; Kaufmann et al. 1991; Saba-El-Leil et al. 1994; Kirch et al. 1995; Matton et al. 1997) or the SI and SC species of *Lycopersicum* (Chetelat and DeVerna 1991; Tsai et al. 1992; Dodds et al. 1993, 1999; Rivers et al. 1993; Royo et al. 1994; Bernatzky et al. 1995; Liedl et al. 1996), where SI and inter-species pre-zygotic barriers have been intensively studied in the past.

Methodologies, not necessarily inspired by the *Nicotiana* approach, could also be established outside the Solanaceae, where material that is neither refractory to transformation nor prone to transgene-induced suppression of the expression of endogenous S genes exists. Their aim would be to search other SI systems for a molecular basis for the involvement of S genes in inter-species isolation. It is hoped that the gene-silencing phenomenon (Sect. 4.2.3.1) will not eliminate *Brassica* species from this research, particularly because, with few exceptions, Dickinson and Hiscock confirmed the application of the SI×SC rule to the Brassicaceae and showed that, in this family, pre-zygotic inter-species isolation could, like SI, be overcome through bud pollination or treatment with a protein inhibitor (Sect. 5.1.3.2).

6.2.6.3 Selection of Bridging Lines

The main objective of the plant breeder interested in inter-species hybridization is to break down the reproductive barrier that separates different species. For this purpose, he has relied, sometimes successfully, on empirical techniques leading (through the screening of inbred populations or the back-crossing of inter-species hybrids to certain pre-selected SC accessions) to the selection of bridging lines (Sect. 5.3.1). Such efforts should continue each time more efficient (and usually more specific) approaches involving in vitro cultures, genetic engineering, the use of metabolic inhibitors or mentor molecules are not available. A compromise between the two strategies has been followed with success by Murfett et al. (1994, 1996), who coped with the difficulty of transforming SC *N. alata* by instead using hybrids between transformed SC *N. langsdorfii* or transformed SC *N. plumbaginifolia* and SC *N. alata* (Sects. 3.5.6.2, 5.1.4.3).

▓ Note added in proofs

The period reviewed by this book ended in December 1999. Essential information was published in the first months of 2000 which relates, in particular, to the isolation, characterization and function of SP11, a member of the PCP family, considered similar to SCR, as pollen determinant in *Brassica campestris* (Takayama et al. 2000; Watanabe et al. 2000).

Current knowledge on the pollen S-gene of *Brassica* is summarized and discussed (Franklin and Franklin 2000). The respective roles of SRK and SLG in the determination of pistil specificity (SRK) and in the full manifestation of the SI response (SLG) appear to have been clearly ascertained (Takasaki et al. 2000). Pollen and coat proteins of *Brassica campestris* which appear to be associated to SLR1 in the control of pollen adhesion have been isolated and characterized (Takayamaz et al. 2000). Recombination analyses showed that the *Brassica* S-locus in a given haplotype (S8) was approximately 50 kb and that recombination did not seem to be suppressed in the flanking regions (Casselman et al. 2000).

The so-called Montreal model for the generation of new S-alleles was intensively discussed in a series of letters which underlined the complexity of the problem and the need for further research (Charlesworth 2000; Uyenoyama and Newbigin 2000; Matton et al. 2000).

References

Abdalla MMF (1970) Inbreeding, heterosis, fertility, plasmon differentiation and *Phytophtora* resistance in *Solanum verrucosum,* and some interspecific crosses in *Solanum.* Agric Res Rep 748:213

Abdalla MMF, Hermsen JGT (1971) Unilateral incompatibility: hypotheses, debate and its implications for plant breeding. Euphytica 21:32–47

Ai Y, Singh A, Coleman CE, Ioerger RT, Kheyr-Pour A, Kao T-H (1990) Self-incompatibility in *Petunia inflata*: isolation and characterization of cDNAs encoding three S-allele-associated proteins. Sex Plant Reprod 3:130–138

Ai Y, Kron E, Kao T-H (1991) S-alleles are retained and expressed in a self-compatible cultivar of *Petunia hybrida*. Mol Gen Genet 230:353–358

Allard RW (1965) Genetic systems associated with colonizing ability in predominantly self-pollinated species. In: Baker HG, Stebbins GL (eds) The genetics of colonizing species. Academic Press, New York, pp 50–76

Allard RW, Jain SK, Workman PL (1968) The genetics of inbreeding populations. Adv Genet 14:55–131

Althausen C (1908) Zur Frage über die Vererbung der langgriffeligen und kurzgriffeligen Blütenform beim Buchweizen und zur Methodik der Veredelung dieser Pflanzen. Zhur Opuitn Agron 9:561–568

Anderson E, de Winton D (1931) The genetic analysis of an unusual relationship between self-sterility and self-fertility in *Nicotiana*. Ann Mo Bot Gard 18:97–116

Anderson MK, Taylor NL, Duncan JF (1974) Self-incompatibility, genotype identification and stability as influenced by inbreeding in red clover (*Trifolium pratense* L.). Euphytica 23:140–148

Anderson MA, Cornish EC, Mau SL, Williams EG, Hoggart R, Atkinson A, Bönig I, Greg B, Simpson RJ, Roche PJ, Haley JD, Penschow JD, Niall HD, Tregear GW, Coghlan JP, Crawford RJ, Clarke AE (1986) Cloning of cDNA for a stylar glycoprotein associated with expression of self-incompatibility in *Nicotiana alata*. Nature 321:38–44

Anderson MA, McFadden GI, Bernatzky R, Atkinson A, Orpin T, Dedman H, Tregear GW, Fernley R, Clarke AE (1989) Sequence variability of three alleles of the self-incompatibility gene of *Nicotiana alata*. Plant Cell 1:483–491

Anderson NO, Liedl BE, Ascher PD, Desborough S (1989) Distinguishing between self-incompatibility and other reproductive barriers in plants using male (MCC) and female (FCC) coefficient of crossability. Sex Plant Reprod 2:116–126

Annerstedt I, Lundqvist A (1967) Genetics of self-incompatibility in *Tradescantia paludosa* (Commelinaceae). Hereditas 58:13

Arasu NN (1968) Self-incompatibility in angiosperms: a review. Genetica 39:1–24

Ascher PD (1966) A gene action model to explain gametophytic self-incompatibility. Euphytica 15:179–183

Ascher PD (1971) The influence of RNA synthesis inhibitors on in vivo pollen tube growth and the self-incompatibility reaction in *Lilium longiflorum* Thunb. Theor Appl Genet 41:75–78

Ascher PD (1974) The self-incompatibility reaction in detached styles of *Lilium longiflorum* Thunb. injected before pollination with 6 methylpurine or puromycin. Inc News 4:57–61

Ascher PD (1986) Incompatibility and incongruity: two mechanisms preventing gene transfer between taxa. In: Mulcahy DL, Mulcahy GB, Ottaviano E (eds) Biotechnology and ecology of pollen. Springer, Berlin Heidelberg New York, pp 251–256

Ascher PD, Drewlow LW (1971) Unilateral interspecific incompatibility in *Lilium*. Yearb N Am Lily Soc 24:70

Ascher PD, Peloquin SJ (1966a) Effect of floral ageing on the growth of compatible and incompatible pollen tubes in *Lilium longiflorum*. Am J Bot 53:99–102

Ascher PD, Peloquin SJ (1966b) Influence of temperature on incompatible and compatible pollen tube growth in *Lilium longiflorum*. Can J Genet Cytol 8:661–664

Ascher PD, Peloquin SJ (1968) Pollen tube growth and incompatibility following intra- and interspecific pollinations in *Lilium longiflorum*. Am J Bot 55:1230–1234

Attia MS (1950) The nature of incompatibility in cabbage. Proc Am Soc Hort Sci 56:369–371

Atwood SS (1942) Genetics of pseudo-self-compatibility and its relation to cross-incompatibility in *Trifolium repens*. J Agric Res 64:699–709

Atwood SS (1944) Oppositional alleles in natural populations of *Trifolium repens*. Genetics 29:428–435

Atwood SS (1947) Cytogenetics and breeding of forage crops. Adv Genet 1:1–67

Atwood SS, Brewbaker JL (1953) Incompatibility in autoploid white clover. Cornell Univ Mem 319:1–47

Austin RB, Flavell RB, Lowe HJB (1986) Molecular biology and crop improvement. Cambridge University Press, Cambridge

Awadalla P, Charlesworth D (1999) Recombination and selection at *Brassica* self-incompatibility loci. Genetics 152(1):413–425

Bagni N, Adamo P, Serafini-Fracassini D (1981) RNA, proteins and polyamines during tube growth in germinating apple pollen. Plant Physiol 68:727–730

Baker HG (1948) Dimorphism and monomorphism in the Plumbaginaceae. I. A survey of the family. Ann Bot 12:207–219

Baker HG (1953) Dimorphism and monomorphism in the Plumbaginaceae. II. Pollen and stigmata in the genus *Limonium*. Ann Bot 17:433–445

Baker HG (1954) Dimorphism and monomorphism in the Plumbaginaceae. 8th Congr Intern Botan Paris Sect, 1954. Rapp Commun 10:133–134

Baker HG (1955) Self-compatibility and establishment after "long-distance, dispersal. Evolution 9:347–348

Baker HG (1966) The evolution, functioning and breakdown of heteromorphic incompatibility systems. I. The Plumbaginaceae. Evolution 20:349–368

Barlow N (1913) Preliminary note on heterostylism in *Oxalis* and *Lythrum*. J Genet 3:53–65

Barlow N (1923) Inheritance of the three forms in trimorphic species. J Genet 13:133–146

Barrett SCH (1988) Evolution of breeding systems in *Eichhornia* (Pontedariaceae): a review. Ann Mo Bot Gard 75:741–760

Barrett SCH (1990) The evolution and adaptive significance of heterostyly. Trends Ecol Evol 5:144–148

Barrett SCH (1992) Heterostylous genetic polymorphism: model systems for evolutionary analysis. In: Barrett SCH (ed) Evolution and function of heterostyly. Springer, Berlin Heidelberg New York, pp 1–24

Barrett SCH (1998) The evolution of mating strategies in flowering plants. Trends Plant Sci 3:325–340

Barrett SCH, Cruzan MB (1994) Incompatibility in heterostylous plants. In: Williams EG, Clarke AE, Knox RB (eds) Genetic control of self-incompatibility and reproductive development in flowering plants. Kluwer, Dordrecht, pp 189–219

Barrett SCH, Cole WW, Arroyo J, Cruzan MB, Lloyd DG (1997) Sexual polymorphism in *Narcissus triandrus* (Amaryllicaceae): is this species tristylous? Heredity 78:135–145

Bartley BGD, Cope FW (1973) Practical aspects of self-incompatibility in *Theobroma cacao* L. In: Moav R (ed) Agricultural genetics. Wiley, New York, pp 109–134

Bateman AJ (1943) Specific differences in *Petunia*. II. Pollen growth. J Genet XLV:236–242

Bateman AJ (1947) Number of S-alleles in a population. Nature 160:3

Bateman AJ (1952) Self-incompatibility systems in angiosperms. I. Theory. Heredity 6:285–310

Bateman AJ (1954) Self-incompatibility systems in angiosperms. II. *Iberis amara*. Heredity 8:305–332

Bateman AJ (1955) Self-incompatibility systems in angiosperms. III. *Cruciferae*. Heredity 9:52–68

Bateson W, Gregory RP (1905) On the inheritance of heterostylism in *Primula*. Proc R Soc Lond B 76:581–586

Baulcombe DC, English JJ (1996) Ectopic pairing of homologous DNA and post-transcriptional gene silencing in transgenic plants. Curr Opin Biotech 7:173–180

Baur E (1911) Einführung in die experimentalle Verberungslehre, 2nd edn. Borntraeger, Berlin

Beardsell DV (1991) The reproductive biology of *Thryptomene calycina* Staph. PhD, University of Melbourne, Melbourne

Beecher B, Murfett J, McClure BA (1998) RNase I from *Escherichia coli* cannot substitute for S-Rnase in rejection of *Nicotiana plumbaginifolia* pollen. Plant Mol Biol 36:553–563

Bell PR (1995) Incompatibility in flowering plants: adaptation of an ancient response. Plant Cell 7:5–16

Bernatzky R, Miller DD (1994) Self-incompatibility is codominant in intraspecific hybrids of self-compatible and self-incompatible *Lycopersicon peruvianum* and *L. hirsutum*. Sex Plant Reprod 7:297–302

Bernatzky R, Tanksley SD (1986) Towards a saturated linkage map in tomato based on isozymes and random cDNA sequences. Genetics 112:887–898

Bernatzky R, Anderson MA, Clarke AE (1988) Molecular genetics of self-incompatibility in flowering plants. Dev Genet 9:1–12

Bernatzky R, Glaven RH, Rivers BA (1995) S-related proteins can be recombined with self-incompatibility in interspecific derivatives of *Lycopersicon*. Biochem Genet 33:215–225

Bertin RI, Barnes C, Guttman SI (1989) Self-sterility and cryptic self-sterility in *Campsis radicans* (Bignoniaceae). Bot Gaz 150:397–403

Bianchi F (1959) Onderzoek naar de erfelijkheid van de bloemvorm bij *Petunia*. University of Amsterdam, Amsterdam

Bokhari MH (1972) A brief review of stigma and pollen tubes types in *Acantholimon* and *Limonium*. Notes R Botan Gard 32:79–84

Bouharmont J (1960) Recherches cytologiques sur la fructification et l'incompatibilité chez *Theobroma cacao* L. INEAC Ser Sci 89:113

Bower MS, Matias D, Fernandes-Carvalho E, Mazzurco M, Gu T, Rothstein SJ, Goring DR (1996) Two members of the thioredoxin-h family interact with the kinase domain of a *Brassica* S-locus receptor kinase. Plant Cell 8:1641–1650

Boyes DC, Nasrallah JB (1993) Physical linkage of the SLG and SRK genes at the self-incompatibility locus of *Brassica oleracea*. Mol Gen Genet 236:369–373

Boyes DC, Nasrallah JB (1995) An anther-specific gene encoded by an S-locus haplotype of *Brassica* produces complementary and differentially regulated transcripts. Plant Cell 7:1283–1294

Boyes DC, Chen C-H, Tantikanjana T, Esch JJ, Nasrallah JB (1991) Isolation of a second S-locus related cDNA from *B. oleracea*: genetic relationship between the S-locus and two related loci. Genetics 127:221–228

Boyes DC, Nasrallah ME, Vrebalov J, Nasrallah JB (1997) The self-incompatibility (S) haplotypes of *Brassica* contain highly divergent and rearranged sequences of ancient origin. Plant Cell 9:237–247

Brace J, Ockendon DJ, King GJ (1993) Development of a method for the identification of S-alleles in *Brassica oleraceae* based on digestion of PCR-amplified DNA with restriction endonucleases. Sex Plant Reprod 6:133–138

Brace J, King GJ, Ockendon DJ (1994) A molecular approach to the identification of S-alleles in *Brassica olearacea*. Sex Plant Reprod 7:203–208

Brandham PE, Owen SJ (1977) The genetic control of self-incompatibility in the genus *Gasteria* (Liliaceae). Heredity 40:155–169

Bredemeyer GMM, Blaas J (1980) Do S-allele specific peroxidase isoenzymes exist in self-incompatible *Nicotiana alata*? Theor Appl Genet 57:119–123

Bredemeyer GMM, Blaas J (1981) S-specific proteins in styles of self-incompatible *Nicotiana alata*. Theor Appl Genet 59:429–434

Brewbaker JL (1954) Incompatibility in autotetraplois *Trifolium repens*. I. Competition and self-compatibility. Genetics 39:307–316

Brewbaker JL (1957) Pollen cytology and incompatibility systems in plants. J Hered 48:217–277

Brewbaker JL (1958) Self-incompatibility in tetraploid strains of *Trifolium hybridum*. Hereditas 44:547–553

Brewbaker JL (1959) Biology of the angiosperm pollen grain. Ind J Genet Plant Breed 19:121–133

Brewbaker JL, Majumder SK (1961) Cultural studies of the pollen population effect and the self-incompatibility inhibition. Am J Bot 48:457–464

Brewbaker JL, Natarajan AT (1960) Centric fragments and pollen-part mutation of incompatibility alleles in *Petunia*. Genetics 45:699–704

Brewer JG (1968) Flowering and seed-setting in *Pyrethrum* (*Chrisantemum cinerariae-folium* Vis). A review. Radiat Bot 9:18–21

Brewer JG, Parlevliet JE (1969) Incompatibility as a new method for identification of *Pyrethrum* clones. Euphytica 18:320–325

Brieger F (1930) Selbsterilität und Kreuzungssterilität. Springer, Berlin, pp 386

Briggs BG (1963) Incompatibility among species of *Darwinia*. Genet Today 1:211

Briggs SP, Koziel M (1998) Engineering new plant strains for commercial markets. Curr Opin Biotech 9:233–235

Brooks RJ, Tobias AM, Lawrence MJ (1996) The population genetics of the self-incompatibility polymorphism in *Papaver rhoeas*. XI. The effects of limited pollen and seed dispersal, overlapping generations and variations in plant size on the variance of S-allele frequencies in populations at equilibrium. Heredity 76:367–376

Brooks RJ, Tobias AM, Lawrence MJ (1997a) A time series analysis of the population genetics of the self-incompatibility polymorphism. 1. Allele frequency distribution of a population with overlapping generations and variation in plant size. Heredity 79:350–360

Brooks RJ, Tobias AM, Lawrence MJ (1997b) A time series analysis of the population genetics of the self-incompatibility polymorphism. 2. Frequency-equivalent population and the number of alleles that can be maintained in a population. Heredity 79:361–364

Broothaerts WJ, Janssens GA, Proost P, Broekaert WF (1995) cDNA cloning and molecular analyses of two self-incompatibility alleles from apple. Plant Mol Biol 27:499–511

Brown G (1983) Cherry stella. Garden J R Hortic Soc 108:277

Brugière N (1999) A structural and functional analysis of the S-locus region in two self-incompatible lines of Brassica *napus*. In: Pollen-stigma interactions. Int Conf (Dept Plant Sci, Univ Oxford) dedicated to the memory of Jack Heslop-Harrison, Oxford, p 23

Bubar JS (1958) An association between variability in ovule, development within ovaries and self-incompatibility in *Lotus* (Leguminosae). Can J Bot 36:65–72

Bullen MR (1960) Interspecific hybridization within the diploid species of *Lotus* (Leguminosae). MSc Thesis. McGill University, Montreal

Burgos L, Pérez-Tornero O, Ballestèr J, Olmas O (1998) Detection and inheritance of stylar ribonucleases associated with self-incompatibility alleles in apricot. Sex Plant Reprod 11:153–158

Burton GW, Hanna WW (1992) Using apomictic tetraploids to make a self-incompatible diploid pensacola bahiagrass clone set seed. J Hered 83:305–306

Cabin RJ, Evans AS, Jennings DL, Marshall DL, Mitchell RJ, Sher AA (1996) Using bud pollinations to avoid self-incompatibility: implications from studies of three mustards. Can J Bot 74:285–289

Cabrillac D, Delorme V, Garin J, Ruffio-Châble V, Giranton JL, Dumas C, Gaude T, Cock JM (1999) The S15 self-incompatibility haplotype in *Brassica oleracea* includes three S-gene family members expressed in stigmas. Plant Cell 11:971–986

Campbell RJ, Ascher PD (1972) High temperature removal of self-incompatibility in *Nemesia strumosa*. Incomp Newslett 1:3–5

Campbell RJ, Ascher PD (1975) Incorporation of radioactive label into nucleic acids of compatible and incompatible pollen tubes of *Lilium longiflorum* Thunb. Theor Appl Genet 46:143–148

Campbell JM, Lawrence MJ (1981) The population genetics of the self-incompatibility polymorphism in *Papaver rhoeas*. II. The number and frequency of S-alleles in a natural population (R106). Heredity 46:81–90

Carluccio F, de Nettancourt D, van Gastel AJG (1974) On a possible involvement of chromosome 3 in the formation of self-compatibility mutations in *Nicotiana alata* Link and Otto. In: IAEA (ed) Polyploidy and induced mutations in plant breeding. IAEA, Vienna, pp 41–50

Carraro L, Lombardo G, Gerola P (1996) Stylar peroxidase and heteromorphic incompatibility reactions in *Primula acaulis* Hil ("thrum" morph). Caryologia 49:101–112

Casselman AL, Vrebalov J, Conner JA, Singai A, Giovanni J, Nasrallah ME, Nasrallah JB (2000) Determining the physical limits of the *Brassica* S-locus by recombination analysis. Plant Cell 12:23–33

Certal AC, Sanchez AM, Kokko H, Broothaerts WJ, Oliveira MM, Feijo JA (1999) S-Rnases in apple are expressed in the pistil along the pollen tube growth path. Sex Plant Reprod 12:94–98

Chapple C, Carpita N (1998) Plant cell walls as targets for biotechnology. Curr Opin Plant Biol 1:179–185

Charlesworth B (1976) Recombination modification in a fluctuating environment. Genetics 83 131–195

Charlesworth D (1979) Some properties of populations with multilocus homomorphic gametophytic systems. Heredity 43:19–25

Charlesworth D (1988) Evolution of homomorphic sporophytic self-incompatibility. Heredity 60:445–453

Charlesworth D, Awadalla P (1998) Flowering plant self-incompatibility: the molecular population genetics of *Brassica* loci. Heredity 81:36404

Charlesworth D (2000) How can two-gene models of self-incompatibility generate new specificities? Plant Cell 12:309–310

Charlesworth B, Charlesworth D (1979 a) The maintenance and breakdown of distyly. Am Nat 114:499–513

Charlesworth D, Charlesworth B (1979 b) The evolution and breakdown of self-incompatibility systems. Heredity 43:41–55

Charlesworth D, Charlesworth B (1990) Inbreeding depression with heterozygote advantage and its effect on selection for modifiers changing the outcrossing rate. Evolution 44(4):870–888

Charlesworth D, Charlesworth B, Strobeck C (1977) Effects of selfing on selection for recombination. Genetics 86:213–226

Chen CH, Nasrallah JB (1990) A new class of S-sequences determined by a pollen recessive self-incompatibility allele of *Brassica*. Mol Gen Genet 222:241–248

Chetelat CB, DeVerna JW (1991) Expression of unilateral incompatibility in pollen of *Lycopersicum penellii* is determined by major loci on chromosomes 1, 6 and 10. Theor Appl Genet 82:704–712

Chichiricco G (1990) Self-incompatibility in *Crocus vernus* sp. *vernus* (Iridaceae). Plant Syst Evol 172:77–82

Christ B (1959) Entwicklungsgeschichtliche und physiologische Untersuchungen über die Selbsterilität von *Cardamine pratensis* L. Z Bot 47:88–112

Chung IK, Lee SY, Lim PO, Oh SA, Kim YS, Nam HG (1999) An S-Rnase gene of *Lycopersicum peruvianum* L. is highly expressed in transgenic tobacco but does not affect self-incompatibility. J Plant Physiol 154:63–70

Clark AG, Kao T-H (1991) Excess non-synonymous substitutions at shared polymorphic sites among self-incompatibility alleles of Solanaceae. Proc Natl Acad Sci USA 88:9823–9827

Clark AG, Kao T-H (1994) Self-incompatibility: theoretical concepts and evolution. In: Williams EG, Clarke AE, Knox RB (eds) Genetic control of self-incompatibility and reproductive development in flowering plants. Kluwer, Dordrecht, pp 220–242

Clark KR, Sims TL (1994) The S-ribonuclease gene of *Petunia hybrida* is expressed in non-stylar tissue, including immature anthers. Plant Physiol 106:25–36

Clark KR, Okuley JJ, Collins PD, Sims TL (1990) Sequence variability and development expression of S-alleles in self-incompatible and pseudo self-compatible *Petunia*. Plant Cell 2:815–826

Clarke AE, Newbigin E (1993) Molecular aspects of self-incompatibility in flowering plants. Annu Rev Genet 27:257–279

Clarke AE, Anderson MA, Bacic T, Harris PJ, Mau SC (1985) Molecular basis of cell recognition during fertilization in higher plants. J Cell Sci 2:261–285

Clarke AE, Anderson MA, Atkinson A, Bacic A, Ebert PR, Jahnen W, Mau SL, Woodward JR (1989) Recent developments in the molecular genetics and biology of self-incompatibility. Plant Mol Biol 13:267–271

Cock JM, Stanchev BS, Delorme V, Croy RRD, Dumas C (1995) SLR3: a modified receptor kinase gene that has been adapted to encode a putative secreted glycoprotein similar to the S-locus glycoprotein. Mol Gen Genet 248:151–161

Coleman CE, Kao T-H (1992) The flanking regions of two *Petunia inflata* S-alleles are heterogenous and contain repetitive sequences. Plant Mol Biol 18:725–737

Compton RH (1913) Phenomena and problems of self-sterility. New Phytol 12:197–206

Conner JA, Tantikanjana T, Stein JC, Kandasamy MK, Nasrallah JB, Nasrallah ME (1997) Transgene-induced silencing of S-locus genes and related genes in *Brassica*. Plant J 11(4):809–823

Conner JA, Conner P, Nasrallah ME, Nasrallah JB (1998) Comparative mapping of the *Brassica* S locus region and its homeolog in *Arabidopsis*: implications for the evolution of mating systems in the Brassicaceae. Plant Cell 10:801–812

Cope FW (1958) Incompatibility in *Theobroma cacao*. Nature 181:279

Cope FW (1962) The mechanism of pollen incompatibility in *Theobroma cacao*. Heredity 17:157–182

Cornish MA, Hayward MD, Lawrence MJ (1979) Self-incompatibility in ryegrass. I. Genetic control in diploid *Lollium perenne*. Heredity 43(1):95–106

Cornish EC, Pettitt JM, Bonig I, Clarke AE (1987) Developmentally-controlled, tissue-specific expression of a gene associated with self-incompatibility in *Nicotiana alata*. Nature 326:99

Correns C (1912) Selbststerilität und Individualstoffe. Festschrift der mathematisch-naturwissenschaftlichen Gesellschaft zur 84. Vers Dtsch Naturforsch Ärzte, pp 1–32

Correns C (1913) Selbststerilität und Individualstoffe. Biol Centr 33:389–423

Crane MB, Brown A (1937) Incompatibility and sterility in the sweet cherry, *Prunus avium* L. J Pomol 15:86–99

Crane MB, Lawrence WJC (1929) Genetical and cytological aspects of incompatibility and sterility in cultivated fruits. J Pomol Hortic Sci 7:276–301

Crane MB, Lewis D (1942) Genetical studies in pears. III Incompatibility and sterility. J Genet 43:31–49

Cresti M, Linskens HF (1999) The discovery of sexual reproduction in higher plants. Acta Biol Cracov 41:19–29

Cresti M, Ciampolini F, Mulcahy DL, Mulcahy GBM (1985) Ultrastructure of *Nicotiana alata* pollen, its germination and early tube formation. Am J Bot 72:719–727

Cresti M, Ciampolini F, Tiezzi A (1986) Ultrastructural studies on *Nicotiana tabacum* *pollen tubes* grown in different culture media (preliminary results). Acta Bot Ned 65:285–292

Crosby JL (1940) High proportions of homostyle plants in populations of *Primula vulgaris*. Nature 145:672–673

Crosby JL (1949) Selection of an unfavourable gene complex. Evolution 3:212–230

Crowe LK (1954) Incompatibility in Cosmos bipinnatus. Heredity 8:1–11

Crowe LK (1955) The evolution of incompatibility in species of *Oenothera*. Heredity 9:293–322

Crowe LK (1964) The evolution of outbreeding in plants. I. The angiosperms. Heredity 19:435–457

Crowe LK (1971) The polygenic control of outbreeding in *Borago officinalis*. Heredity 27(1):111–118

Dahlgren KVO (1916) Eine *acaulis*-Varietät von *Primula officinalis* Jacq. und ihre Erblichkeitsverhältnisse. Svensk Bot Tidskr 10:536

Dahlgren KVO (1922) Vererbung der Heterostylie bei *Fagopyrum* (nebst einigen Bemerkungen über Pulmonaria). Hereditas 3:91–99

Dana MN, Ascher PD (1986a) Sexually localized expression of pseudo self-compatibility (PSC) in *Petunia x hybrida* hort. I. Pollen inactivation. Theor Appl Genet 71:753–577

Dana MN, Ascher PD (1986b) Sexually localized expression of pseudo self-compatibility (PSC) in *Petunia*. Theor Appl Genet 71:578–589

Darlington CD, Mather K (1949) The elements of genetics. Allen and Unwin, London

Darwin C (1862) On the two forms, or dimorphic condition, in the species of *Primula* and on their remarkable sexual relation. J Linn Soc Bot 6:77–96

Darwin C (1868) Animals and plants under domestication, 2nd edn (1875, 2). Appleton, New York

Darwin C (1876) Effects of cross- and self-fertilization in the vegetable kingdom, 2nd edn (1878). Appleton, New York

Darwin C (1879) De la variation des animaux et des plantes à l'état domestique, vol 1, 2nd English edn. Reinwald, Paris

Darwin C (1880) The different forms of flowers on plants of the same species, 2nd edn, 1892 reprint. Murray, London, p XXIV+352

Darwin C (1882 a) Origin of species, 6th edn, 2. Appleton, New York

Darwin C (1982 b) De la variation des animaux et des plantes à l'état domestique, 2nd edn 1980. Vols I and II translated. Murray, London

Darwin C (1890) De la variation des animaux et des plantes à l'état domestique, vol 2, 2nd English edn. Reinwald, Paris

Darwin C (1905) L'origine des espèces au moyen de la sélection naturelle ou la lutte pour l'existence dans la nature. Translation of the final English edition. Schleicher Frères, Paris

Davies DR, Wall ET (1961) Gamma radiation and interspecific incompatibility in plants. Effects of ionizing radiations on seeds. IAEA Symposium, Vienna

Dayton DF (1974) Overcoming self-incompatibility in apple with killed compatible pollen. J Am Soc Hortic Sci 99:190–192

Dearnarley JDW, Clarke KM, Lew RR, Goring DR (1999) Neither compatible nor self-incompatible pollination of *Brassica napus* involve reorganization of the papillar cytoskeleton. New Phytol 141:199–207

Delorme V, Giranton JL, Hatzfeld Y, Friry A, Heizmann P, Ariza MJ, Dumas C, Gaude T, Cock JM (1995) Characterization of the S-locus genes, SLG and SRK, of the *Brassica* S3 haplotypes: identification of a membrane-localized protein encoded by the S-locus receptor kinase gene. Plant J 7:429–440

de Nettancourt D (1969) The gametophytic system of self-incompatibility in higher plants. Theor Appl Genet 39:187–196

de Nettancourt D (1972) Self-incompatibility in basic and applied researches with higher plants. Genet Agraria 26:163–216

de Nettancourt D (1977) Incompatibility in angiosperms. Springer, Berlin Heidelberg New York

de Nettancourt D (1979) Systèmes d'incompatibilité. Bull Soc Bot Fr 2:97–104

de Nettancourt D (1997) Incompatibility in angiosperms. Sex Plant Reprod 10:185–199

de Nettancourt D (1999) Homomorphic self-incompatibility in flowering plants. In: Cresti M, Cai G, Moscatelli A (eds) Fertilization in higher plants. Springer, Berlin Heidelberg New York

de Nettancourt D, Ecochard R (1968) Effects of chronic irradiation upon a self-incompatible clone of *Lycopersicum*. Theor Appl Genet 38:289–293

de Nettancourt D, Grant WF (1963) The cytogenetics of *Lotus* (Leguminosae). II. A diploid interspecific hybrid between *L. tenuis* and *L. filicaulis*. Can J Genet Cytol 5:338–347

de Nettancourt D, Ecochard R, Perquin MDG, Drift van der T, Westerhof M (1971) The generation of new S-alleles at the incompatibility locus of *L. peruvianum* Mill. Theor Appl Genet 41:120–129

de Nettancourt D, Devreux M, Bozzini A, Cresti M, Pacini E, Sarfatti G (1973 a) Ultrastructural aspects of self-incompatibility mechanism in *Lycopersicum peruvianum* Mill. J Cell Sci 12:403–419

de Nettancourt D, Devreux M, Laneri U, Pacini E, Cresti M, Sarfatti G (1973 b) Ultrastructural aspects of unilateral interspecific incompatibility between *Lycopersicum peruvianum* and *L. esculentum*. Caryologia 25:207–217

de Nettancourt D, Devreux M, Laneri U, Cresti M, Pacini E, Sarfatti G (1974 a) Genetical and ultrastructural aspects of self- and cross-incompatibility in interspecific hybrids between self-compatible *L. esculentum* and self-incompatible *L. peruvianum*. Theor Appl Genet 44:278–288 de Nettancourt D, Saccardo F, Laneri U, Capaccio E, Westerhof M, Ecochard R (1974 b) Self-compatibility in a spontaneous tetraploid of *Lycopersicum peruvianum* Mill. In: IAEA (ed) Polyploidy and induced mutations in plant breeding. IAEA, Vienna, pp 77–84

de Nettancourt D, Devreux M, Carluccio F, Laneri U, Cresti M, Pacini E, Sarfatti G, van Gastel AJG (1975) Facts and hypotheses on the origin of S-mutations and on the function of the S-gene in *Nicotiana alata* and *Lycopersicum peruvianum*. Proc R Soc Lond B 188:345–360

Den Nijs APM, Oost EH (1980) Effects of mentor pollen on pollen-pistil incongruities among species of *Cucumis*. Euphytica 29:267–271

Denna DW (1971) The potential use of self-incompatibility for breeding F_1 hybrids of naturally self-pollinated vegetable crops. Euphytica 20:542–548

Denward T (1963) The function of the incompatibility alleles in red clover (*Trifolium pratense* L.). Hereditas 49:189–334

Dereuddre J (1971) Sur la présence de groupes de saccules appartenant au réticulum endo-plasmique dans les cellules des ébauches foliaires en vie ralentie de *Betula verrucosa*. CR Acad Sci 273:2239–2242

Després C, Saba-El-Leil M, Rivard SR, Morse D, Cappadocia M (1994) Molecular cloning of two *Solanum chacoense* S-alleles and a hypothesis concerning their evolution. Sex Plant Reprod 7:169–176

Desrochers AM, Rieseberg LH (1998) Mentor effects in wild species of Helianthus (Asreriaceae). Am J Bot 85(6):770–775

Devey F, Fearon CH, Hayward MD, Lawrence MJ (1994) Self-incompatibility in ryegrass. XI. Number and frequency of alleles in a cultivar of *Lollium perenne*. Heredity 73:262–264

Devreux M, Vallaeys G, Pochet P, Gilles A (1959) Recherches sur l'autostérilité du caféier Robusta (*Coffea canephora* Pierre). Publ INEAC 78:44

de Winton D, Haldane JBC (1933) The genetics of *Primula sinensis*. II. Segretation and interaction of factors in the diploid. J Genet 27:1–44

Dhaliwal AS, Malik CP, Singh MB (1981) Overcoming incompatibility in *Brassica campestris* L. by carbon dioxide and dark fixation of the gas by self and cross pollinated pistils. Ann Bot 48:227–233

Dickinson HG (1976) Common factors in exine deposition. In: Ferguson IK, Muller J (eds) The evolutionary significance of the exine, series 1. Linnean Society Symposium, pp 67–90

Dickinson HG (1994) Simply a social disease? Nature 367:517–518

Dickinson HG (1995) Dry stigmas, water and self-incompatibility in *Brassica*. Sex Plant Reprod 8:1–10

Dickinson HG, Elleman CJ (1985) Pollen-stigma interactions in *Brassica oleracea*. Micron 16:255–270

Dickinson HG, Lewis D (1973 a) The formation of the tryphine coating the pollen grains of *Raphanus*, and its properties relating to the self-incompatibility system. Proc R Soc Lond B 184:148–165

Dickinson HG, Lewis D (1973 b) Cytochemical and ultrastructural differences between intraspecific compatible and incompatible pollinations in *Raphanus*. Proc R Soc Lond B 183:21–28

Dickinson HG, Lewis D (1975) Interaction between the pollen grain coating and the stigmatic surface during compatible and incompatible interspecific pollinations in *Raphanus*. In: Duckett JC, Racey PA (eds) The biology of the male gamete. Biol J Linn Soc 7 [Suppl 1]:165–175

Dickinson HG, Crabbé MJC, Gaude T (1992) Sprorophytic self-incompatibility systems: S-gene products. Int Rev Cytol 140:525–561

Dionne LA (1961) Cytoplasmic sterility in derivates of *Solanum demissum*. Am Potato J 38:117–120

Dobrofsky S, Grant WF (1979) Electrophoretic evidence supporting self-incompatibility in *Lotus corniculatus*. Can J Bot 58:712–716

Dodds PN, Bönig I, Anderson MA, Newbigin E, Clarke AE (1993) The S-Rnase gene of *Nicotiana alata* is expressed in developing pollen. Plant Cell 5:1771–1782

Dodds PN, Clarke AE, Newbigin E (1996) A molecular perspective on pollination in flowering plants. Cell 85:141–144

Dodds PN, Clarke AE, Newbigin E (1997) Molecules involved in self-incompatibility in flowering plants. In: Janick J (ed) Plant breeding reviews, vol 15. Wiley, New York, pp 18–42

Dodds PN, Ferguson C, Clarke AE, Newbigin E (1999) Pollen-expressed S-Rnases are not involved in self-incompatibility in *Lycopersicum peruvianum*. Sex Plant Reprod 12:76–87

Do Rêgo MM, Bruckner CH, da Silva EA, Finger FL, de Siqueira DL, Fernandes AA (1999) Self-incompatibility in Passion fruit: evidence of two loci genetic control. Theor Appl Genet 98 564–568

Doughty J, Hedderson F, McCubbin AG, Dickinson H (1993) Interaction between a coating-borne peptide of the *Brassica* pollen grain and stigmatic S- (self-incompatibility) locus-specific glycoprotein. Proc Natl Acad Sci USA 90:467–471

Doughty J, Dixon S, Hiscock SJ, Willis AC, Parkin IAP, Dickinson HG (1998) PCP-A1, a defensin-like *Brassica* pollen coat protein that binds the S locus glycoprotein, is the product of gametophytic gene expression. Plant Cell 10:1333–1347

Doughty J, Wingett SW, Abott AR, Al-Sady B, Dickinson AG (1999) PCPs and their interactions with stigmatic S-class proteins in *Brassica*: putative roles in SI and pollination. In: Pollen-stigma interactions. Int Conf (Dept Plant Sci Univ Oxford) dedicated to the memory of Jach Heslop-Harrison, Oxford, p 28

Dowrick VPJ (1956) Heterostyly and homostyly in *Primula obconica*. Heredity 10:219–236

Dulberger R (1964) Flower dimorphism and self-incompatibility in *Narcissus tazetta*. Evolution 18:361–373

Dulberger R (1970) Tristyly in *Lytrhum junceum*. New Phytol 69:751–759

Dulberger R (1974) Structural dimorphism of stigmatic papillae in distylous *Linum* species. Am J Bot 61:238–243

Dulberger R (1975) Intermorph structural differences between stigmatic papillae and pollen grains in relation to incompatibility in Plumbaginaceae. Proc R Soc Lond B 188:257–274

Dulberger R (1981) Dimorphic exine sculpturing in three distylous species of *Linum* (Linaceae). Plant Syst Evol 139:113–119

Dulberger R (1987a) Fine structure and cytochemistry of the stigma surface and incompatibility in some distylous *Linum* species. Ann Bot 59:203–217

Dulberger R (1987b) Incompatibility in *Plumbago capensis*. Fine structure and cytochemistry of the reproductive surface and pollen wall. XIV Int Bot Congr, Berlin, Book of Abstracts, p 203

Dulberger R (1992) Floral polymorphisms and their functional significance in the heterostylous syndrome. In: Barrett SCH (ed) Evolution and function of heterostyly. Springer, Berlin Heidelberg New York, pp 41–84

Dulberger R (1999) Incompatibility in distylous *Linum grandiflorum*. In: Pollen-stigma interactions. Int Conf (Dept Plant Sci, Univ Oxford) dedicated to the memory of Jack Heslop-Harrison, Oxford, p 58

Dumas C, Gaude T (1982) Sécrétions et biologie florale. II. Leur rôle dans l'adhésion et la reconnaissance pollen-stigmate. Données récentes, hypothèses et notion d'immunité végétale. Bull Soc Bot Fr 129:89–101

Dumas C, Luu D, Heizmann Ph (1999) Pollen hydration and pollen adhesion on the stigma surface. In: Pollen-stigma interactions. Conf (Dept Plant Sci, Univ Oxford) dedicated to the memory of Jack Heslop-Harrison, Oxford, p 10

Duvick DN (1966) Influence of morphology and sterility on breeding methodology. In: Frey KJ (ed) Proc Plant Breeding Symposium Iowa State University, pp 85–138

Dwyer KG, Balent MA, Nasrallah JB, Nasrallah ME (1991) DNA sequences of self-incompatibility genes from *Brassica campestris* and *B. oleraceae*: polymorphism predating speciation. Plant Mol Biol 16:481–486

Dwyer KG, Kandasamy MK, Mahosky DI, Acciai J, Kudish BI, Miller JE, Nasrallah ME, Nasrallah JB (1994) A superfamily of S-locus related sequences in *Arabidopsis*: diverse structures and expression patterns. Plant Cell 6:1829–1843

Dzelzkalns VA, Nasrallah JB, Nasrallah ME (1992) Cell-cell communication in plants: self-incompatibility in flower development. Dev Biol 153:70–82

East EM (1924) Self-sterility. Bibliogr Genet 5:331–370

East EM (1926) The physiology of self-sterility in plants. Jacques Loeb Mem Vol. J Gen Physiol 8:403–416

East EM (1929) Self-sterility. Bibliographica Genetica 5:331–369

East EM (1932) Studies on self-sterility. IX. The behavior of crosses between self-sterile and self-fertile plants. Genetics 17:175–202

East EM (1934) Norms of pollen tube growth in incompatible matings of self-sterile plants. Proc Natl Acad Sci USA 20:225–230

East EM (1940) The distribution of self-sterility in flowering plants. Proc Am Philos Soc 82:449–518

East EM, Mangelsdorf AJ (1925) A new interpretation of the hereditary behaviour of self-sterile plants. Proc Natl Acad Sci USA 11:166–171

East EM, Park JB (1917) Studies on self-sterility. I. The behavior of self-sterile plants. Genetics 2:505–609

Ebert PR, Anderson MA, Bernatzky R, Altschuler M, Clarke, AE (1989) Genetic polymorphism of self-incompatibility in flowering plants. Cell 56:255–262

Eghis SA (1925) Experiments on the drawing up of a method of buckwheat breeding. Bull Appl Bot Genet Plant Breed 14:235–231

Elleman CJ, Dickinson HG (1986) Pollen-stigma interactions in *Brassica*. IV. Structural reorganisation in the pollen grains during hydratation. Cell Sci 80:141–157

Elleman CJ, Dickinson HG (1990) The role of the exine coating in pollen-stigma interactions. New Phytol 114:511–518

Elleman CJ, Dickinson HG (1994) Pollen-stigma interaction during sporophytic self-incompatibility in *Brassica oleracea*. In: Williams EG, Clarke AE, Knox RB (eds) Genetic control of self-incompatibility and reproductive development in flowering plants. Kluwer, Dordrecht, pp 67–87

Elleman CJ, Dickinson HG (1996) Identification of pollen components regulating pollination-specific responses in the stigmatic papillae of *Brassica oleracea*. New Phytol 133:197–205

Elleman CJ, Wilson CE, Sarker RH, Dickinson H (1988) Interaction between the pollen tube and stigmatic wall following pollination in *Brassica oleracea*. New Phyol 109:111–117

Elleman CJ, Franklin-Tong VE, Dickinson HG (1992) Pollination in species with dry stigma: the nature of the early stigmatic response and the pathway taken by pollen tubes. New Phytol 121:413–424

Ellis MF, Sedgley M, Gardner JA (1991) Interspecific pollen-pistil interaction in *Eucalyptus* (Myrthaceae): the effect of taxonomic distance. Ann Bot 68:185–194

Emerson S (1938) The genetics of self-incompatibility in *Oenothera organensis*. Genetics 23:190–202

Emerson S (1939) A preliminary survey of the *Eonothera organensis* population. Genetics 24:524–537

Emsweller SL, Stuart NW (1948) Use of growth regulating substances to overcome incompatibilities in *Lilium*. Am Soc Hortic Sci 51:581

England FJW (1974) The use of incompatibility for the production of F₁ hybrids in forage grasses. Heredity 32:183–188

Erdtman G (1970) Über Pollendimorphie in Plumbaginaceae- Plumbagineae. (Unter besonderer Berücksichtigung von *Dyerophytum indicum*). Svensk Bot Tidskr 64:184–188

Erdtman G, Dunbar A (1966) Notes on electron micrographs illustrating the pollen morphology in *Armeria maritima* and *A. sibirica*. Grana Palynol 6:338–354

Ernst A (1932) Zur Kenntnis der Heterostylie tropischer Rubiaceen. I. Arch Julius Klaus-Stiftung 7:241–280

Ernst A (1936) Heterostylie-Forschung. Versuche zur genetischen Analyse eines Organisations- und Anpassungsmerkmales. Z Indukt Abstamm Vererbl 71:156–230

Ernst A (1953) „Basic numbers" und Polyploidie und ihre Bedeutung für das Heterostylie-Problem. Arch Julius Klaus-Stiftung 28:1–159

Ernst A (1957) Austausch und Mutation im Komplex -Gen für Blütenplastik und Inkompatibilität bei *Primula*. Z Indukt Abstamm Vererbl 88:517–599

Esser K (1953) Genomverdopplung und Pollenschlauchwachstum bei Heterostylen. Z Indukt Abstamm Vererbl 85:28–50

Eyster WH (1941) The induction of fertility in genetically self-sterile plants. Science 94:144–145

Falque M (1994) Pod and seed development and phenotype of the M1 plants after pollination and fertilization with irradiated pollen in cacao (*Theobroma cacao* L.). Euphytica 75:19–25

Fearon CH, Hayward MD, Lawrence MJ (1984) Self-incompatibility in ryegrass. IX. Cross-compatibility and seed-set in autotetraploid *Lollium perenne*. Heredity 53:423–434

Fearon CH, Cornish MA, Hayward MD, Lawrence MJ (1994) Self-incompatibility in ryegrass. X. Number and frequency of alleles in a natural population of *Lollium perenne*. Heredity 73:262–264

Feijó JA, Certal AC, Boavida L, van Nerum I, Valdiviesso T, Oliveira MM, Broothaerts W (1999) Advances on the study of sexual reproduction in the cork-tree (*Quercus suber*), chestnut (*Castenea sativa*) and in *Rosaceae* (apple and almond). In: Cresti M, Cai G, Moscatelli A (eds) Fertilization in higher plants. Springer-Verlag Berlin Heidelberg New York, pp 377–396

Ferrari TE, Bruns D, Wallace DH (1981) Isolation of a plant glycoprotein involved with control of intercellular recognition. Plant Physiol 67:270–277

Filzer P (1926) Die Selbststerilität von *Veronica syriaca*. Z Indukt Abstamm Vererbl 41:137–197

Finney DJ (1952) The equilibrium of a self-incompatible polymorphic species. Genetica 26:33–64

Fisher RA (1941) The theoretical consequences of polyploid inheritance for the mid style form of *Lythrum salicaria*. Ann Eugen 11:31–38

Fisher RA (1944) The allowance for double reduction in the calculation of genotype frequencies with polyploid inheritance. Ann Eugen 12:169–171

Fisher RA (1949) The theory of inbreeding. Oliver and Boyd, London

Fisher RA (1961) A model for the generation of self-sterility alleles. J Theor Biol 1:411–414

Fisher RA, Martin VC (1948) Genetics of style-length in *Oxalis*. Nature 162:533

Fisher RA, Mather K (1943) The inheritance of style length in *Lythrum salicaria*. Ann Eugen 12:I–II

Flavell RB (1994) Inactivation of gene expression in plants as a consequence of specific sequence duplication. Proc Natl Acad Sci USA 91:3490–3496

Foote HCC, Ride JP, Franklin-Tong VE, Walker EA, Lawrence MJ, Franklin FCH (1994) Cloning and expression of a distinctive class of self-incompatibility (S-) gene from *Papaver rhoeas* L. Proc Natl Acad Sci USA 91:2265–2269

Ford MA, Kay QON (1985) The genetics of incompatibility in *Sinapis arvensis* L. Heredity 54:99–102

Frankel R (1973) Utilization of self-incompatibility as an outbreeding mechanism in hybrid seed production. In: Moav R (ed) Agricultural genetics. Wiley, New York, pp 95–108

Franklin FCH, Franklin-Tong VE, Thorby GJ, Atwall KK, Lawrence MJ (1991) Molecular bases of the self-incompatibility mechanism in *Papaver rhoeas*. Plant Growth Regul 11:5–12

Franklin FCH, Lawrence MJ, Franklin-Tong VE (1995) Cell and molecular biology of self-incompatibility in flowering plants. Int Rev Cytol 158:23377

Franklin TM, Oldknow J, Trick M (1996) SLR1 function is dispensable for both self-incompatible rejection and self-compatible process in *Brassica*. Sex Plant Reprod 9:203–208

Franklin VE, Franklin FCH (2000) Self-incompatibility in *Brassica*:the elusive pollen S-gene is identified! Plant Cell 12:305–308

Franklin-Tong VE, Franklin FCH (1992) Gametophytic self-incompatibility in *Papaver rhoeas* L. Sex Plant Reprod 5:36342

Franklin-Tong VE, Franklin FCH (1993) Gametophytic self-incompatibility: contrasting mechanisms for *Nicotiana* and *Papaver*. Trends Cell Biol 3:340

Franklin-Tong VE, Lawrence MJ, Franklin FCH (1988) An in vitro bioassay for the stigmatic product of the self-incompatibility gene in *Papaver rhoeas* L. New Phytol 110:109–118

Franklin-Tong VE, Lawrence MJ, Franklin FCH (1989) Characterization of a stigmatic component from *Papaver rhoeas* which exhibits the specific activity of a self-incompatibility (SI) gene product. New Phytol 110:319–324

Franklin-Tong VE, Lawrence MJ, Franklin FCH (1990) Self-incompatibility in *Papaver rhoeas*: inhibition of incompatible pollen is dependent on pollen gene expression. New Phytol 116:319–324

Franklin-Tong VE, Ride JP, Read ND, Trewavas AJ, Franklin FCH (1993) The self-incompatibility response in *Papaver rhoeas* is mediated by cytosolic free calcium. Plant J 4:163–177

Franklin-Tong VE, Ride JP, Franklin FCH (1995) Recombinant stigmatic self-incompatibility protein elicits a Ca2+ transient in pollen of *Papaver rhoeas*. Plant J 8:299–307

Franklin-Tong VE, Drobak BK, Allan AC, Watkins PAC, Trewavas AJ (1996) Growth of pollen tubes of *Papaver rhoeas* is regulated by a slow moving calcium wave propagated by inositol 1,4,5-trisphosphate. Plant Cell 8:1305–1321

Franklin-Tong VE, Hackett G, Hepler PK (1997) Ratio-imaging of Ca2+ in the self-incompatibility response in pollen tubes of Papaver rhoeas. Plant J 12(6):1376–1386

Fryxell PA (1957) Mode of reproduction of higher plants. Bot Rev 23:135–233

Fyfe VC (1953) Double reduction at the Mid locus in *Lythrum salicaria*. Heredity 7:285–289

Gadish IZ, Zamir D (1987) Differential zygotic abortion in an interspecific *Lycopersicon* cross. Genome 29:156–159

Gallie DR (1998) Controlling gene expression in transgenics. Curr Opin Plant Biol 1:166–172

Ganders FR (1974) Disassociative pollination in the distylous plant *Jepsonia hetrandra*. Can J Bot 52:2401–2406

Ganders FR (1979) The biology of heterostyly. NZJ Bot 17:607–635

Garber RI, Quisenberry KS (1927) The inheritance of length of style in buckwheat. J Agric Res 34:181–183

Garde NM (1959) Mechanisms of species isolation in tuberous *Solanum*. Agro Lusit 21:19–42

Gaude T, Dumas C (1987) Molecular and cellular events of self-incompatibility. In: Giles KL, Prakash J (eds) Pollen: cytology and development. (International review of cytology, vol 107.) Academic Press, Orlando, pp 333–366

Gaude T, Friry A, Heizmann P, Mariac C, Rougier M, Fobis I, Dumas C (1993) Expression of a self-incompatibility gene in a self-compatible line of *Brassica oleracea*. Plant Cell 5:75–86

Gaude T, Rougier R, Heizmann P, Ockendon DJ, Dumas C (1995) Expression level of the SLG gene is not correlated with the self-incompatibility phenotype in the class II S haplotypes of *Brassica olearacea*. Plant Mol Biol 27:1003–1014

Gause GF (1934) The struggle for existence. Williams and Wilkins, Baltimore

Gebhardt C, Ritter E, Barone A (1991) RFLP maps of potato and their alignment with the homoeologous tomato genome. Theor Appl Genet 83:49–57

Geitmann A (1999) Cell death of self-incompatible pollen tubes: necrosis or apoptosis? In: Cresti M, Cai G, Moscatelli A (eds) Fertilization in higher plants (molecular and cytological aspects). Springer, Berlin Heidelberg New York, pp 113–137

Geitmann A, Cresti M (1998) Ca2+ channels control the rapid expansions in pulsating growth of *Petunia hybrida* pollen tubes. J Plant Physiol 152:439–447

Gelvin SB (1998) The introduction and expression of transgenes in plants. Curr Opin Biol 9:227–232

Gerstel DU (1950) Self-incompatibility studies in Guayule. II Inheritance. Genetics 35:482–506

Gertz A, Wricke G (1989) Linkage between the incompatibility locus Z and a beta-glucosidase locus in rye. Plant Breed 102:255–259

Gibbs PE (1986) Do homomorphic and heteromorphic self-incompatibility systems have the same sporophytic mechanism? Plant Syst Evol 154:285–323

Gibbs PE, Ferguson IK (1987) Correlations between pollen exine sculpturing and angiosperm self-incompatibility system-a reply. Plant Syst Evol 157:143–159

Gilroy S, Bethke PC, Jones RL (1993) Calcium homeostasis in plants. J Cell Sci 106:453–462

Giranton JL, Ariza MJ, Dumas C, Cock JM, Gaude T (1995) The S-locus receptor kinase encodes a soluble glycopotein corresponding to the SRK extracellular domain in *Brassica oleracea*. Plant J 8:827–834

Giranton JL, Cabrillac D, Dumas C, Cock JM, Gaude T (1999a) Characterization of a SRK kinase complex in the plasma membrane of stigmatic cells in *Brassica olearacea* (Abstract). In: Pollen-Stigma Interactions. Int Conf (Dept Plant Sci, Univ Oxford) dedicated to the memory of Jack Heslop-Harrison, Oxford, p 24

Giranton JL, Passelegue E, Dumas C, Cock JM, Gaude T (1999b) Membrane proteins involved in pollen-pistil interactions. Biochimie 81:675–680

Glendinning DR (1960) Selfing of self-incompatible cocoa. Nature 187:170

Golynskaya EL, Bashrikova NV, Tomchuk NN (1976) Phytohemagglutinins of the pistil of *Primula* as possible proteins of generative incompatibility. Soviet Plant Physiol 23:169–176

Golz JF, Clarke AE, Newbigin E, Anderson MA (1998) A relic S-Rnase is expressed in the styles of self-compatible *Nicotana sylvestris*. Plant J 16(5):591–596

Golz JF, Su V, Clarke AE, Newbigin E (1999a) A molecular description of mutations affecting the pollen component of the *Nicotiana* S-locus. Genetics 152:1123–1135

Golz JF, Clarke AE, Newbigin E (1999b) Duplications of the *Nicotiana alata* S-locus affect pollen self-incompatibility (Abstract). In: Pollen Stigma Interactions. Int

Conf (Dept Plant Sci, Univ Oxford) dedicated to the memory of Jack Heslop-Harrison, Oxford, p 19

Gonai H, Hinata K (1971) Effect of temperature on pistil growth and phenotypic expression of self-incompatibility in *Brassica oleracea* L. Jpn J Breed 21:195–198

Goring DR, Rothstein SJ (1992) The S-locus receptor gene in a self-incompatible *Brassica napus* line encodes a functional serine/threonine kinase. Plant Cell 4:1273–1281

Goring DR, Glavin TL, Schafer U, Rothstein SJ (1993) An S-receptor kinase gene in self-compatible *Brassica napus* has a 1 bp deletion. Plant Cell 5:531–539

Gosh S, Shivanna KR (1980) Pollen-pistil interaction in *Linum grandiflorum*. Planta 149:257–261

Gosh S, Shivanna KR (1983) Studies on pollen-pistil interaction in *Linum grandiflorum*. Phytomorphology 32:385–395

Grant V (1949) Pollination systems as isolating mechanisms in angiosperms. Evolution 3:82–97

Grant WF, Bullen MR, de Nettancourt D (1962) The cytogenetics of *Lotus*. I. Embryo-cultured interspecific diploid hybrids closely related to *L. Corniculatus* L. Can J Genet Cytol 4:105–128

Gray JE, McClure BA, Bönig I, Anderson MA, Clarke AE (1991) Action of the style product of the self-incompatibility gene of *Nicotiana alata* (S-Rnase) on in vitro-grown pollen tubes. Plant Cell 3:271–283

Green PJ (1994) The ribonucleases of higher plants. Annu Rev Plant Physiol Plant Mol Biol 45:421–445

Greenberg JT (1996) Programmed cell death: a way of life for plants. Proc Natl Acad Sci USA 93:12094–12097

Gregory RP (1915) Note on the inheritance of heterostylism in *Primula acaulis*. J Genet 4:303

Grun P (1961) Early stages in the formation of internal barriers to gene exchange between diploid species of *Solanum*. Am J Bot 48:78–89

Grun P, Aubertin M (1966) The inheritance and expression of unilateral incompatibility in *Solanum*. Heredity 21:131–138

Grun P, Radlow A (1961) Evolution of barriers to crossing of self-incompatible with self-compatible species of *Solanum*. Heredity 16:137–143

Guilluy CM, Gaude T, Digonnet-Kerhoas C, Chaboud A, Heizmann P, Dumas C (1990) New data and concepts in angiosperm fertiliation. In: Mechanism of fertilization. NATO AS Ser 45:253–270

Guilluy CM, Trick M, Heizmann P, Dumas C (1991) PCR detection of transcripts homologous to the self-incompatibility gene in anthers of *Brassica*. Theor Appl Genet 82:466–472

Günther E, Jüttersonke B (1971) Untersuchungen über die Kreuzungsinkompatibilität zwischen *Lycopersicum peruvianum* (L.) Mill. und *Lycopersicum esculentum* Mill. und den reziproken Bastarden. Biol Zentralbl 90:561–574

Hardon JJ (1967) Unilateral incompatibility betwen *Solanum pennellii* and *Lycopersicum esculentum*. Genetics 57:795–808

Haring VH, Gray JE, McClure BA, Anderson MA, Clarke AE (1990) Self-incompatibility: a self-recognition system in plants. Science 250:937–941

Harlan JR (1945 a) Cleistogamy and chasmogamy in *Bromus carinatus* Hook. Ann. Am J Bot 32:66–72

Harlan JR (1945 b) Natural breeding structure in the *Bromus carinatus* complex as determined by population analyses. Am J Bot 32:142–148

Harrison BJ, Darby L (1955) Unilateral hybridization. Nature 176:982

Harte C (1994) *Oenothera* (contribution of a plant to biology). In: Frankel R, Grossman M, Linskens FH, Maliga P, Riley R (eds) Springer, Berlin Heidelberg New York

Hatakeyama K, Watanabe M, Takasaki T, Ojima K, Hinata K (1998a) Dominance relationships between S-alleles in self-incompatible *Brassica campestris* L. Heredity 80:241–247

Hatakeyama K, Watanabe M, Takasaki T, Hinata K (1998b) High sequence similarity between SLG and the receptor domain of SRK is not necessarily involved in higher dominance relationships in stigma in self-incompatibility. Sex Plant Reprod 11:292–294

Hatakeyama K, Takasaki T, Watanabe M, Hinata K (1998c) Molecular characterization of S-locus genes, SLG anf SRK, in a pollen recessive self-incompatibility haplotype of *Brassica rapa* L. Genetics 149:1587–1597

Hayman DL (1956) The genetic control of incompatibility in *Phalaris coerulescens*. Aust J Biol Sci 9:321–331

Hayman DL, Richter J (1992) Mutations affecting self-incompatibility in *Phalaris coerulescens* Desf. (Poaceae). Heredity 68:495–503

Hearn MJ, Franklin FCH, Ride JP (1996) Identification of a membrane glycoprotein in pollen of *Papaver rhoeas* which binds stigmatic self-incompatibility (S) proteins. Plant J 9(4):467–475

Hecht A (1960) Growth of pollen tubes of *Oenothera organensis* through otherwise incompatible styles. Am J Bot 47:32–36

Hecht A (1964) Partial inactivation of an incompatibility substance in the stigmas and styles of *Oenothera*. In: Linskens HF (ed) Pollen physiology and fertilization. North-Holland, Amsterdam, pp 237–243

Heizmann P (1992) Sporophytic self-incompatibility. In: Dattée Y, Dumas C, Gallais A (eds) Reproductive biology and plant breeding. Springer, Berlin Heidelberg New York, pp 153–162

Henny RJ, Ascher PD (1973) Effect of auxin (3-indoleacetic acid) on in vivo compatible and incompatible pollen tube growth in detached styles of *Lilium longiflorum* Thunb. Incomp Newslett 3:14–17

Hermsen JGT, Sawicka E (1979) Incompatibility and incongruity in tuber-bearing *Solanum* species. In: Hawkes JG, Lester RN, Skelding AD (eds) The biology and taxonomy of the Solanaceae, series 7. Linnean Society Symposium, London. Academic Press, London, pp 445–454

Heslop-Harrison J (1967) Ribosome sites and S-gene action. Nature 218:230–237

Heslop-Harrison J (1968) Pollen wall development. Science 161:230–237

Heslop-Harrison J (1975) Incompatibility and the pollen stigma interaction. Annu Rev Plant Physiol 26:403–425

Heslop-Harrison J (1979) An interpretation of the hydrodynamics of pollen. Am J Bot 66:737–743

Heslop-Harrison J (1982) Pollen stigma interaction and cross-incompatibility in the grasses. Science 215:1358–1364

Heslop-Harrison J, Heslop-Harrison Y (1975) Enzymic removal of proteinaceous pellicle of the stigma papilla prevents pollen tube entry in the Caryophyllaceae. Ann Bot 39:163–165

Heslop-Harrison J, Heslop-Harrison Y, Knox RB, Howlett BJ (1973) Pollen wall proteins: gametophytic and sporophytic fraction in the pollen wall of the Malvaceae. Ann Bot 37:403–412

Heslop-Harrison J, Knox RB, Heslop-Harrison Y (1974) Pollen-wall proteins: exine-held fractions associated with the incompatibility response in Cruciferae. Theor Appl Genet 44:133–137

Heslop-Harrison J, Knox RB, Heslop-Harrison Y, Mattsson O (1975) Pollen-wall proteins: emission and role in incompatibility responses. In: Duckett JG, Racey PA (eds) The biology of the male gamete. Biol J Linn Soc 7 [Suppl 1]:189–202

Heslop-Harrison Y, Shivanna KR (1977) The receptive surface of the angiosperms stigma. Ann Bot 41:1233–1258

Heslop-Harrison Y, Heslop-Harrison J, Shivanna KR (1981) Heterostyly in *Primula*. 1. Fine structural and cytochemical features of the stigma and style in *Primula vulgaris* Huds. Protoplasma 107:171–187

Hildebrand F (1863) cited in Darwin 1879–1880 (French translations of Darwin 1863)

Hildebrand F (1896) Über Selbststerilität bei einigen Cruciferen. Ber Dtsch Botan Ges 14:324–331

Hinata K, Nishio T (1981) Con A-peroxidase method: an improved procedure for staining S-glycoproteins in cellulose-acetate electrofocusing in crucifers. Theor Appl Genet 60:281–283

Hinata K, Okazaki K (1986) Role of stigma in the expression of self-incompatibility in crucifers in view of genetic analysis. In: Mulcahy DL, Mulcahy GBM, Ottaviano E (eds) Biotechnology and ecology of pollen. Springer, Berlin Heidelberg New York, pp 185–190

Hinata K, Nishio T, Kimura J (1982) Comparative studies on S-glycoproteins purified from different S-genotypes in self-incompatible *Brassica* species. II. Immunological specificities. Genetics 100:649–657

Hinata K, Watanabe M, Toriyama K, Isogai A (1993) A review of recent studies on homomorphic self-incompatibility. Int Rev Cytol 143:257–296

Hinata K, Isogai A, Isuzugawa K (1994) Manipulation of sporophytic self-incompatibility in plant breeding. In: Williams EG, Clarke AE, Knox RB (eds) Genetic control of self-incompatibility and reproductive development in flowering plants. Kluwer, Dordrecht, pp 102–115

Hinata K, Watanabe M, Yamakawa S, Satta Y, Isogai A (1995) Evolutionary aspects of the S-related genes of the *Brassica* self-incompatibility system: synonymous and nonsynonymous base substitutions. Genetics 140:1099–1104

Hiratsuka S (1992a) Detection and inheritance of a stylar protein associated with a self-incompatibility genotype in Japanese pear. Euphytica 61:55–59

Hiratsuka S (1992b) Characterization of an S-allele associated protein in Japanese pear. Euphytica 62:103–110

Hiscock SJ, Dickinson HG (1993) Unilateral incompatibility within the Brassicaceae: further evidence for the involvement of the self-incompatibility (S) locus. Theor Appl Genet 86:744–753

Hiscock SJ, Dewey FM, Coleman JOD, Dickinson HG (1994) Identification and localization of an active cutinase in the pollen of *Brassica napus* L. Planta 193:377–384

Hiscock SJ, Doughty J, Willis AC, Dickinson HG (1995) A 7 kDa pollen coating-borne peptide from *Brassica napus* interacts with S-locus glycoprotein and S-locus related glycoprotein. Planta 196:367–374

Hitz B (1999) Economic aspects of transgenic crops which produce novel products. Curr Opin Plant Biol 2:135–138

Hoffmann M (1966) Bestimmung der Selbstinkompatibilitätsallele und Inkompatibilitätsreaktion bei *Lycopersicum peruvianum* (L.) Mill. Diplomarb Greifswald

Hogenboom NG (1972a) Breaking breeding barriers in *Lycopersicon*. 1. The genus *Lycopersicon* its breeding barriers and the importance of breaking these barriers. Euphytica 21:221–227

Hogenboom NG (1972b) Breaking breeding barriers in *Lycopersicon*. 2: Breakdown of self-incompatibility in *L peruvianum* (L) Mill. Euphytica 21:228–243

Hogenboom NG (1972c) Breaking breeding barriers in *Lycopersicon*. 3. Inheritance of self-compatibility in *L. peruvianum* (L) Mill. Euphytica 21:244–256

Hogenboom NG (1972d) Breaking breeding barriers in *Lycopersicon*. 4. Breakdown of unilateral incompatibility between *L. peruvianum* (L.) Mill. and *L. esculentum* Mill. Euphytica 21:397–404

Hogenboom NG (1972 e) Breaking breeding barriers in *Lycopersicon*. 5. The inheritance of the unilateral incompatibility between *L. peruvianum* (L.) Mill. and *L. esculentum* Mill. and the genetics of its breakdown. Euphytica 21:405–414

Hogenboom NG (1973) A model for incongruity in intimate partner relationships. Euphytica 22:219–233

Hogenboom NG (1975) Incompatibility and incongruity: two different mechanisms for the non-functioning of intimate partner relationships. Proc R Soc Lond B 188:361–375

Hogenboom NG (1984) Incongruity: non-functioning of intercellular and intracellular partner relationships through non-matching information. In: Linskens HF, Heslop-Harrison J (eds) Encyclopedia of plant physiology. Springer, Berlin Heidelberg New York, pp 640–654

Holmgreen A (1989) Thioredoxin and glutaredoxin systems. J Biol Chem 264:13963–13966

Hopper JE, Peloquin SJ (1968) X-ray inactivation of the stylar component of the self-incompatibility reaction in *Lilium longiflorum*. Can J Genet Cytol 10:941–944

Hopper JE, Ascher PD, Peloquin SJ (1967) Inactivation of self-incompatibility following temperature pretreatments of styles in *Lilium longiflorum*. Euphytica 16:215–220

Howard HW (1942) Self-incompatibility in polyploid forms of *Brassica* and *Raphanus*. Nature 149:302

Howlett BJ, Knox RB, Heslop-Harrison J (1973) Pollen-wall proteins: release of the allergen antigen E from intine and exine sites in pollen grains of ragweed and Cosmos. J Cell Sci 13:603–619

Huang S, Lee HS, Karunanandaa B, Kao T-H (1994) Ribonuclease activity of *Petunia inflata* S-proteins is essential for rejection of self-pollen. Plant Cell 6:1021–1028

Hughes MB, Babcock EB (1950) Self-incompatibility in *Crepis foetida* L. stubsp. Rhoedaifolia. Genetics 35:570–588

Hülskamp M, Kopczak SD, Horejsi TF, Kihl B, Pruitt RE (1995) Identification of genes required for pollen-stigma recognition in *Arabidopsis thaliana*. Plant J 8(5):703–714

Ikeda S, Nasrallah JB, Dixit R, Press S, Nasrallah ME (1997) An aquaporin-like gene required for the *Brassica* self-incompatibility response. Science 276:1564–1566

Imanishi S (1988) Efficient ovule culture for the hybridization of *Lycopersicum esculentum* and *L. peruvianum*, *L. glandulosum*. Jpn J Breed 39:36404

Imanishi S, Tanaka H, Harada S, Hishumira R, Takahashi S, Takashina T, Oumura S (1996) Development of interspecific hybrids between *Lycopersicum esculentum* and *L. peruvianum* var *humifusum* and introgression of wild type invertase gene into *L. esculentum*. Breeding Sci 46(4):355–359

Imrie BC, Kirkman CT, Ross DR (1972) Computer simulation of a sporophytic self-incompatibility breeding system. Aust J Biol Sci 25:343–349

Ioerger TR, Clark AG, Kao T-H (1990) Polymorphism of the self-incompatibility locus predates speciation. Proc Natl Acad Sci USA 87:9732–9735

Ioerger TR, Gohlke JR, Xu B, Kao T-H (1991) Primary structural features of the self-incompatibility protein in Solanaceae. Sex Plant Reprod 4:81–87

Ishimizu T, Norioka S, Kanai M, Clarke AE, Sakiyama F (1996) Location of cysteine and cysteine residues in S-ribonucleases associated with gametophytic self-incompatibility. Eur J Biochem 242:627–635

Ishimizu T, Endo T, Yamaguchi-Kabata Y, Nakamura KT, Sakiyama F, Norioka S (1998 a) Identification of regions in which positive selection may operate in S-Rnase of Rosaceae: implication for S-allele-specific recognition S-Rnase. FEBS Lett 440: 337–342

Ishimizu T, Shinkawa T, Sakiyama F, Norioka S (1998b) Primary structural features of rosaceous S-Rnases associated with gametophytic self-incompatibility. Plant Mol Biol 37:931–941

Ishimizu T, Mitsukami Y, Shinkawa T, Natsuka S, Hase S, Miyagi M, Sakiyama F, Norioka S (1999) Presence of asparagine-linked N-acetylglucosamine and chitobiose in *Pyrus pyrifolia* S-RNases associated with gametophytic self-incompatibility. Eur J Biochem 236:624–634

Isogai A, Takayama S, Tsukamoto H, Kanabara C, Hinata K, Okazaki K, Susuki A (1988) Existence of a common glycoprotein homologous to S-glycoproteins in two self-incompatible homozygotes of *Brassica campestris*. Plant Cell Physiol 29:1321–1336

Iversen J (1940) Blutenbiologische Studien. I. Dimorphie und Monomorphie bei *Armeria*. K Danse Vidensk Selsk Skr 15:1–39

Jackson JF, Linskens HF (1990) Bioassays for incompatibility. Sex Plant Reprod 3:207–212

Jahnen W, Lush WM, Clarke AE (1989) Inhibition of in vitro pollen tube growth by isolated S-glycoproteins of *Nicotiana alata*. Plant Cell 1:501–510

Jakowitsch J, Papp I, Moscone EA, van der Winden J, Matzke M, Matzke AJM (1999) Molecular and cytogenetic characterization of a transgene locus that induces silencing and methylation of homologous promoters in *trans*. Plant J 17(2):131–140

Jeffrey EC (1916) The anatomy of woody plants. Univ Press, Chicago

Johnson AG (1972) Some causes of variations in the proportion of selfed seed present in F_1 hybrid seed lot of Brussels sprouts. Euphytica 21:309–316

Jordan ND, Kakeda K, Conner A, Ride JP, Franklin-Tong VE, Franklin FCH (1999) S-protein mutants indicate a functional role for SBP in the self-incompatibility reaction of *Papaver rhoeas*. Plant J 20(1):119–125

Jost L (1907) Über die Selbststerilität einiger Blüten. Bot Z 65:77–117

Kakeda K, Kowyama Y (1996) Sequences of *Ipomea trifida* cDNAs related to the *Brassica* S-locus genes. Sex Plant Reprod 9:309–310

Kakeda K, Jordan ND, Conner A, Ride JP, Franklin-Tong VE, Franklin FCH (1998) Identification of residues in a hydrophilic loop of the *Papaver rhoeas* S-protein that play a crucial role in recognition of incompatible pollen. Plant Cell 10:1723–1731

Kakizaki Y (1930) Studies on the genetics and physiology of self- and cross-incompatibility in the common cabbage. Jpn J Bot 5:133–208

Kandasamy MK, Paolillo CD, Faraday JB, Nasrallah JB, Nasrallah ME (1989) The S-locus specific glycoproteins of *Brassica* accumulate in the cell wall of developing stigma papillae. Dev Biol 134:462–472

Kandasamy MK, Dwyer KG, Paolillo DJ, Doney RC, Nasrallah JB, Nasrallah ME (1990) *Brassica* S-proteins accumulate in the intercellular matrix along the path of pollen tubes in transgenic tobacco pistils. Plant Cell 2:39–49

Kandasamy MK, Nasrallah JB, Nasrallah ME (1994) Pollen-pistil interactions and developmental regulation of pollen tube growth in *Arabidopsis*. Development 120:3405–3418

Kao T-H, Huang S (1994) Gametophytic self-incompatibility: a mechanism for self/nonself discrimination during sexual reproduction. Plant Physiol 105:461–466

Kao T-H, McCubbin AG (1996) How flowering plants discriminate between self and non-self pollen to prevent inbreeding. Proc Natl Acad Sci USA 93:12059–12065

Kao T-H, McCubbin AG (1997) Molecular and biochemical bases of gametophytic self-incompatibility in Solanaceae. Plant Physiol Biochem 35(3):171

Kao T-H, McCubbin AG, Verica JA, Wang X (1999) Searching for the missing links in the mechanism of Solanaceous type self-incompatibility. In: Pollen-stigma interactions. Int Conf (Dept Pl Sci, Univ Oxford) dedicated to the memory of J. Heslop-Harrison, p 18

Karunanandaa B, Huang S, Kao T-H (1994) Carbohydrate moiety of the *Petunia inflata* S3 protein is not required for self-incompatibility interactions between pollen and pistil. Plant Cell 6:1933–1940

Kaufmann H, Kirch HH, Wenner T, Thompson R (1991) The relationship of major pistil proteins of *Solanum tuberosum* to self-incompatibility. In: Hallick RB (ed) 3rd international congress of plant molecular biology. Abstract 37. International Society for Plant Molecular Biology, Tucson, USA

Kendall WA, Taylor NL (1969) Effects of temperature on pseudo-compatibility in *Trifolium pratense*. Theor Appl Genet 39:123–126

Kenrick J, Kaul V, Williams EJ (1986) Self-incompatibility in *Acacia retinodes*: site of pollen tube arrest is in the nucellus. Planta 169:245–250

Kheyr-Pour A, Bintrim SB, Ioerger TR, Remy R, Hammond SA, Kao T-H (1990) Sequence diversity of pistil S-proteins associated with gametophytic self-incompatibility in *Nicotiana alata*. Sex Plant Reprod 3:88–97

Kinman ML (1963) Current status of sunflower production and research, and the possibilities of this crop as an oilseed in the United States. Meeting Oilseed and Peanut Research and Marketing Advisory Committee, Peoria, vol I (11)

Kirch HH, Li Y-Q, Seul U, Thompson RD (1995) The expression of a potato (*Solanum tuberosum*) S-RNase gene in *Nicotiana tabacum* pollen. Sex Plant Reprod 8:77–84

Kishi-Nichizawa N, Isogai A, Watanabe M, Hinata K, Yamakawa S, Shojima S, Susuki A (1990) Ultrastructure of papillar cells in *Brassica campestris* revealed by liquid helium rapid freezing and substitution-fixation method. Plant Cell Physiol 31(8): 1207–1219

Klekowski EJJ (1988) Mutation, developmental selection and plant evolution. Columbia University Press, New York

Kleman-Mariac C, Rougier M, Cock JM, Gaude T, Dumas C (1995) S-locus glycoproteins are expressed along the path of pollen tubes in *Brassica* pistils. Planta 196: 614–621

Knight R, Rogers HH (1953) Sterility in *Theobroma cacao* L. Nature 172:164

Knight R, Rogers HH (1955) Incompatibility in *Theobroma cacao*. Heredity 9:69–77

Knox RB, Kenrick J (1983) Polyad function in relation to the breeding system of *Acacia*. In: Mulcahy DL, Ottaviano E (eds) Pollen: biology and implications for plant-breeding. Elsevier Biomedical, New York, pp 411–417

Knox RB, Willing R, Ashford AE (1972) Role of pollen-wall proteins as recognition substances in interspecific incompatibility in poplars. Nature 237:381–383

Kölreuter JG (1764) Vorläufige Nachricht von einigen das Geschlecht der Pflanzen betreffenden Versuchen und Beobachtung, nebst Fortsetzungen. Ostwald's Klassiker Nr 41. Engelmann, Leipzig, 1, 2 u. 3, pp 266

Koltin Y, Stamberg J (1973) Is there a universal genetic structure of incompatibility systems? Incomp Newslett 2:40–41

Komari T, Hiei Y, Ishida YTK, Kubo T (1998) Advances in cereal gene transfer. Curr Opin Plant Biol 1:161–165

Konar RN, Linskens HF (1966) Chemistry of the stigmatic fluid of *Petunia hybrida*. Planta 71:372–387

Korban SS (1986) Interspecific hybridization in *Malus*. Hortic Sci 21:41–48

Kovaleva LV, Milyaeva EL, Chailakyan MKH (1978) Overcoming self-incompatibility by inhibitors of nucleic acid and protein metabolism Phytomorphology 28:445–449

Kowyama Y, Takahashi H, Murakoa K, Tani T, Hara K, Shiotani I (1994) The number, frequency and dominance relationships of S-alleles in diploid *Ipomea trifida*. Heredity 73:275–283

Kowyama Y, Kakeda K, Nakano R, Hattori T (1995) SLG/SRK-like genes are expressed in the reproductive tissues of *Ipomea trifida*. Sex Plant Reprod 8:333–338

Kowyama Y, Kakeda K, Kondo K, Imada T, Hattori T (1996) A putative receptor protein gene in *Ipomea trifida*. Plant Cell Physiol 7(5):681–685

Kowyama Y, Kakeda K, Tsuchiya T (1999) Molecular analysis of sporophytic self-incompatibility in Ipomea trifida (Abstract). In: Pollen-stigma interactions. Int Conf (Dept Plant Sci, Univ Oxford) dedicated to the memory of Jack Heslop-Harrison, p 31

Kroh M (1966) Reaction of pollen after transfer from one stigma to another (contribution to the character of the incompatibility mechanism in *Cruciferae*). Züchter 36:185–189

Kuboyama T, Chung CS, Takeda G (1994) The diversity of interspecific pollen-pistil incongruity in *Nicotiana*. Sex Plant Reprod 7:250–258

Kucer V, Polak L (1975) The serological specificity of S-alleles of homozygous incompatible lines of the marrow-stem kale (*Brassica oleracea* var *acephala* DC.). Biol Plant 17:50–54

Kunishige M, Hirata Y (1978) Studies on crosses of lillies: on the mixed pollen method. Jpn Bull Hortic Res Sta 47:80–85

Kunz C, Chang A, Faure JD, Clarke AE, Poly GM, Anderson MA (1996) Phosphorylation of style S-Rnases by Ca2+-dependent protein kinases from pollen tubes. Sex Plant Reprod 9:25–34

Kurian V, Richards V (1997) A new recombinant in the heteromorphy "S" supergene in *Primula*. Heredity 78:383–390

Kurup S, Ride JP, Jordan ND, Fletcher G, Franklin-Tong VE, Franklin FCH (1998) Identification and cloning of related self-incompatibility S-genes in *Papaver rhoeas* and *Papaver nudicaule*. Sex Plant Reprod 11:192–198

Kusaba M, Nishio T (1999a) Comparative analyses of S-haplotypes with very similar SLG alleles in *Brassica rapa* and *Brassica oleraceae*. Plant J 17(1):83–91

Kusaba M, Nishio T (1999b) The molecular mechanism of self-recognition in *Brassica* self-incompatibility. Plant Biotechnol 16:93–102

Kusaba M, Nishio T, Satta Y, Hinata K, Ockendon DJ (1997) Striking sequence similarity in inter- and intraspecific comparisons of class I SLG alleles from *Brassica oleracea* and *Brassica campestris*: implications for the evolution and recognition mechanism. Proc Natl Acad Sci USA 94:7673–7678

Kusaba M, Susuki T, Okazaki K, Nishio T (1999) Is SLG essential for self-incompatibility in *Brassica*? (Abstract). In: Pollen-stigma interactions. Int Conf (Dept Plant Sci, Univ Oxford) dedicated to the memory of Jack Heslop-Harrison, p 25

Labarca C, Kroh M, Loewus F (1970) The composition of stigmatic exudate from *Lillium longiflorum*. Labelling studies with myo-insitol, D-glucose and L-proline. Plant Physiol 46:150–156

Labroche P, Poirier-Hamon S, Pernes J (1983) Inheritance of leaf peroxidase isoenzyme in *Nicotiana alata* and linkage with the self-incompatibility locus. Theor Appl Genet 65:163–170

Lalonde BA, Nasrallah ME, Dwyer KG, Chen KG, Barlow B, Nasrallah JB (1989) A highly conserved gene with homology to the S-locus specific glycoprotein structural gene. Plant Cell 1:249–258

Lamm R (1950) Self-incompatibility in *Lycopersicum peruvianum* Mill. Hereditas 36:509–511

Lande R, Schemske DW (1985) The evolution of self-fertilization and inbreeding depression in plants.1. Genetic models. Evolution 39:24–40

Lane MD, Lawrence MJ (1993) The population genetics of the self-incompatibility polymorphism in *Papaver rhoeas*. Heredity 71:596–602

Langridge P, Baumann U, Juttner J (1999) Revisiting and revising the self-incompatibility genetics of *Phalaris coerulescens*. Plant Cell 11(10):1826

Larsen J, Larsen K, Lundqvist A, Osterbye U (1973) Complex self-incompatibility systems within the angiosperms and the possibility of reconstructing a self-incompatibility system from different forms within a self-fertile species. Incomp Newslett 3:79–80

Larsen K (1977) Self-incompatibility in *Beta vulgaris* L. I. Four gametophytic, complementary S-loci in sugar beet. Hereditas 85:227–248

Larsen K (1978) Oligoallelism in the multigenic incompatibility system of *Beta vulgaris* L. Incomp Newslett 10:23–28

Larsen K (1986) The complex S-gene system for control of self-incompatibility in the buttercup genus *Ranunculus*. Hereditas 113:29–46

Lashermes P, Couturon E, Moreau N, Paillard M, Louarn J (1996) Ineritance and genetic mapping in *Coffea canephora* Pierre. Theor Appl Genet 93:458–462

Lawrence MJ (1975) The genetics of self-incompatibility in *Papaver rhoeas*. Proc R Soc Lond B 188:275–285

Lawrence MJ (1996) Number of incompatibility alleles in clover and others species. Heredity 76:610–615

Lawrence MJ, Franklin-Tong VE (1994) The population genetics of the self-incompatibility polymorphism in *Papaver rhoeas*. IX. Evidence of an extra effect of selection acting on the S-locus. Heredity 72:353–364

Lawrence MJ, O'Donnell S (1981) The population genetics of the self-incompatibility polymorphism in *Papaver rhoeas*. III. The number and frequency of S-alleles in two further natural populations (R102 and R104). Heredity 47:53–61

Lawrence MJ, Afzal M, Kenrick J (1978) The genetical control of self-incompatibility in Papaver rhoeas. Heredity 40(2):239–253

Lawrence MJ, Marshall DF, Curtis VE, Fearon CH (1985) Gametophytic self-incompatibility re-examined: a reply. Heredity 54:131–138

Lawrence MJ, Lane MD, O'Donnell S, Franklin-Tong VE (1993) The population genetics of the self-incompatibility polymorphism in *Papaver rhoeas*. V. Cross-classification of the S-alleles of samples from three natural populations. Heredity 71:581–590

Lawrence MJ, O'Donnell S, Lane MD, Marshall DF (1994) The population genetics of the self-incompatibility polymorphism in *Papaver rhoeas*. VIII. Sampling effects as a possible cause of unequal allele frequencies. Heredity 72:345–352

Lee HS, Singh A, Kao T-H (1992) RNase X2, a pistil-specific ribonuclease from *Petunia inflata*, shares sequence similarity with solanaceous S proteins. Plant Mol Biol 20:1131–1141

Lee HS, Huang S, Kao T-H (1994) S-proteins control rejection of incompatible pollen in *Petunia inflata*. Nature 367:560–563

Lee HS, Chung YY, Karunanandaa B (1997) Embryo sac development is affected in Petunia inflata plants with an antisense gene encoding the extracellular domain of receptor kinase PRKl. Sex Plant Reprod 10:341–350

Leffel RC (1963) Pseudo-self-compatibility and segregation of gametophytic self-incompatibility alleles in red clover, *Trifolium pratense* L. Crop Sci 3:377–380

Lehman E (1926) The heredity of self-strility in *Veronica syriaca*. Mem Hortic Soc NY 3:313–320

Lewis D (1942) The physiology of incompatibility in plants. I. The effect of temperature. Proc R Soc Lond B 131:13–27

Lewis D (1943) The physiology of incompatibility in plants. II. *Linum grandiflorum*. Ann Bot 7:115–122

Lewis D (1947) Competition and dominance of incompatibility in diploid pollen. Heredity 1:85–108

Lewis D (1948) Structure of the incompatibility gene. I. Spontaneous mutation rate. Heredity 2:219–236

Lewis D (1949a) Incompatibility in flowering plants. Biol Rev 24:472–496

Lewis D (1949b) Structure of the incompatibility gene. II. Induced mutation rate. Heredity 3:339–355

Lewis D (1951) Structure of the incompatibility gene. III. Types of spontaneous and induced mutation. Heredity 5:399–414

Lewis D (1952) Serological reactions of pollen incompatibility substances. Proc R Soc Lond B 140:127–135

Lewis D (1954) Comparative incompatibility in angiosperms and fungi. Adv Genet 6:235–285

Lewis D (1956) Genetics and plant breeding. Brookh Symp Biol 9:89–100

Lewis D (1960) Genetic control of specificity and activity of the S antigen in plants. Proc R Soc Lond B 151:468–477

Lewis D (1961) Chromosome fragments and mutations of the incompatibility gene. Nature 190:990–991

Lewis D (1965) A protein dimer hypothesis on incompatibility. In: Geerts SJ (ed) Proc 11th Int Congr Genet, The Hague, 1963.Genet Today 3:656–663

Lewis D (1975) Heteromorphic incompatibility system under disruptive solution. Proc R Soc Lond B 188:247–256

Lewis D (1977) Sporophytic incompatibility with 2 and 3 genes. Proc R Soc Lond B 196:161–170

Lewis D (1982) Incompatibility, stamen movement and pollen economy in a hetero-styled tropical forest tree, *Cratoxylum formosum* (Guttiferae). Proc R Soc Lond B 214:273–283

Lewis D (1994) Gametophytic-sporophytic incompatibility. In: Williams EG, Clarke AE, Knox RB (eds) Genetic control of self-incompatibility and reproductive development in flowering plants. Kluwer, Dordrecht, pp 88–101

Lewis D, Crowe LK (1954) Structure of the incompatibility gene. IV. Types of mutation in *Prunus avium* L. Heredity 8:357–363

Lewis D, Crowe LK (1958) Unilateral incompatibility in flowering plants. Heredity 12:233–256

Lewis D, Jones DA (1992) The genetics of heterostyly. In: Barrett SCH (ed) Evolution and function of heterostyly. Springer, Berlin Heidelberg New York, pp 129–150

Lewis D, Rao AN (1971) Evolution of dimorphism and population polymorphism in *Pemphis acidula* Forst. Proc R Soc B 178:79–94

Lewis D, Burrace S, Walls D (1967) Immunological reactions of single pollen grains, electrophoresis and enzymology of pollen protein exudates. J Exp Bot 18:371–378

Lewis D, Verma SC, Zuberi MI (1988) Gametophytic-sporophytic in the Cruciffereae – *Raphanus sativus*. Heredity 61:355–366

Li YQ, Moscatelli A, Cai G, Cresti M (1997) Functional interactions among cytoskeleton, membranes and cell wall in the pollen tube of flowering plants. Int Rev Cytol 176:133–199

Li X, Niedl J, Hayman D, Langridge P (1994) Cloning a putative self-incompatibility gene from the pollen of the grass *Phalaris coerulescens*. Plant Cell 6:1923–1932

Li X, Niedl J, Hayman D, Langridge P (1995) Thioredoxin activity in the C terminus of *Phalaris* S-proteins. Plant J 8:133–138

Li X, Niedl J, Hayman D, Langridge P (1996) A self-fertile mutant of Phalaris produces an S protein with reduced thioredoxin activity. Plant J 10(3):505–513

Li X, Paech N, Niedl J, Hayman D, Langridge P (1997) Self-incompatibility in the grasse: evolutionary relationship of the S-gene from *Phalaris coerulescens*. Plant Mol Biol 34:223–232

Liedl BE, Liu SC, Esposito D, Mutschler MA (1993) Identification and mapping of the S-gene in a self-compatible F2 population of *L. esculentum* x *L. penellii*. Rep Tomato Genet Coop 43:33

Liedl BE, McCormick S, Mutschler MA (1996) Unilateral incongruity in crosses involving *Lycopersicon penellii* and *L. esculentum* is distinct from self-incompatibility in expression, timing and location. Sex Plant Reprod 9:299–308

Linskens FH (1954) Biochemical studies about the incompatibility reaction in the style of *Petunia*. Congr Int Bot Paris, Sect 9, pp 146–147

Linskens FH (1955) Physiologische Untersuchungen der Pollenschlausch-Hemmung selbst-steriler Petunien. Z Bot 43:1–44

Linskens FH (1958) Zur Frage der Entstehung der Abwehrkörper bei der Inkompatibilitätsreaktion von *Petunia*. I. Mitteilung: Versuche zur Markierung der Griffel mit P^{32}- und C^{14}-Verbindungen. Ber Dtsch Bot Ges 71:3–10

Linskens FH (1960) Zur Frage der Entstehung der Abwehr-Körper bei der Inkompatibilitäts-Reaktion von *Petunia*. III. Mitteilung: Serologische Teste. Z Bot 48:126–135

Linskens FH (1961) Biochemical aspects of incompatibility. Rec Adv Bot 2:1500–1503

Linskens FH (1962) Die Anwendung der „Clonal Selection Theory" auf Erscheinungen der Selbstinkompatibilität bei der Befruchtung der Blütenpflanzen. Portug Acta Biol Ser A 6:231–238

Linskens HF (1964) The influence of castration on pollen tube growth after self-pollination. In: Linskens HF (ed) Pollen physiology and fertilization. North Holland, Amsterdam, pp 230–236

Linskens HF (1965) Biochemistry of incompatibility. In: Geerts SJ (ed) Proc 11th Int Congr Genet, The Hague, 1963. Genet Today 3:621–636

Linskens HF (1975) Incompatibility in *Petunia*. Proc R Soc Lond B 188:299–311

Linskens FH (1977) Incompatibility reactions during the flowering period of several *Petunia* clones. Acta Bot Neerl 26(5):411–415

Linskens FH (1986) Recognition during the progamic phase. In: Cresti M, Dallai R (eds) Symp Biology of reproduction and cell motility in plants and animals. Univ di Siena, Siena, pp 21–31

Linskens HF (1988) Present status and future propects of sexual reproduction research in higher plants. In: Cresti M, Gori P, Pacini E (eds) Sexual reproduction in higher plants. Proc 10th Int Symp. Springer, Berlin Heidelberg New York, pp 451–458

Linskens HF, Esser KL (1957) Über eine spezifische Anfärbung der Pollenschläuche im Griffel und die Zahl der Kallosepfropfen nach Selbstdung und Fremddung. Naturwissenschaften 44:1–2

Linskens HF, Heinen W (1962) Cultinase-Nachweis in Pollen. Z Bot 50:338–347

Linskens HF, Kroh M (1967) Inkompatibilität der Phanerogamen. In: Ruhland W (ed) Encyclopedia of plant physiology, vol 18. Springer, Berlin Heidelberg New York, pp 506–530

Linskens HF, Schrauwen JAM, van der Donk M (1960) Überwindung der Selbstinkompatibilität durch Röntgenbestrahlung des Griffels. Naturwissenschaften 46:547

Liu B, Kusaba M, Li J-H, Newbigin EJ (1997) Molecular characterization of 48 A, a pollen-expressed gene closely linked to the *Nicotiana alata* S-Locus. Annual Report of Plant Cell Biology Center, School of Botany, University of Melbourne, Parkville, Victoria, Australia, p 27

Lloyd DG, Webb CJ (1992 a) The evolution of heterostyly. In: Barrett SCH (ed) Evolution and function of heterostyly. Springer, Berlin Heidelberg New York, pp 151–178

Lloyd DG, Webb CJ (1992 b) The selection of heterostyly. In: Barrett SCH (ed) Evolution and function of heterostyly. Springer, Berlin Heidelberg New York, pp 179–207

Lord JM (1985) Precursors of ricin and *Ricinus communis* agglutinin. Glycosylation and processing during synthesis and intracellular transport. Eur J Biochem 146:411–416

Lundqvist A (1954) Studies on self-sterility in rye, *Secale cereale* L. Hereditas 40:278–294

Lundqvist A (1955) Genetics of self-incompatibility in *Festuca pratensis* Huds. Hereditas 40:278–294

Lundqvist A (1956) Self-incompatibility in rye. I. Genetic control in the diploid. Hereditas 42:293–348

Lundqvist A (1961) A rapid method for the analysis of incompatibilities in grasses. Hereditas 47:705–707

Lundqvist A (1962a) The nature of the two loci incompatibility system in grasses. I. The hypothesis of a duplicative origin. Hereditas 48:153–168

Lundqvist A (1962b) Self-incompatibility in diploid *Hordeum bulbosum* L. Hereditas 48:138–152

Lundqvist A (1964) The nature of the two-loci incompatibility system in grasses. IV. Interaction between the loci in relation to pseudo-compatibility in *Festuca pratensis* Huds. Hereditas 52:221–234

Lundqvist A (1965) The genetics of incompatibility. In: Geerts SJ (ed) Proc 11th Int Congr Genet, The Hague, 1963. Genet Today 3:637–647

Lundqvist A (1968) The mode of origin of self-fertility in grasses. Hereditas 59:413–426

Lundqvist A (1969) Auto-incompatibility and breeding. Genet Agraria 23:365–380

Lundqvist A (1975) Complex self-incompatibility systems in angiosperms. Proc R Soc Lond B 188:235–245

Lundqvist A (1990a) Variability within and among populations in the 4-gene system for control of self-incompatibility in *Ranunculus polyanthemos*. Hereditas 113:47–61

Lundqvist A (1990b) One locus sporophytic S-gene system with traces of gametophytic pollen control in *Cerastium arvense* ssp *strictum* (Caryophyllaceae). Hereditas 113:203–215

Lundqvist A (1990c) The complex S-gene system for control of self-incompatibility in the buttercup genus *Ranunculus*. Hereditas 113:29–46

Lundqvist A (1991) Four-locus S-gene control of self-incompatibility made probable in *Lilium martagon*. Hereditas 114:57–63

Lundqvist A (1992) The self-incompatibility system in *Caltha palustris* (Ranunculaceae). Hereditas 117:145–151

Lundqvist A (1993) The self-incompatibility system in *Lotus tenuis* (Fabaceae). Hereditas 119:59–66

Lundqvist A (1994a) "Slow" and "quick" alleles without dominance interaction in the sporophytic one-locus self-incompatibility system of *Stellaria holostea* (Caryophyllaceae). Hereditas 120:151–157

Lundqvist A (1994b) The self-incompatibility system in *Ranunculus repens* (Ranunculaceae). Hereditas 120:191–202

Lundqvist A, Osterbye U, Larsen K, Linde-Laursen I (1973) Complex self-incompatibility systems in *Ranunculus acris* L. and *Beta vulgaris* L. Hereditas 74:161–168

Lush WM, Clarke AE (1997) Observations of pollen tube growth in *Nicotiana alata* and their implications for the mechanism of self-incompatibility. Sex Plant Reprod 10:27–35

Lush WM, Opat AS, Nie F, Clarke AE (1997) An in vitro assay for assessing the effects of growth factors on Nicotiana pollen tubes. Sex Plant Reprod 10:351–357

Lush WM, Grieser F, Wolters-Arts M (1998) Directional guidance of *Nicotiana alata* pollen tubes in vitro and on the stigma. Plant Physiol 118:733–741

Luu DT, Heizmann P, Dumas C, Trick M, Cappadocia M (1997a) Involvement of SLR1 genes in pollen adhesion to the stigma surface in Brassicaceae. Sex Plant Reprod 10:227–235

Luu DT, Heizmann P, Dumas C (1997b) Pollen-stigma adhesion in kale is not dependent on the self-incompatibility genotype. Plant Physiol 115:1221–1230

Luu DT, Marty-Mazars D, Trick M, Dumas C, Heizmann P (1999) Pollen-stigma adhesion in *Brassica* spp involves SLG and SLR1 glycoproteins. Plant Cell 11:251–262

Mackenzie A, Heslop-Harrison J, Dickinson HG (1967) Elimination of ribosomes during meiotic prophase. Nature 215:997–999

Makinen YLA, Lewis D (1962) Immunological analysis of incompatibility (S) proteins and of cross-reacting material in a self-compatible mutant of *Oenothera organensis*. Genet Res 3:352–363

Malhó R, Read ND, Pais MS, Trewavas AJ (1994) Role of cytosolic free calcium in the reorientation of pollen tube growth. Plant J 5:331–341

Marcucci MC, Visser T (1987) Pollen tube growth in apple and pear styles in relation to self-incompatibility, incongruity and pistil load. Adv Hortic Sci 1:90–94

Martin FW (1958) Staining and observing pollen tubes in the style by means of fluorescence. Stain Technol 34:125–128

Martin FW (1961) The inheritance of self-incompatibility in hybrids of *Lycopersicon esculentum* Mill. x *L. chilense* Dun. Genetics 46:1443–1454

Martin FW (1964) The inheritance of unilateral incompatibility in *Lycopersicon hirsutum*. Genetics 50:459–469

Martin FW (1967) The genetic control of unilateral incompatibility between two tomato species. Genetics 56:391–398

Martin FW (1968) The behavior of *Lycopersicon* incompatibility alleles in an alien genetic milieu. Genetics 60:101–109

Martin FW (1973) Fertility, incompatibility and morphological abnormalities of sweet potato inbreds. Incomp Newslett 2:29–31

Mascarhenas JP (1990) Gene activity during pollen development. Annu Rev Plant Physiol 41:317–338

Mascarhenas JP (1993) Molecular mechanisms of pollen tube growth and differentiation. Plant Cell 5:1303–1314

Mather K (1943) Specific differences in *Petunia*. I. Incompatibility. J Genet 45:215–235

Mather K (1944) Genetical control of incompatibility in angiosperms and fungi. Nature 153:392–394

Mather K (1950) The genetical architecture of heterostyly in *Primula sinensis*. Evolution 4:340–352

Mather K (1953) The genetical structure of populations. Evolution 7:66–95

Mather K (1975) Discussion of Hogenboom, N.G. A discussion on incompatibility in flowering plants. Proc R Soc Lond B 188:233–375

Mather K, de Winton D (1941) Adaptation and counteradaptation of the breeding system *in Primula*. Ann Bot 5:297–311

Matsubara S (1980) Overcoming self-incompatibility in *Raphanus sativus* with high temperature. J Am Soc Hortic Sci 105:842–846

Matsushita M, Watanabe M, Yamakawa S, Takayama S, Isogai A, Hinata K (1996) The SLGs corresponding to the same S24-haplotype are perfectly conserved in three different self-incompatible *Brassica campestris* L. Genes Genet Syst 71:255–258

Matton DP, Mau SL, Okamoto S, Clarke AE, Newbigin E (1995) The S-locus of *Nicotiana alata*: genomic organization and sequence analysis of two S-RNase alleles. Plant Mol Biol 28:847–858

Matton DP, Maes O, Laublin G, Xike Q, Morse D, Cappadocia M (1997) Hypervariable domains of self-incompatibility ribonucleases mediate allele)specific pollen recognition. Plant Cell 9:1757–1766

Matton DP, Morse D, Cappadocia M (1998) Reply to Zurek et al (1998). Plant Cell 10:316–317

Matton DP, Luu DT, Xike Q, Laublin G, O'Brien M, Maes O, Morse D, Cappadocia M (1999) Production of an S-RNase with dual specificity suggests a novel hypothesis for the generation of new S-alleles. Plant Cell 11:2087–2097

Matton DP, Luu DT, Morse D, Cappadocia M (2000) Establishing a paradigm for the generation of new S-alleles. Plant Cell 12: 313–315

Mattsson O (1983) The significance of exine oils in the initial reaction between pollen and stigma in *Armeria maritima*. In: Mulcahy DL, Ottaviano E (eds) Pollen: biology and implications for plant-breeding. Elsevier Biomedical, New York, pp 257–264

Mattsson O, Knox RB, Heslop-Harrison J, Heslop-Harrison Y (1974) Protein pellicle of stigmatic papillae as a probable recognition site in incompatibility reactions. Nature 247:298–300

Mau SL, Raff JW, Clarke AE (1982) Isolation and partial characterization of components of *Prunus avium* L. styles including an antigenic glycoprotein associated with a self-incompatibility genotype. Planta 156:505–516

Mau SL, Williams EG, Atkinson A, Anderson MA, Cornish EC, Grego B, Simpson RJ, Kheyr-Pour A, Clarke AE (1986) Style proteins of a wild tomato (*Lycopersicum peruvianum*) associated with expression of self-incompatibility. Planta 169:184–191

Mayo O, Hayman DL (1968) The maintenance of two loci systems of gametophytically determined self-incompatibility. In: Oshima C (ed) Proc 12th Int Congr Genet, Oshima, vol 1, p 331

Mayo O, Hayman DL (1973) The stability of systems of gametophytically determined self-incompatibility. Incomp Newslett 2:15–18

McCarthy EM, Rasmussen MA, Anderson WW (1995) A theoretical assessment of speciation. Heredity 74:502–509

McClure BA, Haring VH, Ebert PR, Anderson MA, Simpson RJ, Sakiyama F, Clarke AE (1989) Style self-incompatibility gene products of *Nicotiana alata* are ribonucleases. Nature 342:955–957

McClure BA, Gray JE, Anderson MA, Clarke AE (1990) Self-incompatibility in *Nicotiana alata* involves degradation of pollen ribosomal RNA. Nature 347:757–760

McClure AG, Du H, Liu YH, Clarke AE (1993) S-locus products in *Nicotiana alata* pistils are subject to organ-specific post-transcriptional processing but not post-translational processing. Plant Mol Biol 22:177–181

McCubbin AG, Kao T-H (1996) Molecular mechanisms of self-incompatibility. Curr Opin Biotech 7:55–160

McCubbin AG, Kao T-H (1999) The emerging complexity of self-incompatibility (S) loci. Sex Plant Reprod 12:1–5

McCubbin AG, Chung YY, Kao T-H (1997) A mutant S Rnase of *Petunia inflata* lacking RNase activity has an S-allele dominant negative effect on self-incompatibility interactions. Plant Cell 9:85–95

McGuire DC, Rick CM (1954) Self-incompatibility in species of *Lycopersicon* sect. *Eriopersicon* and hybrids with *L. esculentum*. Hilgardia 23:101–124

Melton B (1970) Effects of clones, generation of inbreeding and years on self-fertility in alfalfa. Crop Sci 10:497–500

Meyer P, Saedler H (1996) Homology-dependent gene silencing in plants. Annu Rev Plant Phys 47:23–48

Miri RK, Bubar JS (1966) Self-incompatibility as an out-crossing mechanism in birdsfoot trefoil (*Lotus corniculatus*). Can J Plant Sci 46:411–418

Modlibowska I (1945) Pollen tube growth and embro-sac development in apples and pears. J Pomol Hortic Sci 45:57–89

Moore HM, Nasrallah JB (1990) A *Brassica* self-incompatibility gene is expressed in the stylar transmitting tissue of transgenic tobacco. Plant Cell 2:29–38

Mu JH, Lee HS, Kao T-H (1994) Characterization of a pollen-expressed receptor-like kinase gene of *Petunia inflata* and the activity of its encoded kinase. Plant Cell 6:709–721

Mulcahy DL, Mulcahy GB (1983) Gametophytic self-incompatibility reexamined. Science 220:1247–1251

Müller F (1868) Notizen über die Geschlechtsverhältnisse brasilianischer Pflanzen. Bot Z 26:113–116

Munro R (1868) On the reproduction and cross-fertilization of *Passifloras*. Bot Soc 9:399–402

Murfett J, McClure BA (1998) Expressing foreign genes in the pistil: a comparison of S-RNase constructs in different *Nicotiana* backgrounds. Plant Mol Biol 37:561–569

Murfett J, Cornish EC, Ebert PR, Bönig I, McClure BA, Clarke AE (1992) Expression of a self-incompatibility glycoprotein (S2-ribonuclease) from *Nicotiana alata* in transgenic *Nicotiana tabacum*. Plant Cell 4:1063–1074

Murfett J, Atherton T, Mou B, Gasser C, McClure BA (1994) S-RNase expressed in transgenic *Nicotiana* causes S-allele-specific pollen rejection. Nature 367:563–566

Murfett J, Bourque JE, McClure BA (1995) Antisense suppression of S-Rnases using RNA polymerases II- and III-transcribed gene constructs. Plant Mol Biol 29:201–212

Murfett J, Strabala TJ, Zurek DM, Mou B, Beecher B, McClure BA (1996) S-RNase and interspecific pollen rejection in the genus *Nicotiana*: multiple pollen – rejection pathways contribute to unilateral incompatibility between self-incompatible and self-compatible species. Plant Cell 8:943–958

Murray BG (1974) Breeding systems and floral biology in the genus *Briza*. Heredity 33:285–292

Murray BG (1986) Floral biology and self-incompatibility in *Linum*. Bot Gaz 147:327–333

Mutschler MA, Liedl BE (1994) Interspecific barriers crossing barriers in *Lycopersicon* and their relationship to self-incompatibility. In: Williams EG, Clarke AE, Knox RB (eds) Genetic control of self-incompatibility and reproductive development in flowering plants. Kluwer, Dordrecht, pp 164–188

Naarborgh AT, Willemse MTM (1992) The ovular incompatibility system in *Gasteria verrucosa* (Mill.) H. Duval. Euphytica 58:231–240

Nagylaki T (1975) The deterministic behavior of self-incompatibility alleles. Genetics 79:545–550

Nakanishii T, Hinata K (1973) An effective time for CO_2 gas treatment in overcoming self-incompatibility in *Brassica*. Plant Cell Physiol 14:873–879

Nakanishii T, Sawano M (1989) Changes in pollen tube behaviour induced by carbon dioxide and their role in overcoming self-incompatibility in *Brassica*. Sex Plant Reprod 2:109–115

Nakanishii T, Esashi Y, Hinata K (1969) Control of self-incompatibility by CO_2 gas in *Brassica*. Plant Cell Physiol 10:925–927

Nasrallah JB (1997) Evolution of the *Brassica* self-incompatibility locus: a look into S-gene polymorphism. Proc Natl Acad Sci USA 94:9516–9519

Nasrallah JB (1999) Identifying the genes for pollen-stigma recognition in *Brassica* (Abstract). In: Pollen-stigma interactions. Int Conf (Dept Plant Sci, Univ Oxford) dedicated to the memory of Jack Heslop-Harrison, Oxford, p 22

Nasrallah JB, Nasrallah ME (1989) The molecular genetics of self-incompatibilityin *Brassica*. Annu Rev Genet 23:121–139

Nasrallah JB, Nasrallah ME (1993) Pollen-stigma signaling in the sporophytic self-incompatibility response. Plant Cell 5:1325–1335

Nasrallah JB, Doney RC, Nasrallah ME (1985a) Biosynthesis of glycoproteins involved in the pollen-stigma interaction of incompatibility in developing flowers of *Brassica oleracea*. Planta 165:100–107

Nasrallah JB, Kao T-H, Goldberg ML, Nasrallah ME (1985b) A cDNA clone encoding an S-specific glycoprotein from *Brassica oleracea*. Nature 318:263–267

Nasrallah JB, Kao T-H, Chen CH, Goldberg ML, Nasrallah ME (1987) Amino-acid sequence of glycoproteins encoded by three alleles of the S-locus of *Brassica oleracea*. Nature 326:617–619

Nasrallah JB, Yu SM, Nasrallah ME (1988) Self-incompatibility genes of *Brassica oleracea*: expression, isolation and structure. Proc Natl Acad Sci USA 85:5551–5555

Nasrallah JB, Nishio T, Nasrallah ME (1991) The self-incompatibility genes of *Brassica*: expression and use in genetic ablation of floral tissues. Annu Rev Plant Phys 42:393–422

Nasrallah JB, Rundle SJ, Nasrallah ME (1994) Genetic evidence for the requirement of the *Brassica* S-locus receptor kinase gene in the self-incompatibility response. Plant J 5(3):373–384

Nasrallah ME (1974) Genetic control of quantitative variation in self-incompatibility proteins detected by immunodiffusion. Experientia 40:279–281

Nasrallah ME, Nasrallah JB (1986) Immunodetection of S-gene products on nitrocellulase electroblots. In: Mulcahy DL, Mulcahy GBM (eds) Biotechnology and ecology of pollen. Springer, Berlin Heidelberg New York, pp 251–256

Nasrallah ME, Wallace DH (1967a) Immunogenetics of self-incompatability in *Brassica oleracea* L. Heredity 22:519–527

Nasrallah ME, Wallace DH (1967b) Immunochemical detection of antigens in self-incompatibility genotype of cabbage. Nature 213:700–701

Nasrallah ME, Wallace DH (1968) The influence of modifier genes on the intensity and stability of self-incompatibility in cabbage. Euphytica 17:495–503

Nasrallah ME, Barber JT, Wallace DH (1970) Self-incompatibility proteins in plants: detection, genetics and possible mode of action. Heredity 25:23–27

Nasrallah ME, Wallace DH, Savo RM (1972) Genotype, protein, phenotype relationships in self-incompatibility of *Brassica*. Genet Res 20:151

Nasrallah ME, Kandasamy MK, Nasrallah JB (1992) A genetically defined trans-acting locus regulates S-locus function in *Brassica*. Plant J 2:497–506

Nass N, Clarke AE, Newbigin E (1997) Reproductive barriers in flowering plants. In: Geneve RL, Preece JE, Merkle SA (eds) Biotechnology of ornamental plants. CAB International, New York, pp 121–138

Nassar NMA, Carvalho CG, Veira C (1996) Overcoming crossing barriers between cassava *Manihot esculentum* Crantz and a wild relative *M pohli* Warva. Brazilian J Bot 19:617–620

Newbigin E (1996) The evolution of self-incompatibility: a molecular voyeur's perspective. Sex Plant Reprod 9:357–361

Newton DL, Kendall WA, Taylor NL (1970) Hybridization of some *Trifolium* species through stylar temperature treatments. Theor Appl Genet 40:59–62

Nishio T, Hinata K (1977) Analysis of S-specific proteins in stigma of *Brassica oleracea* by isoelectric focusing. Heredity 38:391–396

Nishio T, Hinata K (1978) Stigma proteins in self-incompatible *Brassica campestris* and self-compatible relatives, with special reference to S-allele specificity. Jpn J Genet 53:197–205

Nishio T, Hinata K (1979) Purification of an S-specific glycoprotein in self-incompatible *Brassica campestris*. Jpn J Genet 54:307–311

Nishio T, Hinata K (1980) Rapid detection of S-glycoproteins of self-incompatible crucifers using Con-A reaction. Euphytica 29:217–221

Nishio T, Kusaba M (1999) Polymorphism of SRKs and self-incompatible strains without SLG in *Brassica*. In: Pollen-stigma interactions. Int Conf (Dept Plant Sci, Univ Oxford) dedicated to the memory of Jack Heslop-Harrison, Oxford, p 25

Nishio T, Sakamoto K (1993) PCR-RFLP of S-locus for the identification of breeding lines in *Brassica*. Jpn J Breed [Suppl] 1:274

Nishio T, Toriyama K, Sato T, Kandasamy MK, Paolillo DJ, Nasrallah JB, Nasrallah ME (1992) Expression of S locus glycoprotein genes from *Brassica oleracea* and *B. campestris* in transgenic plants of self-compatible *B. napus* cv Westar. Sex Plant Reprod 5:101–109

Nishio T, Sakamoto K, Yamaguchi J (1994) PCR-RFLP of S-locus for identification of breeding lines in cruciferous vegetables. Plant Cell Rep 13:546–550

Nishio T, Kusaba M, Watanabe M, Hinata K (1996) Registration of S-alleles in *Brassica campestris* by the restriction fragment sizes of SLGs. Theor Appl Genet 92:388–394

Nishio T, Kusaba M, Sakamoto K, Ockendon DJ (1997) Polymorphism of the kinase domain of the S-locus receptor kinase gene (SRK) in *Brassica oleraceae*. Theor Appl Genet 95:335–342

Nou IS, Watanabe M, Isuzugawa K, Isogai A, Hinata K (1993a) Isolation of S-alleles from a wild population of *Brassica campestris* L. at Balceme, Turkey and their characterization by S-glycoproteins. Sex Plant Reprod 6:71–78

Nou IS, Watanabe M, Isogai A, Hinata K (1993b) Comparison of S-alleles and S-glycoproteins between two wild populations of *Brassica campestris* in Turkey and Japan. Sex Plant Reprod 6:79–86

Ockendon DJ (1974) Distribution of self-incompatibility alleles and breeding structure of open-pollinated cultivars of Brussels sprouts. Heredity 33:159–171

Ockendon DJ (1975) Dominance relationships between S-alleles in the stigmas of Brussels sprouts of *Brassica oleracea* var. *gemmifera*. Euphytica 24:165–172

Ockendon DJ (1985) Genetics and physiology of self-incompatibility in *Brassica*. In: Sussex I, Ellingboe A, Crouch M, Malmberg R (eds) Plant cell/cell interactions (Current communications in molecular biology series). Cold Spring Harbour Press, New York, pp 1–6

Odland ML, Noll CJ (1950) The utilization of cross-compatability and self-incompatibility in the production of F_1 hybrid cabbage. Proc Am Soc Hortic Sci 55:391–402

O'Donnell S, Lane MD, Lawrence MJ (1993) The population genetics of the self-incompatibility polymorphism in *Papaver rhoeas*. VI. Estimation of the overlap between the allelic complements of a pair of populations. Heredity 71:159–171

Oehlkers F (1943) Über die Aufhebung des Gonen-und Zygotenausfalls bei *Oenothera*. I. Teil. Untersuchungen an *Oenothera Cockerelli*. Flora NF 36:106–119

Okazaki K, Hinata K (1987) Repressing the expression of self-incompatibility in crucifers by short-term high temperature treatment. Theor Appl Genet 73:496–500

Oldknow J, Trick M (1995) Genomic sequence of an SRK-like gene linked to the S-locus of a self-incompatible *Brassica oleracea*. Sex Plant Reprod 8:247–253

Oliver SG, Ward JM (1985) A dictionary of genetic engineering. Cambridge University Press, Cambridge

O'Neil P, Singh MB, Neales TF, Knox RB, Williams FG (1984) Carbon dioxide blocks the stigma callose response following incompatible pollination in *Brassica*. Plant Cell Environ 7:285–288

Opeke L, Jacob VJ (1969) Studies on methods of overcoming self-incompatibility in *T. cacao* L. 2nd Int Cacao Res Conf, Bahia, Brazil, pp 356–358

Ornduff R (1970) Incompatibility and the pollen economy of *Jepsonia parryi*. Am J Bot 57:1036–1041

Ornduff R (1971) The reproductive system of *Jepsonia heterandra*. Evolution 25:300–311

Ornduff R (1972) The breakdown of trimorphic incompatibility in *Oxalis* section Corniculatae. Evolution 26:52–65

Ornduff R (1975) Heterostyly and pollen flow in *Hypericum aegypticum* (Guttiferae). Bot J Linn Soc 71:51–57

Ornduff R (1979) Heterostyly in *Oplonia* (Acanthaceae). J Arnold Arbor 60:382–385

Osterbye U (1975) Self-incompatibility in *Ranunculus acris* L. Genetics interpretation and evolutionary aspects. Hereditas 80:91–112

Osterbye U (1977) Self-incompatibility in *Ranunculus acris* L.: four S-loci in a german population. Hereditas 87:174–178

Osterbye U (1986) Self-incompatibility in *Ranunculus acris* L. III. S-loci numbers and allelic identities. Hereditas 104:61–73

Oxley D, Bacic A (1996) Disulphide bonding in a stylar self-incompatibility ribonuclease of *Nicotiana alata*. Eur J Biochem 242:75–80

Oxley D, Munro SLA, Craik DJ, Bacic A (1996) Structure of the N-glycans on the S3 and S6 allele stylar self-incompatibility ribonucleases of *Nicotiana alata*. Glycobiology 6:611–618

Oxley D, Munro SLA, Craik DJ, Bacic A (1998) Structure and distribution of N-glycans on the S-allele stylar self-incompatibility ribonuclease of Nicotiana alata. J Biochem 123:978–983

Pandey KK (1956 a) Incompatibility in autotetraploid *Trifolium pratense*. Genetics 41:353–366

Pandey KK (1956 b) Mutations of self-incompatibility alleles in *Trifolium pratense* and *T. repens*. Genetics 41:327–343

Pandey KK (1957) Genetics of incompatibility in *Physalis ixocarpa* Brot. A new system. Am J Bot 44:879–887

Pandey KK (1958) Time of S-allele action. Nature 181:1220–1221

Pandey KK (1959) Mutations of the self-incompatibility gene (S) and pseudo-compatibility in angiosperms. Lloydia 22:222–234

Pandey KK (1960) Incompatibility in *Abitulon hybridum*. Am J Bot 47:877–883

Pandey KK (1962 a) Interspecific incompatibility in *Solanum* species. Am J Bot 49:874–882

Pandey KK (1962b) A theory of S-gene structure. Nature 196:236–238

Pandey KK (1964) Elements of the S-gene complex. I. The SF1 alleles in *Nicotiana*. Genet Res 2:397–409

Pandey KK (1965) Centric chromosome fragments and pollen-part mutation of the incompatibility gene in *Nicotiana alata*. Nature 206:792–795

Pandey KK (1967) Elements of the S-gene complex. II. Mutation and complementation at the SI locus in *Nicotiana alata*. Heredity 22:255–283

Pandey KK (1968) Colchicine-induced changes in the self-incompatibility behavior of *Nicotiana*. Genetica 39:257–271

Pandey KK (1969) Elements of the S-gene complex. V. Interspecific cross-compatibility relationships and therory of the evolution of the S complex. Genetica 40:447–474

Pandey KK (1970 a) Time and site of the S-gene action, breeding systems and relationships in incompatibility. Euphytica 19:364–372

Pandey KK (1970 b) Elements of the S-gene complex. VI. Mutations of the self-incompatibility gene, pseudo-compatibility and origin of new incompatibility alleles. Genetica 41:477–516

Pandey KK (1970 c) Self-incompatibility alleles produced through inbreeding. Nature 227:689–690

Pandey KK (1972) Origin of genetic variation: regulation of genetic recombination in the higher organisms – a theory. Theor Appl Genet 42:250–260

Pandey KK (1973) Theory and practice of induced androgenesis. New Phytol 72:1129–1140

Pandey KK (1974) Overcoming interspecific pollen incompatibility through the use of ionising radiation. Heredity 33:179

Pandey KK (1977) Origin of complementary incompatibility systems in flowering plants. Theor Appl Genet 49:101–109

Pandey KK (1978) Proposed causal mechanisms of the "mentor" pollen effects. Incomp Newslett 10:87–93

Pandey KK (1979) The genus *Nicotiana*: evolution of incompatibility in flowering plants. In: Hawkes JG, Lester RN, Skelding AD (eds) The biology and taxonomy of the Solanaceae, series 7. Linnean Society Symposium, London. Academic Press, New York, pp 421–434

Pandey KK (1980) Evolution of incompatibility systems in plants: origin of independent and complementary control of incompatibility in angiosperms. New Phytol 84:381–400

Parry SK, Liu YH, Clarke AE, Newbigin E (1997a) S-RNases and other plant extracellular ribonucleases. In: D'Alessio G, Riordan JF (eds) Ribonucleases structures and functions. Academic Press, New York, pp 191–211

Parry S, Newbigin E, Currie G, Bacic A, Oxley D (1997b) Identification of active-site histidine rezsidues of a self-incompatibility ribonuclease froùm a wild tomato. Plant Physiol 115:1421–1429

Parry S, Newbigin E, Craik D, Nakamura KT, Bacic A, Oxley D (1998) Structural analysis and molecular model of a self-incompatibility Rnase from wild tomato. Plant Physiol 116:463–469

Pastuglia M, Roby D, Dumas C, Cock JM (1997) Rapid induction by wounding and bacterial infection òf an S I gene family receptor-like kinase gene in *Brassica oleracea*. Plant Cell 9:49–60

Pelletier G (1993) Somatic hybridization. In: Hayward MD, Bosemark NO, Romagosa I (eds) Plant breeding (principles and prospects). Chapman and Hall, London, pp 93–106

Pellew C (1928) Annual report John Innes Institute, annual report. John Innes Horticultural Institution, Norwich, United Kingdom

Phillipp M, Schou O (1981) An unusual heteromorphic incompatibility system: distyly, self-incompatbility, pollen load, and fecundity in *Anchusa officinalis* (Boraginaceae). New Phytol 89:693–703

Picard J, Demarly Y (1952) Autofertilité chez la luzerne. Son conditionnement biologique. Proc 6th Int Grasslands Congr, 1952, vol 1, pp 260–266

Pickersgill B (1993) Interspecific hybridization by sexual means. In: Hayward MD, Bosemark NO, Romagosa I (eds) Plant breeding (principles and prospects). Chapman and Hall, London, pp 63–78

Pierson ES, Cresti M (1992) Cytoskeleton and cytoplasmic organization of pollen and pollen tubes. Int Rev Cytol 140:73–125

Pimienta E, Polito VS (1983) Embryo-sac development in almond (*Prunus dulcis*) as affected by cross-, self- and non-pollination. Ann Bot 51:469–479

Polya GM, Micucci V, Rae AL, Harris PJ, Clarke AE (1986) Ca2+-dependent protein phosphorylation in germinated pollen of *Nicotiana alata*, an ornemental tobacco. Physiol Plant 67:151–157

Potrykus I (1993) Gene transfer to plants: approaches and available techniques. In: Hayward MD, Bosemark NO, Romagosa I (eds) Plant breeding (principles and prospects). Chapman and Hall, London, pp 126–137

Prabha K, Sood R, Gupta SC (1982) High temperature-induced inactivation of sporophytic self-incompatibility in *Ipomea fistulosa*. New Phytol 92:115–122

Prakash G (1975) Plant regulators with antagonistic and synergistic effects in petiolar abscission of *Catharanthus roseus* (L.). Incomp Newslett 6:122–125

Prell H (1921) Das Problem der Unfruchtbarkeit. Naturw Wochenschr N F 20:440–446

Preuss D, Lemieux B, Yen G, Davis RWA (1993) A conditional sterile mutation eliminates surface components from *Arabidopsis* pollen and disrupts cell signalling during fertilization. Genes Dev 7:974–985

Pushkarnath M (1942) Studies on sterility in potatoes. I. The genetics of self- and cross-incompatibilities. Ind J Genet 2:11–19

Pushkarnath M (1953) Studies on sterility in potatoes. IV. Genetics of incompatibility in *Solanum araccpapa*. Euphytica 2:49–58

Quiros C, Ochoa O, Douches D (1986) *L. peruvianum x L. pennellii* sexual hybrids. Rep Tomato Genet Coop 36:31–32

Raff JW, Knox RB, Clarke AE (1981) Style antigens of *Prunus avium* L. Planta 153:125–129

Ramulu SK (1982 a) Failure of obligate inbreeding to produce new S-alleles in *Lycopersicum peruvianum* Mill. Incomp Newslett 14:103–110

Ramulu SK (1982 b) Genetic instability at the S-locus of *Lycopersicum peruvianum* plants regenerated from in vitro culture of anthers: generation of new specificities and S-allele reversions. Heredity 49:319–330

Rangaswamy NS, Shivanna KR (1971) Overcoming self-incompatibility in *Petunia axillaris* (Lam.) B. S. P. II. Placenta pollination in vitro. J Ind Bot Soc 50A:286–296

Rangaswamy NS, Shivanna KR (1972) Overcoming self-incompatibility in *Petunia axillaris*. III. Two site pollinations in vitro. Phytomorphology 21:284–289

Rao TS (1970) Effects of acute gamma irradiation on self-incompatibility and related characters in brown sarson. Radiat Bot 10:569–575

Read M, Bacic A, Clarke AE (1992) Pollen tube growth in culture. I. Control of morphology and generative nucleus division in cultured pollen tubes of *Nicotiana*. In: Ottaviano E, Mulcahy DL, Sari Gorla M, Mulcahy GB (eds) angiosperm pollen and ovules. Springer, Berlin Heidelberg New York, pp 162–167

Read SM, Newbigin E, Clarke AE, McClure BA, Kao T-H (1995) Disputed ancestry: comments on a model for the origin of incompatibility in flowering plants. Plant Cell 7:661–665

Reimann-Phillipp R (1965) The application of incompatibility in plant breeding. In: Geerts SJ (ed) Proc 11th Int Congr Genet, The Hague, 1963. Genet Today 3:656–663

Richards AJ (1986) Plant breeding systems. Allen and Unwin, London

Richards AJ (1997) Plant breeding systems, 2nd edn. Chapman and Hall, London

Richards AJ, Mitchell J (1990) The control of incompatibility in dystilous *Pulmonaria affinis*(Boraginaceae). Bot J Linn Soc 104:369–380

Richards RA, Thurling N (1973 a) The genetics of self-incompatibility in *Brassica campestris* L. ssp. oleifera. Characteristics of S-locus. Control of self-incompatibility. Genetica 44:428–438

Richards RA, Thurling N (1973 b) The genetics of self-incompatibility in *Brassica campestris* L. ssp. oleifera metzg. Genotypic and environmental modification of S-locus control. Genetica 44:439–453

Richman AD, Kohn JR (1996) Learning from rejection: the evolution biology of single locus incompatibility. Trends Ecol Evol 11:497–502

Richman AD, Kohn JR (1999) Evolution of S-allele diversity in the Solanaceae (Abstract). In: Pollen-stigma interactions. Int Conf (Dept Plant Sci, Univ Oxford) dedicated to the memory of Jack Heslop-Harrison, Oxford, p 37

Richman AJ, Kao T-H, Schaeffer SW, Uyenoyama M (1995) S-allele sequence diversity in natural populations of *Solanum carolinense* (Horsenettle). Heredity 75:405–415

Richman AD, Uyenoyama MK, Kohn JR (1996 a) S-allele diversity in a natural population of *Physalis crassifolia* (Solanaceae) (ground cherry) assessed by RT-PCR. Heredity 76:497–505

Richman AD, Uyenoyama MK, Kohn JR (1996 b) Allelic diversity and gene genealogy at the self-incompatibility locus in the Solanaceae. Allen and Unwin, London

Richman AD, Broothaerts W, Kohn JR (1997) Self-incompatibility R-Nases from three plant families: homology or convergence? Am J Bot 84(7):912–917

Rick CM (1960) Hybridization between *Lycopersicum penellii*: phylogenetic and cyto-genetic significance. Proc Natl Acad Sci USA 46:78–82

Rick CM (1982) Genetic relationships between self-incompatibility and floral traits in the tomato species. Biol Zentralbl 101:185–198

Rick CM (1986) Reproductive isolation of the *Lycopersicum peruvianum* complex. In: D'Arcy WG (ed) Solanaceae: biology and systematics. Academic Press, London, pp 667–677

Rick CM, Chetelat RT (1991) The breakdown of self-incompatibility in *Lycopersicum hirsutum*. In: Hawkes L, Nee E (eds) Solanaceae. III. Taxonomy, chemistry, evolution. Royal Botanic Gardens, Kew, Linnean Society of London, London, pp 253–256

Ride JP, Davies EM, Franklin FCH, Marshall DF (1999) Analysis of Arabidopsis genome reveals a large new gene family in plants. Plant Mol Biol 39:927–932

Rieseberg LH, Desrochers AM, Youn SJS (1995) Interspecific pollen competition as a reproductive barrier between sympatric species of *Helianthus* (Asteraceae). Am J Bot 82:515–519

Riley HP (1932) Self-sterility in Shepherds'purse. Genetics 17:231–295

Riley HP (1936) The genetics and physiology of self-sterility in the genus *Capsella*. Genetics 21:24–39

Rivers BA, Bernatzky R (1994) Protein expression of a self-compatible allele from *Lycopersicum peruvianum*: introgression and behavior in a self-incompatible background. Sex Plant Reprod 7:357–362

Rivers BA, Bernatzky R, Robinson SJ, Jahnen-Dechent W (1993) Molecular diversity at the self-incompatibility locus is a salient feature in natural populations of wild tomato (*Lycopersicum peruvianum*). Mol Gen Genet 238:419–427

Robbins TP, Harbord RM, Sonneveld T, Clarke K (1999) Molecular genetics of self-incompatibility in *Petunia hybrida*. In: Pollen-stigma interactions. Int Conf (Dept Plant Sci, Univ Oxford) dedicated to the memory of Jack Heslop-Harrtison, Oxford, p 20

Roberts IN, Stead AD, Ockendon DJ, Dickinson HG (1980) Pollen-stigma interactions in *Brassica oleracea*. Theor Appl Genet 58:241–246

Robert LS, Allard S, Franklin TM, Trick M (1994) Sequence and expression of endogenous S-locus glycoprotein genes in self-compatible *Brassica napus*. Mol Gen Genet 242:209–216

Roggen HP, van Dijk AJ (1972) Breaking incompatibility in *Brassica oleracea* L. Euphytica 21:48–51

Ronald WG, Ascher PD (1975) Self-compatibility in garden Chrysanthemum: occurrence, inheritance and breeding potential. Theor Appl Genet 46:45–54

Rosen WG (1971) Pistil-pollen interactions. In: Heslop-Harrison J (ed) Pollen: development and physiology. Butterworths, London, pp 239–254

Rowlands DG (1958) The nature of the breeding system in the field bean (*Vicia faba*) and its relationship to breeding for yield. Heredity 12:113–126

Rowlands DG (1964) Self-incompatibility in sexually propagated cultivated plants. Euphytica 13:157–162

Royo J, Kunz C, Kowyama Y, Anderson MA, Clarke AE, Newbigin E (1994) Loss of a histidine residue at the active site of S-locus ribonuclease is associated with self-incompatibility in *Lycopersicum peruvianum*. Proc Natl Acad Sci USA 91:6511–6514

Rudd JJ, Franklin FCH, Lord JM, Franklin-Tong VE (1996) Increased phosphorylation of a 26-kD pollen protein is induced by the self-incompatibility response in *Papaver rhoeas*. Plant Cell 8:713–724

Rudd JJ, Franklin FCH, Lord JM, Franklin-Tong VE (1997) Ca2+-independent phosphorylation of a 68-kD protein is stimulated by the self-incompatibility response. Plant J 12:507–514

Ruiter RK, Mettenmeyer T, van Laarhoven D, van Eldik GJ, Doughty J, van Herpen MMA, Schrauwen JAM, Dickinson HG, Willems GJ (1997) Proteins of the pollen coat of *Brassica oleracea*. J Plant Physiol 150:85–91

Rundle SJ, Nasrallah ME, Nasrallah JB (1993) Effects of inhibitors of protein serine/ thronine phosphatase on pollination in *Brassica*. Plant Physiol 103:1165–1171

Saba-El-Leil RK, Rivard S, Morse D, Cappadocia M (1994) The S11 and S13 self-incompatibility alleles in *Solanum chacoense* Bitt. are remarkably similar. Plant Mol Biol 24:571–583

Sage TL, Williams AG (1991) Self-incompatibility in *Asclepias*. Plant Cell Incomp Newslett 23:55–57

Sage TL, Bertin RI, Williams EG (1994) Ovarian and other late-acting self-incompatibility systems. In: Williams AG, Clarke AE, Knox RB (eds) Genetic control of self-incompatibility and reproductive development in flowering plants. Kluwer, Dordrecht, pp 116–140

Sage TL, Strumas F, Cole WW, Barrett SCH (1999) Differential ovule development following self-and cross-pollinations: the basis of self-fertility in *Narcissus triandrus* (Amaryllidaceae). Am J Bot 86(6):855–870

Salamini F, Motto M (1993) The role of gene technology in plant breeding. In: Hayward MD, Bosemark NO, Romagosa I (eds) Plant breeding (principles and prospects). Chapman and Hall, London, pp 138–159

Sampson DR (1957) The genetics of self-incompatibility in the radish. J Heredity 48:26–29

Sampson DR (1958) The genetics of self-incompatibility in *Lesquerella densipila* and in the hybrid *L. densipila* x *L. lescurit*. Can J Bot 36:39–56

Sampson DR (1960) An hypothesis of gene interaction at the S-locus in self-incompatibility systems of angiosperms. Am Nat 94:283–292

Sampson DR (1962) Pollen-stigma incompatibility in *Cruciferea*. Can J Genet Cytol 4:38–49

Sampson DR (1967) Frequency and distribution of self-incompatibility alleles in *Raphanus raphanistrum*. Genetics 56:241–251

Sampson DR (1974) Equilibrium frequency of sporophytic self-incompatibility alleles. Can J Genet Cytol 16:611–668

Sareen PK, Kakar SN (1977) Immunological studies on self-incompatibility in *Brassica campestris* L. Var brown sarson. Z Pflanzenzucht 79:324–330

Sarfatti G, Ciampolini F, Pacini E, Cresti M (1974) Effects of actinomycin on *Lycopersicum peruvianum* pollen tube growth and self-incompatibility reaction. In: Linskens HF (ed) Fertilization in higher plants. Nijmegen, Netherlands

Sarker RH, Elleman CJ, Dickinson HG (1988) Control of pollen hydration in *Brassica* requires continued protein synthesis, and glycosylation is necessary for intraspecific incompatibility. Proc Natl Acad Sci USA 85:4340–4344

Sasaki Y, Iwano M, Matsuda N, Susuki G, Watanabe M, Isogai A, Toriyama K (1998) Localisation of an SLG protein expressed under the regulation of a tapetum-specific promoter in anthers of transgenic *Brassica napus*. Sex Plant Reprod 11:245–250

Sassa H, Hirano H, Ikehashi H (1992) Self-incompatibility related RNases in styles of Japanese pear (*Pyrus serotina* Rehd). Plant Cell Physiol 33:811–814

Sassa H, Hirano H, Ikehashi H (1993) Identification and characterization of sytlar glycoproteins associated with self-incompatibility genes of Japanese pear (*Pyrus serotina*) Rehd. Mol Gen Genet 241:17–25

Sassa H, Mase N, Hirano H, Ikeashi H (1994) Identification of self-incompatibility-related glycoproteins in styles of apples (*Malus domestica*). Theor Appl Genet 89:201–205

Sassa H, Nishio T, Kowyama Y, Hirano H, Koba T, Ikehashi H (1996) Self-incompatibility (S) alleles of the Rosaceae encode members of a distinct class of the T2/S ribonuclease superfamily. Mol Gen Genet 250:547–557

Sassa H, Hirano H, Nishio T, Koba T (1997) Style-specific self-compatible mutation caused by deletion of the S-Rnase gene in Japanese pear (*Pyrus serotina*). Plant J 12(1):223–227

Sato T, Thorsness MK, Kandasamy MK, Nishio T, Hirai M, Nasrallah JB, Nasrallah ME (1991) Activity of an S-locus gene promoter in pistil and anthers of transgenic *Brassica*. Plant Cell 3:867–876

Sawyer S (1989) Statistical tests for detecting gene conversion. Mol Biol Evol 6 526–538

Schibilla H, Louant B (1975) The recovery of compatibility mutations at the S-locus of *Lycopersicum peruvianum*. Incomp Newslett 5:89–91

Schierup MH, Vekemans X, Christiansen FB (1997) Evolutionary dynamics of sporophytic self-incompatibility alleles in plants. Genetics 147:835–846

Schopfer CR, Nasrallah ME, Nasrallah JB (1999) The male determinant of self-incompatibility in *Brassica*. Science 286:1697–1700

Scott J (1865) On the individual sterility and cross impregnation of certain species of *Oncidium*. J Linn Soc Lond 8:163–167

Schou O (1984) The dry and wet stigmas of *Primula obconica*: ultrastructural and cytochemical dimorphisms. Protoplasma 121:99–113

Schou O, Philipp M (1984) An unusual heteromorphic incompatibility system. III. On the genetic control of distyly and self-incompatibility in *Anchusa officinalis* L. (Boraginaceae). Theor Appl Genet 68:139–144

Scribailo RW, Barrett SCH (1991) Pollen-pistil interactions in tristylous *Pontederia sagitata* (Pontederiaceae). II. Patterns of pollen tube growth. Am J Bot 78:1662–1682

Scutt CP, Gates PJ, Gatehouse JA, Boulter D, Croy RRD (1990) A cDNA-encoding an S-locus specific glycoprotein from *Brassica oleracea* plants containing the S5 self-incompatibility allele. Mol Gen Genet 220:409–413

Scutt CP, Fordham-Skelton AP, Croy RRD (1993) Ocadaic acid causes breakdown of self-incompatibility in *Brassica oleracea*. Evidence for the involvement of protein phosphatase in the incompatibility response. Sex Plant Reprod 33:412–416

Sears ER (1937) Cytological phenomena connected with self-sterility in the flowering plants. Genetics 22:130–181

Seavey SR, Bawa KS (1986) Late-acting self-incompatibility in angiosperms. Bot Rev 52:195–219

Sedgley M (1974) Assessment of serological techniques for S-allele identification in *Brassica oleracea*. Euphytica 23:543–552

Sedgley M (1994) Self-incompatibility in woody horticultural species. In: Williams AG, Clarke AE, Knox RB (eds) Genetic control of self-incompatibility and reproductive development in flowering plants. Kluwer, Dordrecht, pp 141–163

Sharma KD, Boyes JW (1961) Modified incompatibility of Buckwheat following irradiation. Can J Bot 39:1241

Shiba H, Hinata K, Susuki A, Isogai A (1995) Breakdown of self-incompatibility in *Brassica* by the antisense RNA of the SLG gene. Proc Jpn Acad 71:81–83

Shih CY, Rappaport L (1971) Regulation of bud rest in tubers of potato, *Solanum tuberosum* L. VIII. Early effects of gibberellin A3 and abscissic acid on ultrastructures. Plant Physiol 48:31–35

Shivanna KR, Rangaswamy NS (1969) Overcoming self-incompatibility in *Petunia axillaris*. I. Delayed pollination, pollination with stored pollen and bud pollination. Phytomorphology 19:372–380

Shivanna KR, Heslop-Harrison J, Heslop-Harrison Y (1981) Heterostyly in *Primula*. 2. Sites of pollen inhibition, and effects of pistil constituents on compatible and incompatible pollen tube growth. Protoplasma 107:319–337

Shivanna KR, Heslop-Harrison J, Heslop-Harrison Y (1983) Heterostyly in *Primula*. 3. Pollen water economy: a factor in the intramorph-incompatibility response. Protoplasma 117:175–184

Sims TL, Ordanic M, Fjellstrom RGM, Collins PD (1999) Molecular genetics of gametophytic self-incompatibility in *Petunia hybrida* and *Trifolium pratense*(Abstract). In: Pollen-stigma interactions. Int Conf (Dept Plant Sci, Univ Oxford) dedicated to the memory of Jack Heslop-Harrison, Oxford, p 52

Singh D (1958) Rape and mustard. ICOC, Hyderarabad (India), pp 47–51

Singh A, Kao T-H (1992) Gametophytic self-incompatibility: biochemical, molecular,genetic and evolutionary aspects. Int Rev Cytol 140:449–483

Singh A, Paolillo DJ (1990) Role of calcium in the callose response of self-pollinated *Brassica* stigma. Am J Bot 77:128–133

Song KM, Susuki JY, Slocum MMK, Williams PH, Osborn TC (1991) A linkage map of *Brassica rapa* (syn. *campestris*) based on restriction fragment length polymorphism loci. Theor Appl Genet 82:296–304

Stahl RJ, Arnoldo MA, Glavin TL, Goring DR, Rothstein SJ (1998) The self-incompatibility phenotype in *Brassica* is altered by the transformation of a mutant S-locus receptor kinase. Plant Cell 10:209–218

Stanchev BS, Doughty J, Scutt CP, Dickinson HG, Croy RRD (1996) Cloning of PCP1, a member of a family of pollen coat protein (PCP) genes from *Brassica oleracea* encoding novel cysteine-rich proteins involved in pollen-stigma interactions. Plant J 10(2):303–313

Stebbins GL (1950) Variation and evolution in plants. Columbia Univ Press, New York

Stebbins GL (1957) Self-fertilization and population variability in the higher plants. Am Nat 91:337–354

Stein JC, Nasrallah JB (1993) A plant receptor-like gene, the S-locus receptor kinase of *Brassica oleracea*, encodes a functional serine/threonine kinase. Plant Physiol 101:1103–1106

Stein JC, Howlett BJ, Boyes DC, Nasrallah ME, Nasrallah JB (1991) Molecular cloning of a putative receptor protein kinase of *Brassica oleracea*. Natl Acad Sci USA 88:8816–8820

Stein JC, Dixit R, Nasrallah ME, Nasrallah JB (1996) SRK, the stigma-specific S-locus receptor kinase of *Brassica* is targeted to the plasma membrane in transgenic tobacco. Plant Cell 8:429–445

Steinbachs JE, Holsinger KE (1999) Pollen transfer dynamics and the evolution of gametophytic self incompatibility. J Evol Biol 12:770–778

Steiner E (1956) New aspects of the balanced lethal mechanism in *Oenothera*. Genetics 41:486–500

Stephens JC (1985) Statistical methods of DNA sequence analysis:.detection of intragenic recombination or gene conversion. Mol Biol Evol 2:539–556

Stephenson AG, Doughty J, Dixon S, Elleman CJ, Hiscock SJ, Dickinson HG (1997) The male determinant of self-incompatibility in *Brassica oleraceae* is located in the pollen coating. Plant J 12(6):1351–1359

Stettler RF (1968) Irradiated mentor pollen: its use in remote hybridization of black cottonwood. Nature 219:746–747

Stettler RF, Ager AA (1984) Mentor effects in pollen interactions. In: Linskens HF, Heslop-Harrison J (eds) Encyclopedia of plant physiology: cellular interactions, vol 17. Springer, Berlin Heidelberg New York, p 609

Stettler RF, Guries RP (1976) The mentor pollen phenomenum in black cottonwood. Can J Bot 54:820–830

Stettler RF, Koster R, Steenackers V (1980) Interspecific crossability studies in poplars. Theor Appl Genet 58:273–282

Stevens JP, Kay QON (1989) The number, dominance relationships and frequencies of self-incompatibility alleles in a natural population of *Sinapis arvensis* L. in South Wales. Heredity 62:199–205

Stevens VAM, Murray BG (1982) Studies on heteromorphic self-incompatibility systems: physiological aspects of the incompatibility system in *Primula obconica*. Theor Appl Genet 61:245–256

Stone D (1957) Studies in population differentiation and variation in *Myosorus* of the Ranunculaceae. PhD Thesis, Univ California Berkeley

Stone SL, Arnoldo MA, Goring DR (1999) A breakdown of *Brassica* self-incompatibility in ARC1 antisense transgenic plants. Science 286:1729–1731

Stout AB (1917) Fertility in *Cichorium Intybus*: the sporadic occurrence of self-fertile plants among the progeny of self-sterile plants. Am J Bot 4:375–395

Stout AB (1920) Further experimental studies on self-incompatibility in hermaphroditic plants. J Genet 9:85–129

Stout AB (1925) Studies of *Lythrum salicaria*. II. A new form of flower in the species. Bull Torrey Bot Club 52:81–85

Stout AB (1952) Reproduction in *Petunia*. Mem Torrey Bot Club 20:1–202

Stout AB, Chandler C (1933) Pollen behaviour in *Hemerocallis* with special reference to incompatibilities. Bull Torrey Botan Club 60:397–416

Stout AB, Chandler C (1942) Hereditary transmission of induced tetraploidy and compatibility in fertilisation. Science 96:257

Straub J (1946) Zur Entwicklungsphysiologie der Selbststerilität von *Petunia*. Z Naturforsch 1:287–291

Straub J (1947) Zur Entwicklungsphysiologie der Selbststerilität von *Petunia*. II. Das Prinzip des Hemmungs-Mechanismus. Z Naturforsch 2b:433–444

Sulaman W, Arnoldo MA, Yu KF, Tulsieram L, Rothstein SJ, Goring DR (1997) Loss of callose in the stigma does not affect the *Brassica* self-incompatibility phenotype. Planta 203:327–331

Susuki G, Watanabe M, Toriyama K, Isogai A, Hinata K (1996) Expression of SLG9 and SRK9 in transgenic tobacco. Plant Cell Physiol 37:866–869

Susuki G, Watanabe M, Kai N, Matsuda N, Toriyama K, Takayama S, Isogai A, Hinata K (1997a) Three members of the S-multigene family are linked to the S-locus in *Brassica*. Mol Gene Genet 256:257–264

Susuki G, Watanabe M, Toriyama K, Isogai A, Hinata K (1997b) Direct cloning of the *Brassica* S-locus by using a P1-derived arificial chromosome (PAC) vector. Gene 199:133–137

Susuki G, Kai N, Hirose T, Fukui K, Nishio T, Takayama T, Isogai A, Watanabe M, Hinata K (1999) Genomic organization at the S-locus: identification and characterization of genes in SLG/SRK region of S9 Haplotype of *Brassica campestris* (syn. *rapa*). Genetics 153:391–400

Sutton I (1918) Report on tests of self-sterility in plums, cherries and apples at the John Innes Horticultural Institution. J Genet 7:281–300

Takasaki T, Hatakeyama K, Watanabe M, Isogai A, Hinata K (1999) Introduction of SLG (S-locus glycoprotein) alters the phenotype of endogenous S haplotype, but confers no new S-haplotype specificity in *Brassica rapa* L. Plant Mol Biol 40:659–668

Takasaki T, Hatakeyama K, Suzuki G, Watanabe M, Isogai A, Hinata K (2000) The S-receptor kinase determines self-incompatibility in *Brassica* stigma. Nature 403:913–916

Takashina T, Imanishi S, Egashira H (1997) Evaluation of the cross-incompatibility of "*peruvianum*-complex" lines with *Lycopersicon esculentum* Mill. by the ovule selection method. Breed Sci 47:33–37

Takayama S, Isogai A, Tsukomoto C, Ueda Y, Hinata K, Okazaki K, Susuki A (1986) Isolation and some characterization of S-locus-specific glycoproteins associated with self-incompatibility in *Brassica campestris*. Agric Biol Chem 50:1673–1676

Takayama S, Isogai A, Tsukamoto C, Ueda Y, Hinata K, Susuki A (1987) Sequences of S-glycoproteins, products of the *Brassica campestris* self-incompatibility locus. Nature 326:102–105

Takayama S, Shiba H, Iwano M, Shimosato H, Che F-S, Kai N, Watanabe M, Susuki G, Hinata K, Isogai A (2000) The pollen determinants of self-incompatibility in *Brassica campestris*. Proc Natl Acad Sci USA 97:1920–1925

Takayama S, Shiba H, Iwano M, Asano K, Che F-S, Watanabe M, Hinata K, Isogai A (2000) Isolation and characterization of pollen coat proteins of *Brassica campestris* that interact with S locus-related glycoprotein 1 involved in pollen-stigma adhesion. Proc Natl Acad Sci USA 97(7):3765–3770

Takhtajan A (English edn 1969) Flowering plants (origin and dispersal). First published 1961. Oliver and Boyd, Edinburgh

Talmage DW (1959) Mechanism of the antibody response. In: Zirkle RE (ed) A symposium on molecular biology. Chicago Univ, pp 91–101

Tanksley SD, Loaiza-Figueroa F (1985) Gametophytic self-incompatibility is controlled by a single major locus on chromosome 1 in *Lycopersicum peruvianum*. Proc Natl Acad Sci USA 82:5093–5096

Tantikanjana T, Nasrallah ME, Stein JC, Chen CH, Nasrallah JB (1993) An alternative transcript of the S-locus glycoprotein gene in a class II pollen-recessive self-incompatibility haplotype of *Brassica oleracea* encodes a membrane-anchored protein. Plant Cell 5:657–666

Tao R, Yamane H, Sassa H, Mori H, Gradziel TM, Dandekar AM, Sugiura A (1997) Identification of stylar RNases associated with gametophytic self-incompatibility in almond (*Prunus dulcis*). Plant Cell Physiol 38(3):304–311

ten Hoopen R, Harbord RM, Maes T, Nanninga N, Robbins TP (1998) The self-incompatibility (S) locus in *Petunia hybrida* is located on chromosome III in a region syntenic for the Solanaceae. Plant J 16(6):729–734

Thompson KF (1957) Self-incompatibility in marrow-stem kale, *Brassica oleracea* var. acephala. Demonstration of sporophytic system. J Genet 55:45–60

Thompson KF (1967) Breeding problems in kale (*Brassica oleracea*) with particular reference to marrow-stem kale. Rep Plant Breed Inst 7:34

Thompson KF (1972) Competitive interaction between two-alleles in a sporophytically controlled incompatibility system. Heredity 28:1–8

Thompson KF, Taylor JP (1965) Identical S-alleles in different botanical varieties of *Brassica oleracea*. Nature 208:306–307

Thompson KF, Taylor JP (1966) The breakdown of self-incompatibility in cultivars of *Brassica oleracea*. Heredity 21:637–648

Thompson KF, Taylor JP (1971) Self-compatibility in kale. Heredity 27:459–471

Thompson RD, Uhrig H, Hermsen JGT, Sazlamini F, Kaufmann H (1991) Investigation of a self-compatible mutation in *Solanum tuberosum* clones inhibiting S-allele activity in pollen differentially. Mol Gen Gent 226:283–288

Thorn EC (1991) Crossability barriers between barley (*Hordeum vulgare* L.) and *Hordeum bulbosum* L. Hereditas 114:213–218

Thorsness MK, Kandasamy MK, Nasrallah ME, Nasrallah JB (1991) A *Brassica* S-locus gene promoter targets toxic gene expression and cell death to the pistil and pollen of transgenic *Nicotiana*. Dev Biol 143:173–184

Tobias CM, Nasrallah JB (1996) An S-locus related gene in *Arabidopsis* encodes a functional kinase and produce two classes of transcripts. Plant J 10:523–531

Tobias CM, Howlett BJ, Nasrallah JB (1992) An *Arabidopsis thaliana* gene with sequence similarity to the S-locus receptor kinase of *Brassica oleracea*. Plant Physiol 99:284–290

Tobutt KR, Boskovi R, Sonneveld T (1999) Detection of incompatible alleles in cherry and almond by the analysis of stylar ribonucleases. In: Pollen-stigma interactions. Int Conf (Dept Plant Sci, Univ Oxford, Oxford) dedicated to the memory of J. Heslop-Harrison, Oxford, p 53

Tomkova J (1959) Problems of autosterility in *Nicotiana alata* Link et Otto (Formation of the inhibitory factor in the pistil). Biol Plant 1:328–329

Toryama K, Thorsness MK, Nasrallah JB, Nasrallah ME (1991a) A *Brassica* S-locus gene promoter directs sporophytic expression in the anther tapetum of transgenic *Arabidopsis*. Dev Biol 143:427–431

Toriyama K, Stein JC, Nasrallah ME, Nasrallah JB (1991b) Transformation of *Brassica olearacea* with an S-locus gene from *B. campestris* changes the self-incompatibility phenotype. Theor Appl Genet 81:769–776

Townsend CE (1965) Seasonal and temperature effects on self-compatibility in tetraploid alsike clover, *Trifolium hybridum* L. Crop Sci 5:329–332

Townsend CE (1966) Self-compatibility response to temperature and the inheritance of the response in tetraploid alsike clover, *Trifolium hybridum* L. Crop Sci 6:415–419

Townsend CE (1968) Self-compatibility studies with diploid alsike clover, *Trifolium hybridum* L. Crop Sci 8:269–272

Townsend CE (1969) Self-compatibility studies with diploid alsike clover, *Trifolium hybridum* L. IV. Inheritance of type II self-compatibility in different genetic backgrounds. Crop Sci 9:443–446

Townsend CE (1971) Further studies on the inheritance of a self-compatibility response to temperature and the segregation of S-alleles in diploid alsike clover. Crop Sci 11:860–863

Trewavas AJ, Gilroy S (1991) Signal transduction in plant cells. Trends Genet 7:356–361

Trick M, Heizmann P (1992) Sporophytic self-incompatibility systems: *Brassica* S-gene family. Int Rev Cytol 140 485–524

Tsai DS, Lee HS, Post LC, Kreiling KM, Kao T-H (1992) Sequence of an S-protein of *Lycopersicum peruvianum* and comparison with other solanaceous S-proteins. Sex Plant Reprod 5:256–263

Tupý J (1959) Callose formation in pollen tubes and incompatibility. Biol Plant 1:192–198

Tupý J (1961) Investigation of free amino-acids in cross-, self- and non pollinated pistils of *Nicotiana alata*. Biol Plant 3(1):47–64

Tupý J, Hrabetova E, Balatkova V (1977) Evidence for ribosomal RNA synthesis in pollen tubes in culture. Biol Plant 19:226–230

Twell D, Wing RA, Yamaguchi J, McCormick S (1990) Pollen-specific gene expression in transgenic plants: coordonate regulation of two different tomato gene promoters during microsporogenesis. Development 109:705–713

Umbach AL, Lalonde BA, Kandasamy MK, Nasrallah JB, Nasrallah ME (1990) Immunodetection of protein glycoforms encoded by two independent genes of the self-incompatibility multigene family of *Brassica*. Plant Physiol 93:739–747

Ünal M (1986) A comparative cytological study on compatible and incompatible tubes of *Petunia hybrida*. First Univ Fen Fak Mec (Türkite) Ser B51:36495

Ushijima K, Sassa H, Tao R, Yamane H, Dandekar AM, Gradziel TM, Hirano H (1998a) Cloning and characterization of cDNAs encoding S-Rnases from almond

(*Prunus dulcis*): primary structural features and sequence diversity of the the S-Rnases in Rosaceae. Mol Gen Genet 260:261–268

Ushijima K, Sassa H, Hirano H (1998b) Characterization of the flanking regions of the S-Rnase genes of Japaneses pear (*Pyrus serotina*) and apple (*Malus domesticus*). Gene 211:159–167

Uyenoyama MK (1988) On the evolution of genetic incompatibility systems. II. Initial increase of strong gametophytic self-incompatibility under partial selfing and half-sib mating. Am Nat 131:700–722

Uyenoyama MK (1991) On the evolution of genetic incompatibility systems. VI. A three-locus modifier model for the origin of gametophytic self-incompatibility. Genetics 128:453–469

Uyenoyama MK (1995) A generalized least-squares estimate for the origin of sporophytic self-incompatibility. Genetics 139:975–992

Uyenoyama MK (1997) Genealogical structure among alleles regulating self-incompatibility in natural populations of flowering plants. Genetics 147:1389–1400

Uyenoyama MK (1999) Evolutionary processes under self-incompatibility (Abstract). In: Pollen-stigma interactions. Int Conf (Dept Plant Sci, Univ Oxford) dedicated to the memory of Jack Heslop-Harrison, Oxford, p 38

Uyenoyama MK, Newbigin E (2000) Evolutionary dynamics of dual-specificity self-incompatibility alleles. Plant Cell 12:310–312

van der Donk JAW (1974a) Differential synthesis of RNA in self- and cross-pollinated styles of *Petunia hybrida* L. Mol Gen Genet 131:1–8

van der Donk JAW (1974b) Synthesis of RNA and protein as a function of time and type of pollen tube-style interaction in *Petunia hybrida* L. Mol Gen Genet 134:93–98

van der Donk JAW (1975a) Translation of plant messengers in egg cells of *Xenopus laevis*. Nature 256:674–675

van der Donk JAW (1975b) Recognition and gene expression during the incompatibility reaction in *Petunia hybrida* L. Mol Gen Genet 141:305–316

van der Pluijm J, Linskens HF (1966) Feinstruktur der Pollen-Schläuche im Griffel von *Petunia*. Züchter 36:220–224

van Gastel AJG (1976) Mutability of the self-incompatibility locus and identification of the S-bearing chromosome in *Nicotiana alata*. Doctoral, Agric Res Rep 852. Landbouwhogeschool, Wageningen

van Gastel AJG, de Nettancourt D (1974) The effects of different mutagens on self-incompatibility in *Nicotiana alata* Link and Otto. I. Chronic gamma irradiation. Radiat Bot 14:43–50

van Gastel AJG, de Nettancourt D (1975) The effects of different mutagens on self-incompatibility in *Nicotiana alata* Link and Otto. II. Acute irradiations with X-rays and fast neutrons. Heredity 34:381–392

van Nerum I, Incerti F, Keulemans J, Broothaerts WJ (1999) Analysis of transgenic apple trees containing a sense or antisense gene construct (Abstract). In: Pollen-stigma interactions. Int Conf (Dept Plant Sci, Univ Oxford) dedicated to the memory of Jack Heslop-Harrison, Oxford, p 53

Vantuyl JM, Marcucci MC, Visser T (1982) Pollen and pollination experiments. VII. The effect of pollen treatment and application method on incompatibility and incongruity in *Lilium*. Euphytica 31:613–619

Vargas Eyre J, Smith G (1916) Some notes on the Linaceae. J Genet 5:189–197

Vaucheret H, Béclin C, Elmayan T, Feuerbach F, Godon C, Morel J-B, Mourrain P, Palauqui J-C, Vernhettes S (1998) Transgene-induced gene silencing in plants. Plant J 16(6):651–659

Vekemans X, Slatkin M (1994) Gene and allelic genealogies at a gametophytic self-incompatibility locus. Genetics 137:1157–1165

Verica JA, McCubbin AG, Kao T-H (1998) Are the hypervariable regions of S-RNases sufficient for allelic-specific recognition of pollen? Letter to the editor. Plant Cell 10:314–316

Verma S.C, Malik R, Dhir I (1977) Genetics of the incompatibility system in the crucifer *Eruca sativa*. Proc R Soc Lond B 96:131–159

Visser T (1981) Pollen and pollination experiments. IV. Mentor pollen and pioneer pollen techniques regarding incompatibility and incongruity in apple and pear. Euphytica 30:363–369

Visser T, Marcucci MC (1986) The performance of double or mixed pollinations with compatible and self-incompatible or incongruous pollen of pear and apple. Euphytica 31:305–312

Visser T, Oost EH (1982) Pollen and pollination experiments. V. An empirical basis for a mentor pollen effect observed on the growth of incompatible pollen in pear. Euphytica 31:305–312

Visser T, Verhaegh JJ (1980) Pollen and pollination experiments. II. The influence of the first pollination on the effectiveness of the second one in apple. Euphytica 29:385–390

Vithanage HIMV, Gleeson PA, Clarke AE (1980) The nature of callose produced during self-pollination in *Secale cereale*. Planta 148:498–509

von Übisch G (1921) Zur Genetic der trimorphen Heterostylie sowie einige Bemerkungen zur dimorphen Heterostylie. Biol Zentralbl 41:88–96

Vuilleumier BS (1967) The origin and evolutionary development of heterostyly in the angiosperms. Evolution 21:210–226

Walker EA (1994) Cloning, characterization and expression in *E. coli* of S-(self-incompatibility) alleles from *Papaver rhoeas*. PhD Thesis, Univ of Birmingham, Birmingham

Walker EA, Ride JP, Kurup S, Franklin-Tong VE, Franklin FCH (1996) Molecular analysis of two functional homologues of the *Papaver rhoeas* self-incompatibility gene isolated from different populations. Plant Mol Biol 30:983–994

Walker JC, Zhang R (1990) Relationship of a putative receptor protein kinase from maize to the S-locus glycoproteins of *Brassica*. Nature 345(6277):743–746

Walker JC (1993) Receptor-like protein kinase genes of *Arabidopsis thaliana*. Plant J 3:451–456

Walles B, Han SP (1998) Ribosomes in incompatible pollen tubes in the Solanacea. Physiol Plant 103:461–465

Waser NM, Price MV (1991) Reproductive costs of self-pollination in *Ipomopsis aggregata* (Polemoniaceae): are ovules usurped? Am J Bot 78:1036–1043

Watanabe M, Hinata K (1999) Self-incompatibility. In: Gomez-Campo C (ed) Biology of *Brassica coenospecies*. Dev Plant Genet Breed 4. Elsevier, Amsterdam

Watanabe M, Shiozawa H, Isogai A, Susuki A, Taheuchi T, Hinata K (1991) Existence of S-glycoprotein-like proteins in anthers of self-incompatible species of *Brassica*. Plant Cell Physiol 32:1039–1047

Watanabe M, Nou IS, Takayama S, Yamakawa S, Isogai A, Susuki A, Takeuchi T, Hinata K (1992) Variations in and inheritance of NS-glycoprotein in self-incompatible *Brassica campestris* L. Plant Cell Physiol 33:343–351

Watanabe M, Takasaki T, Toriyama K, Yamakawa S, Isogai A, Susuki A, Hinata K (1994) A high degree of homology exists between the protein encoded by SLG and the S-receptor domain encoded by SRK in self-incompatible *Brassica campestris* L. Plant Cell Physiol 35:1221–1229

Watanabe M, Susuki G, Kai N, Takasaki T, Nishio T, Takayama S, Isogai A, Hinata K (1999a) Genomic organization of SLG/SRK region of an S9 haplotype of *Brassica campestris* L. (Abstract). In: Pollen-stigma interactions. Int Conf (Dept Plant Sci, Univ Oxford) dedicated to the memory of Jack Heslop-Harrison, Oxford, p 27

Watanabe M, Susuki G, Toriyama K, Takayama K, Isogai A, Hinata K (1999b) Two anther-expressed genes downstream of SLG9: identification of a novel S-linked gene specifically expressed in anthers at the uninucleate stage of *Brassica campestris* (syn *rapa* L.). Sex Plant Reprod 12:127–134

Watanabe M, Susuki G, Hatakeyama K, Isogai A, Hinata K (1999c) Molecular Biology of self-incompatibility in *Brassica* species. Plant Biotechnol 16:263–272

Watanabe M, Ito A, Takada Y, Ninomiya C, Kakizaki T, Takahata Y, Hatakeyama K, Hinata K, Susuki G, Takasaki T, Satta Y, Shiba H, Takayama S, Isogai A (2000) Highly divergent sequences of the pollen self-incompatibility (S) gene in class-I S haplotypes of *Brassica campestris* (syn. *rapa*) L. FEBS Lett 23647:1–6

Wedzony M, Filek M (1998) Changes of electric potential in pistils of *Petunia hybrida* and *Brassica napus* L. during pollination. Acta Physiol Plant 20:291–297

Wehling P, Hackauf B, Wricke G (1994a) Phosphorylation of pollen grains in relation to self-incompatibility in rye (*Secale cereale* L.). Sex Plant Reprod 7:67–75

Wehling P, Hackauf B, Wricke G (1994b) Identification of self-incompatibility related PCR fragments in rye (*Secale cereale* L.) by DGGE. Plant J 5:891–893

Wehling P, Hackauf B, Wricke G (1995) Characterization of the two factors self-incompatibility system in *Secale Cereale* L. In: Kuck G, Wricke G (eds) Genetic mechanisms for hybrid breeding. Adv Plant Breed 18:149–161

Wexelsen H (1945) Studies on the fertility, inbreeding and heterosis in red clover (*Trifolium pratense* L.) Norske Videnskapsakad. Oslo. Mat-Nat Kl

Whitehead WL, Davis RL (1954) Self and cross-compatibility in alfalfa (*Medicago sativa*). Agron J 46:452–456

Whitehouse HLK (1950) Multiple-allelomorph incompatibility of pollen and style in the evolution of the angiosperms. Ann Bot 14:198–216

Wiering D (1958) Artificial pollination of cabbage plants. Euphytica 7:223–227

Willemse MTM (1999) Pollen coat signals with respect to pistil activation and ovula penetration in Gasteria verrucosa (Mill.) H. Duval. In: Cresti M, Cai G, Moscatelli A (eds) Fertilization in higher plants (molecular and cytological aspects). Springer, Berlin Heidelberg New York, pp 145–156

Willemse MTM, Franssen-Verheijen MAW (1988) Pollen tube growth and its pathway in *Gasteria verrucosa* (Mill.) Duval. Phytomorphology 38:127–132

Williams EG, Ramm-Anderson S, Dumas C, Mau SL, Clarke AE (1982) The effects of isolated components of *Prunus avium* styles on in vitro growth of pollen tubes. Planta 156:517–519

Williams W (1947) Genetics of red clover (*Trifolium pratense* L.) compatibility. III. J Genet 48:69–79

Williams W (1951) Genetics of incompatibility in alsike clover, *Trifolium hybridum* L. Heredity 5:51–72

Wingett SW, Doughty J, Abbott AR, Dickinson FG (1999) The role of a pollen coat protein, PCP-A1, in pollen-stigma interactions. In: Pollen-stigma interactions. Int Conf (Dept Plant Sciene, Oxford) dedicated to the memory of Jack Heslop-Harrison, p 84

Wolters-Arts M, Lush WM, Mariani C (1998) Lipids are required for directional pollen-tube growth. Nature 392:818–821

Wong KC, Watanabe M, Hinata K (1994a) Protein profiles in pin and thrum floral organs of distylous *Averrhoa carambola* L. Sex Plant Reprod 7:107–115

Wong KC, Watanabe M, Hinata K (1994b) Fluorescence and scanning electron microscopic study on self-incompatibility in distylous *Averrhoa carambola* L. Sex Plant Reprod 7: 116–121

Wricke G (1974) Seed set in rye after selfing under high temperature conditions. Incomp Newslett 4:23–27

Wricke G (1984) Hybridzüchtung beim Roggen mit Hilfe der Inkompatibilität. Vortr Pflanzenzüchtg 5:43–54

Wricke G, Wehling P (1985) Linkage between an incompatibility locus and a peroxidase locus (*Prx* 7) in rye. Theor Appl Genet 71:289–291

Wright S (1939) The distribution of self-sterility alleles in populations. Genetics 24:538–552

Wright S (1964) The distribution of self-sterility alleles in populations. Evolution 18:609–619

Xonocostle-Cázares B, Lozoya-Gloria E, Herrero-Estrella L (1993) Gene cloning and identification. In: Hayward MO, Bosemark NO, Romagosa I (eds) Plant Breeding (principles and prospects). Chapman and Hall, London, pp 107–122

Xu B, Mu J, Nevinsq DL, Grun P, Kao T-H (1990a) Cloning and sequencing of cDNAs encoding two self-incompatibility associated proteins in *Solanum chacoense*. Mol Gen Genet 224:341–346

Xu B, Grun P, Kheyr-Pour A, Kao T-H (1990b) Identification of pistil-specific proteins associated with three self-incompatibility alleles in *Solanum chacoense*. Sex Plant Reprod 3:54–60

Xue Y, Carpenter R, Dickinson HG, Coen ES (1996) Origin of allelic diversity in *Antirrhinum* S-locus Rnases. Plant Cell 8:805–814

Yamakawa K (1971) Effects of chronic gamma radiation on hybridization between *Lycopersicon esculentum* and *L. peruvianum*.Gamma Field Symp, Japan, 1971. Inst Rad Breed 10:11–31

Yamakawa S, Watanabe M, Hinata K, Susuki A, Isogai A (1995) The sequences of S-receptor kinases (SRK) involved in self-incompatibility and their homologies to S-locus glycoproteins of *Brassica campestris*. Biosci Biotech Biochem 59(1):161–162

Yamashita K, Oda K, Nakamura N (1990) Seed development in self-pollination of 4 x *Hyuganatsu* and reciprocal crosses between 2 x and 4 x *Hyuganatsu*, and overcoming the self-incompatibility of 2 x *Hyuganatsu* using pollen of 4 x *Hyuganatsu*. J Jpn Soc Hortic Sci 59:23–28

Yasuda S (1934) Physiological research on self-incompatibility in *Petunia violacea*. Bull Coll Agric For Morioka 20:1

Yu K, Schafer U, Glavin TL, Goring DR, Rothstein SJ (1996) Molecular characterization of the S-locus in two self-incompatible *Brassica napus* lines. Plant Cell 8:2369–2380

Zavada MS (1984) The relation between pollen exine sculpturing and self-incompatibility mechanisms. Plant Syst Evol 147:63–78

Zavada MS (1990) Correlations between pollen exine sculpturing and angiosperm self-incompatibility systems – a rebuttal. Taxon 39(3):442–447

Zuberi MI, Lewis D (1988) Gametophytic-sporophytic incompatibility in the Cruciferae-*Brassica campestris*. Heredity 61:367–377

Zuberi MI, Zuberi S, Lewis D (1981) The genetics of incompatibility in *Brassica*. I. Inheritance of self-compatibility in *Brassica campestris* L. var *toria*. Heredity 46:175–190

Zurek DM, Mou B, Beecher B, McClure BA (1997) Exchanging sequence domains between S-Rnases from *Nicotiana alata* disrupts pollen recognition. Plant J 11(4):797–808

Subject Index